# POWERTRAIN SYSTEMS FOR NET-ZERO TRANSPORT

PROCEEDINGS OF THE 2021 POWERTRAIN SYSTEMS FOR NET-ZERO TRANSPORT CONFERENCE, LONDON, UK, 7-8 DECEMBER 2021

# POWERTRAIN SYSTEMS FOR NET-ZERO TRANSPORT

*Edited by*

**Institution of Mechanical Engineers**

CRC Press is an imprint of the
Taylor & Francis Group, an **informa** business

A BALKEMA BOOK

*CRC Press/Balkema is an imprint of the Taylor & Francis Group, an informa business*



*Typeset by Integra Software Services Pvt. Ltd., Pondicherry, India*



*Library of Congress Cataloging-in-Publication Data*

A catalog record has been requested for this book

Published by: CRC Press/Balkema
Schipholweg 107C, 2316 XC Leiden, The Netherlands
e-mail: enquiries@taylorandfrancis.com
www.routledge.com – www.taylorandfrancis.com

ISBN: 978-1-032-11281-7 (Hbk)
ISBN: 978-1-032-11283-1 (pbk)
ISBN: 978-1-003-21921-7 (eBook)
DOI: 10.1201/9781003219217

Table of Contents

## SESSION 3: ENGINES WITH SUSTAINABLE FUELS (E.G HYDROGEN, E-FUELS, BIOMETHANE)



## SESSION 4: SIMULATION, MODELLING AND EXPERIMENTAL TECHNIQUES

## SESSION 5: REAL-WORLD DRIVING EMISSION (RDE) AND EMISSIONS ANALYSIS



## SESSION 6: REAL-WORLD DRIVING EMISSION (RDE) AND EMISSIONS CONTROL SYSTEMS



## SESSION 7: POWERTRAIN DEVELOPMENT SYSTEMS AND ANALYSIS

**Organising Committee**

Powertrain Systems and Fuels Group

**Member Credits:**

| | |
|---|---|
| Hua Zhao | Brunel University London |
| Frank Atzler | TU Dresden |
| Choongsik Bae | KAIST (Korea Advanced Institute of Science and Technology) |
| Ramanarayanan Balachandran | University College London |
| Hugh Blaxill | MAHLE Powertrain LLC |
| Ralph Clague | Jaguar Land Rover |
| Roger Cracknell | Shell Global Solutions |
| Colin Garner | Loughborough University |
| Sean Harman | Ford Motor Company |
| David Heaton | CAT |
| Bengt Johansson | Chalmers |
| Richard Osborne | Ricardo |
| Steve Sapsford | SCE Ltd. |
| Alan Tolley | JCB |
| Khizer Tufail | Ford Motor Company |
| Jamie Turner | Bath University |
| Carsten Weber | Ford Motor Company |
| Steve Whelan | HORIBA MIRA |
| Anna Wise | Innovate UK |
| Mingfa Yao | Tianjin University |

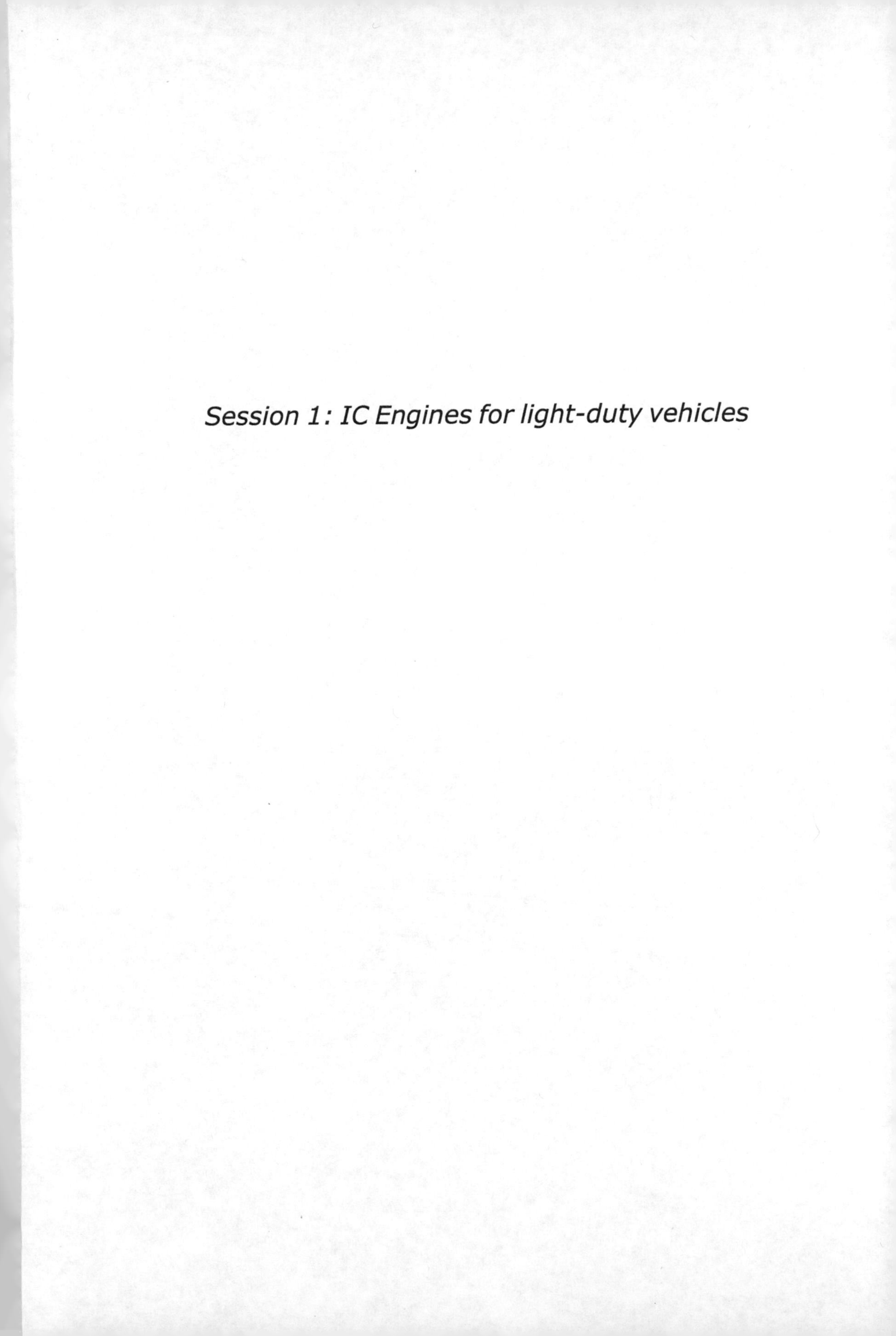

*Session 1: IC Engines for light-duty vehicles*

# Combustion sensitivity to charge motion in a dilute jet ignition engine

**M.P. Bunce[1,2], A. Cairns[2], S.K.P. Subramanyam[1], N.D. Peters[1], H.R. Blaxill[1]**
[1]MAHLE Powertrain LLC, USA
[2]University of Nottingham, UK

**ABSTRACT**

Though there are multiple viable powertrain options available for the automotive sector, those that contain internal combustion engines will continue to account for the majority of global sales for the next several decades. It is therefore imperative to continue the pursuit of novel combustion concepts that produce efficiency levels significantly higher than those of current engines. Introducing high levels of dilution in spark ignited (SI) engines has consistently proven to produce an efficiency benefit compared to conventional stoichiometric engine operation. However, this combustion mode can present challenges for the ignition system. Pre-chamber jet ignition enables stable, highly dilute combustion by both increasing the ignition energy present in the system and distributing it throughout the combustion chamber. Previous work by the authors have shown that jet ignition produces 15-25% increases in thermal efficiency over baseline SI engines with only relatively minor changes to engine architecture.

Lean combustion in general and jet ignition in particular represent fundamentally different engine operating modes compared to those of conventional stoichiometric SI engines. Therefore, there are some system sensitivities not present in stoichiometric engines that must be investigated in order to fully optimize the jet ignition system. Differing types and magnitudes of charge motion are incorporated in SI engines to aid with mixture preparation but the influence of charge motion over lean combustion performance, particularly in jet ignition engines, is less well understood. This study analyzes the impact that charge motion has on both pre-chamber and main chamber combustion. A 1.5L 3-cylinder gasoline engine is outfitted with multiple intake port configurations producing varying magnitudes and types of charge motion. Pre-chamber and main chamber combustion stability and other burn parameter responses are analyzed at part-load conditions. The results show that there is combustion sensitivity to charge motion, resulting in >1 percentage point spread in peak thermal efficiency for the configurations tested, and that this sensitivity manifests most significantly under low ignitability conditions such as heavy dilution. These results provide guidance for future system optimization of jet ignition engines.

## 1 INTRODUCTION

### 1.1 Background

The perpetual desire to conserve fuel is being coupled with an increasing modern awareness of the deleterious environmental impact of tailpipe emissions from the transportation sector. In response, increasingly stringent global legislation of greenhouse gas emissions will require a step change in internal combustion engine (ICE)

DOI: 10.1201/9781003219217-1

efficiency. A method being increasingly explored to accomplish this goal is dilute gasoline combustion [1-8]. The major limitation in developing dilute combustion systems is the less favorable ignition quality of the mixture. This has necessitated the development of higher energy ignition sources [9,10]. A pre-chamber jet igniter application is one such technology, having been researched extensively [11-15]. Pre-chamber combustion concepts have demonstrated the potential for stable main chamber combustion at higher levels of dilution than are allowable in typical SI engines [16].

With a pre-chamber combustor, products from the combustion event inside of the pre-chamber are pushed into the main combustion chamber through a nozzle, which subsequently ignite the main chamber contents. This creates a stronger, more distributed ignition source within the main combustion chamber than would be provided by a standard single-point spark plug. Active pre-chamber concepts contain an auxiliary fueling source in the pre-chamber, enabling de-coupled control over air-fuel ratio in each chamber. This allows the highly reactive jets from a conventional near-stoichiometric combustion event in the pre-chamber to serve as the ignition source for an homogeneous ultra-lean main chamber. MAHLE Jet Ignition® (MJI) is an auxiliary fueled pre-chamber concept that has been under development for several years [17-19]. MJI is designed to be a low-cost, practical ultra-lean combustion enabling technology. A rendering of pre-chamber placement in a typical cylinder head is shown in Figure 1.

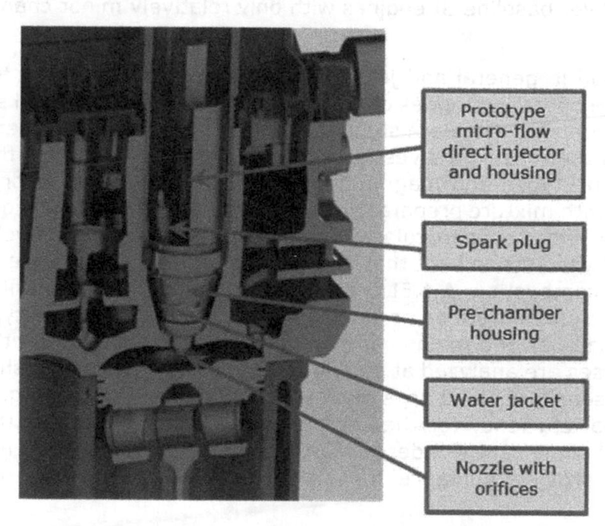


**Figure 1. CAD model rendering of a partial cutaway of the pre-chamber assembly in a cylinder head.**

Jet ignition concepts generally and MJI specifically possess numerous parameters than can be optimized in order to increase BTE, minimize engine-out emissions, or aid practical engine operation. While many of these parameters have been studied extensively by the authors [2] and others [5,12,14], one parameter for which there is minimal published data on its effect on jet ignition combustion is charge motion.

Charge motion in SI engines is typically used to drive or enhance mixture preparation in the cylinder. With the advent of DI SI engines, the role of charge motion in mixture preparation has become especially critical to ensuring successful combustion and low emissions. The pervasive type of charge motion used in SI engines is tumble, which typically interacts with the bulk of the injector spray. Tumble requires certain length scales and tends to degrade as the piston nears top-dead center (TDC) [20-22], though this effect is highly dependent on combustion chamber geometry, especially compression ratio and stroke-to-bore ratio. It devolves into a general non-ordered turbulent kinetic energy (TKE) with high velocity but no uniform flow field. As such tumble motion tends to not contribute strongly to combustion in and of itself, but high levels of TKE present during the combustion process can increase turbulent flame speed, thereby increasing combustion burn rate. This effect is particularly useful for lean engines, as it helps compensate for the reduction in laminar flame speed inherent in the colder lean combustion environment. High levels of turbulence can, however, have the detrimental effect of stretching the spark kernel, resulting in misfires, and also increase in-cylinder heat loss.

Swirl motion is generally not purposefully used in production SI engines as it provides little mixture preparation benefit. It does not degrade near TDC to nearly the same extent as tumble and therefore it is a potentially useful form of charge motion for lean combustion concepts as it exists during the combustion process. Literature [23-25] and previous simulations performed by MAHLE Powertrain have shown contradictory effects of swirl on lean combustion.

Quader (et al) demonstrated that charge motion has a competing influence on kernel formation and flame front propagation in homogeneous lean combustion SI engines [3]. High levels of charge motion, regardless of type, can have the effect of stretching the flame kernel resulting in misfires. Contrarily, high levels of charge motion prove beneficial to increasing flame speed as the flame slowly consumes the lean charge. Stratified lean combustion with targeted mixture preparation to ensure an Ignitable mixture near the spark plug is one potential solution that has been proposed to mitigate the kernel formation challenge of high tumble dilute engines [26,27]. Alternatively, pre-chamber concepts have the potential to effectively separate and quarantine the spark plug from the majority of the main combustion chamber flow field. This could potentially lead to high levels of TKE in the main chamber being beneficial to reducing burn duration and increasing enleanment while reducing the risk of kernel stretching.

### 1.2 Objective

The objective of this study is to understand the impact of charge motion level and type on jet ignition combustion performance comprehensively throughout the cycle and to quantify the thermal efficiency potential of optimized charge motion in a jet ignition engine.

## 2 APPROACH

A multi-cylinder jet ignition engine is used for this study, with an intake system configurable for a range of charge motion types and levels. Data is taken at a part-load non-knock limited condition, and results are evaluated across a sweep of λ at constant speed and load.

## 3 EXPERIMENT

The MAHLE DI3 Downsizing demonstrator engine is selected as the basis of the dedicated Jet Ignition engine due to the authors' familiarity with the engine, access to the design and underlying analyses, and manufacturer agnostic nature of the platform. The engine is repurposed here for a boosted ultra-lean application, with a target peak BMEP of approximately 15 bar. Development of the DI3 engine is well documented [28,29]. The development of the jet ignition variant of this engine was previously published [30]. Table 1 lists the specifications of the MJI DI3 engine. The MJI DI3 is depicted in Figure 2.

**Table 1. Engine specifications.**

| | |
|---|---|
| Configuration | In-line 3 cylinder |
| Displaced volume | 1500 cm$^3$ |
| Stroke; Bore | 92.4 mm; 83 mm |
| Compression Ratio | 15:1 for this study |
| Injection | PFI main chamber, DI pre-chamber |
| Fuel | Pump Grade Premium Gasoline |
| Boost System | Variable-geometry turbocharger |

**Figure 2. MJI DI3 engine.**

An Air Flow Rig was used to evaluate tumble ratio and swirl number of several charge motion variants of the engine. The Air Flow Rig forces air to flow through the cylinder head, while valve lift is adjusted statically in 1 mm increments. Tumble ratio and swirl number are calculated by integrating the area under the resulting non-dimensional tumble vs. lift and swirl vs. lift curves, respectively.

Four charge motion cases were evaluated: baseline, increased tumble, introduction of swirl, and a combination of swirl and tumble (denoted as "swumble" in subsequent sections). Charge motion differences from the baseline were induced through the use of plate inserts into each of the intake ports (Figures 3 and 4). The baseline configuration used no inserts and represents a moderate tumble engine consistent with tumble levels in modern DI SI engines (a tumble ratio of approximately 3). For the tumble variant, a plate insert was used that directed flow to exit past the valve in a more severe tumble motion. For the swirl variant, a splitter plate was used with a slight incline across the diameter of the port to induce swirl. The swumble variant used the swirl plate and tumble plate in series. The relative change in tumble ratio and swirl number with respect to the baseline (no inserts) port are listed in Table 2.

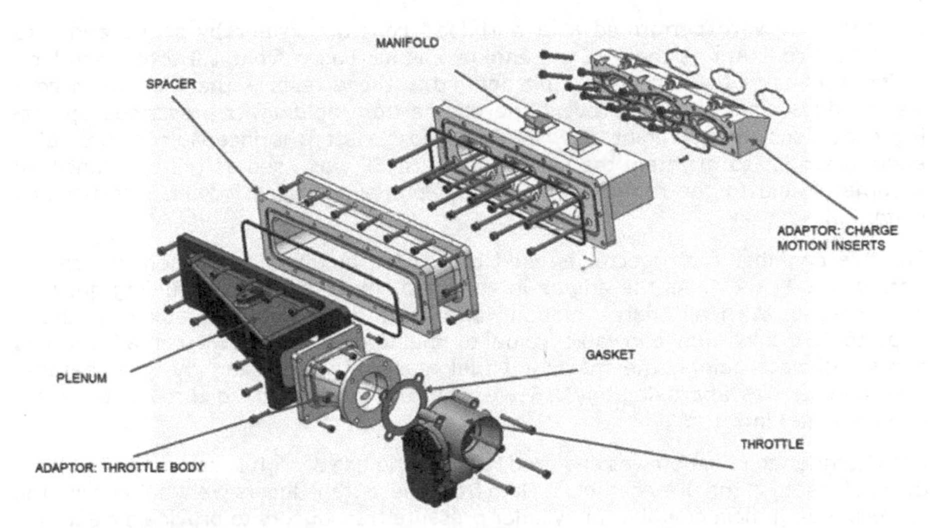


**Figure 3. MJI DI3 engine air intake system – exploded view.**

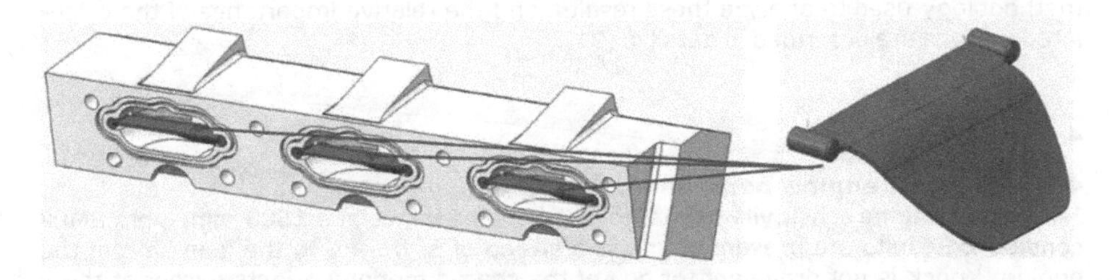

**Figure 4. Charge motion insert adaptor with tumble insert installed.**

**Table 2. Relative tumble ratio and swirl number of charge motion configurations evaluated in this study.**

| Configuration | Relative Tumble Ratio | Relative Swirl Number |
|---|---|---|
| Baseline | - | - |
| Tumble | +13% | -25% |
| Swirl | -39% | +1075% |
| Swirl+Tumble | +13% | +75% |

Sweeps of λ were performed at a part load condition, whereby speed and load were held constant as the λ of the engine was increased from 1.0 to its lean limit in increments of 0.1. The lean limit defined in these tests is the λ at which consistent detectable misfires prevent the engine from holding its proscribed operating conditions, or the point at which the boost system is incapable of providing enough airflow to maintain the desired load. BMEP was used as the constant load parameter due to the reduced influence of pumping losses for this non-boosted condition.

The pre-chamber fuel injector is used to provide auxiliary fuel when the engine achieves a λ = 1.4. As the engine is enleaned, the pre-chamber fueling quantity is increased. With all charge motion variants, the pre-chamber auxiliary fuel is kept to the minimum allowable value to maintain COV < 3%. For all variants across all data points, the maximum fuel mass injected using the pre-chamber fuel injector was approximately 1.5% of the fuel mass injected through the main chamber fuel injector.

Pre-chamber combustion was analyzed through the use of high speed pressure transducers located in the pre-chambers. Data from these transducers were paired with the corresponding main chamber in-cylinder pressure transducers to provide a clear perspective on intra-chamber pressure-based behavior. For this study, high speed pre-chamber and main chamber results from one of the three cylinders are presented in order to avoid the use of corrections for minor cylinder-to-cylinder differences. Each presented high speed data point represents an analysis of 300 consecutive cycles. The methodology used to analyze these results, and the relative importance of the calculated metrics are described in detail in [31].

## 4 RESULTS

### 4.1 Overall engine performance

Jet ignition engine sensitivity to charge motion is examined at a 1500 rpm, 6 bar BMEP condition. Results are presented across a sweep of λ, from 1 to the lean limit of the engine. Knock is not prevalent for any of the charge motion variants except at the λ values closest to 1. Figure 5 shows the two relevant combustion stability metrics, COV and LNV. The acceptable limits are depicted by the red dashed lines.

In Figure 5, it is evident that the tumble variant maintains acceptable stability throughout the range of λ from 1.0 to 2.0, without any partial burn events. The baseline variant performs similarly but with increased instability from λ 1.5 and a stability limit at λ 1.9. There is also more pronounced deterioration in LNV in this lean λ range. The swirl and swumble variants perform measurably poorer, with stability limits reached between λ 1.3 and 1.6.

In this study, the process for determining pre-chamber injection quantity for the baseline variant involved increasing injection pulsewidth to allow the minimum quantity of fuel required to maintain main chamber COV ≤ 3%. This combustion stability requirement was used as the primary criterion for determining pre-chamber injection quantity, with adherence to the baseline variant's fuel flow-λ relationship as the secondary criterion. As can be witnessed in Figure 5, this resulted in relatively consistent quantity-λ relationships amongst the baseline, tumble, and swumble charge motion variants at this speed-load condition. Some minor discrepancies exist between pulsewidth used and resulting pre-chamber fuel mass flow amongst these variants. This is likely due to minor differences in background pressure at time of injection. The swirl variant required approximately twice the pre-chamber fuel quantity that the other variants required. The swirl variant also required pre-chamber auxiliary fuel injection to begin at an earlier λ value in the sweep (beginning at 1.2 versus 1.4 for the other variants). Despite this increased quantity requirement, the swirl variant consistently demonstrated inferior combustion stability behavior to the other variants across the full sweep of λ values. This indicates both a high level of instability in the system induced by swirl motion and that the instability originates in mixture preparation in the pre-chamber, or at least it cannot be adequately mitigated through the traditional means of increasing pre-chamber fuel injection quantity. The poor stability results of the swirl charge motion variant propagate to other metrics as well, including burn durations and efficiency as will be demonstrated.

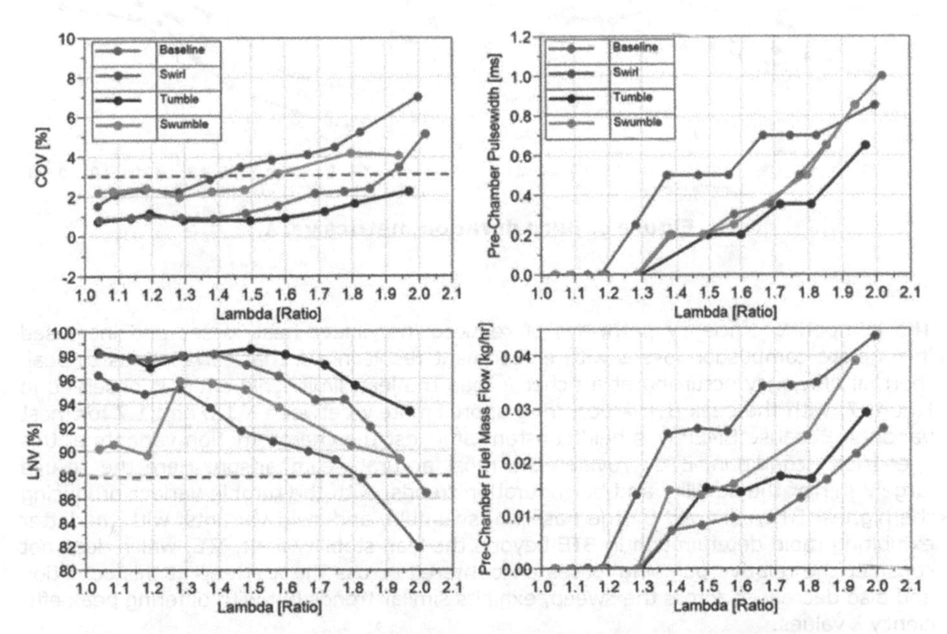


**Figure 5. Pre-chamber combustion stability (left) and fuel injection parameters (right) vs. λ.**

Figure 6 shows the CA50 for the charge motion variants, confirming that light knock may be present near λ 1 but is absent for all variants from λ 1.2. The instability in CA50 in the near-lean region (λ = 1.0-1.3) is due to cylinder-to-cylinder variation that manifests under lean conditions but is mitigated by the addition of pre-chamber auxiliary fuel starting at λ = 1.4 for most variants. An examination of the burn duration segments shows that the two variants that include increased tumble motion (tumble and swumble variants) produce faster overall combustion duration. The difference in burn duration amongst the variants becomes prominent under lean conditions, with minimal separation at λ 1.

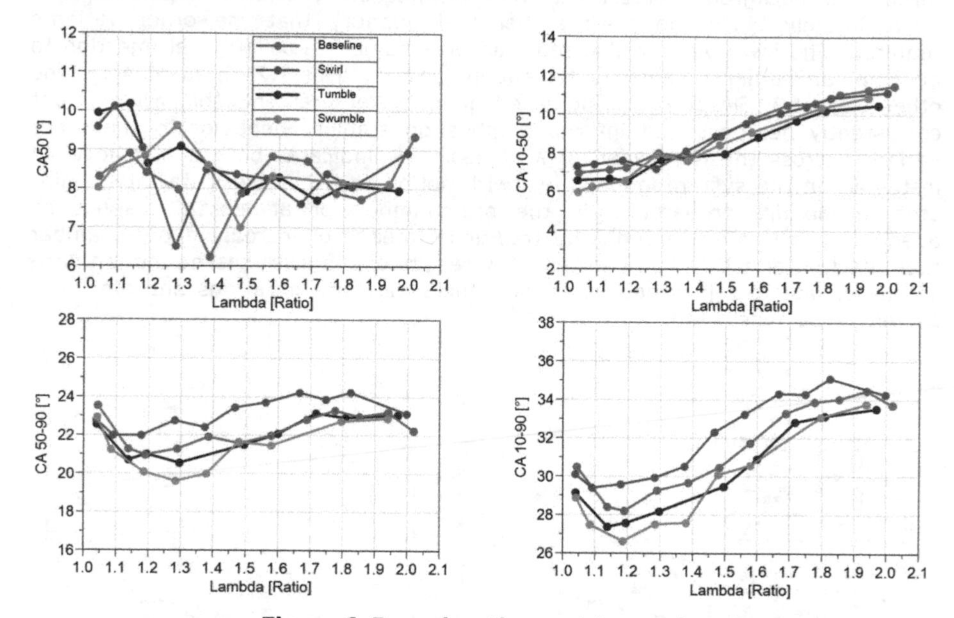


**Figure 6. Burn duration metrics vs. λ.**

The competing efficiency pathways of reduced in-cylinder heat losses and increased incomplete combustion losses with enleanment result in a λ that corresponds to peak thermal efficiency occurring at a richer λ than the lean limit. This effect is observed in Figure 7, with the peak BTE λ occurring approximately between λ 1.6 and 1.7 for most variants. Because BMEP was held constant amongst the charge motion variants at this speed/load condition, BTE provides the most accurate comparison. Here the results largely mirror the stability and burn duration trends, with the tumble variant producing the highest BTE, followed by the baseline, swumble, and swirl variants, with the latter exhibiting rapid deterioration in BTE beyond the lean stability limit. ITE, which does not consider the relative pumping losses encountered across the λ sweep at this condition and also decreases across the sweep, exhibits similar trends but with differing peak efficiency λ values.

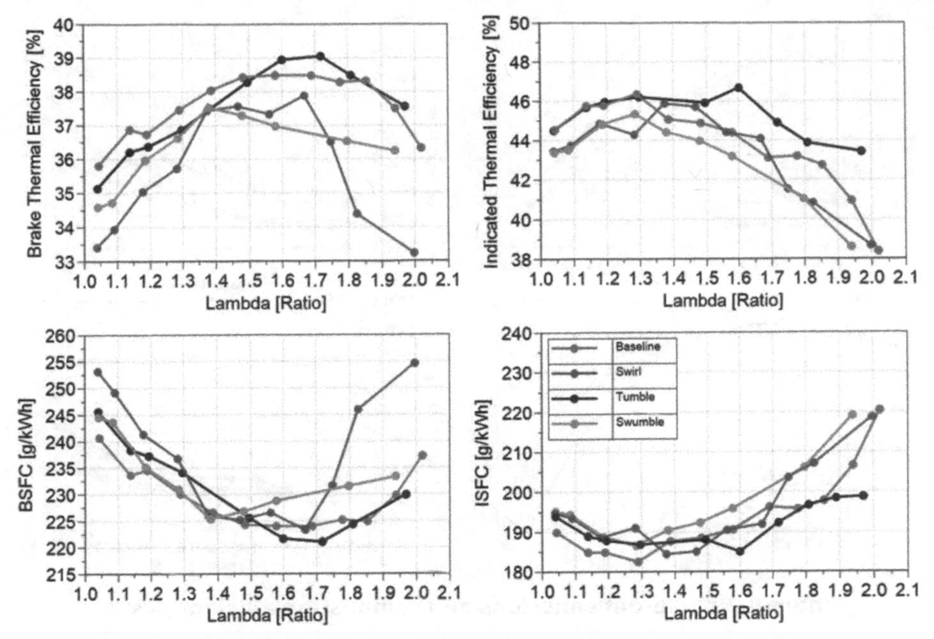


**Figure 7. Efficiency and fuel consumption metrics vs. λ.**

Analysis of $NO_x$ emissions trends versus λ in Figure 8 show relative parity amongst the charge motion variants from λ = 1-1.6. The erratic trends in the range beyond λ = 1.6 do not appear to mirror any other major parameter's trend, and are likely the result of increasingly unstable combustion in this region, particularly in the swirl and swumble variant data. Therefore, it does not appear that charge motion has any noticeable impact on $NO_x$ formation at this condition. However, the comparison of Figures 5 and 8 demonstrates the benefit of enhanced combustion stability in the ultra-lean region, namely the ability to further reduce $NO_x$ emissions by operating at stably leaner λ values.

An examination of the CO emissions trend shows a significant drop in emissions from λ 1 to the near-lean region. CO then slowly increases with further enleanment. THC emissions decrease slightly in the near-lean region but then increase with enleanment, severely in the case of poor stability variants such as swirl.

The CO and THC results translate well to the combustion efficiency trend. While combustion efficiency reduces with increasing enleanment, the swirl variant produces depressed combustion efficiency versus the other charge motion variants across the λ range starting from λ 1.2. With late burning performance having a prominent impact on combustion efficiency, this swirl variant performance is expected. Conversely, the tumble variant produces the highest relative combustion efficiencies under lean conditions. Note that the combustion efficiencies depicted in Figure 8, especially under lean conditions, are lower than would be expected for this type of combustion system. This is due to the relatively high CR for an SI engine coupled with the homogeneous mixture leading to a relatively greater crevice volume fuel percentage of total fuel than would be found in production engines. Also note that the piston and ring combination used for this study are not production-intent and are not based on any existing production designs, and are therefore not optimized for the purposes of this combustion system.

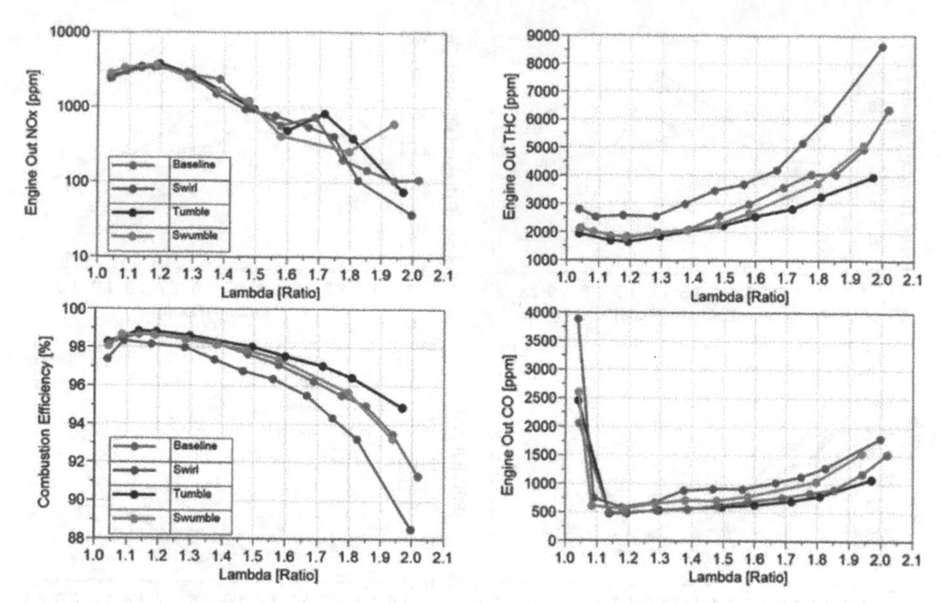


**Figure 8. Engine-out emissions and combustion efficiency vs. λ.**

The results presented so far demonstrate clearly superior performance when additional tumble is introduced in the engine, and clearly inferior performance when swirl is introduced. While there is parity in the results at stoichiometric conditions, the relative difference amongst the variants grows as the engine is enleaned. The engine is therefore most sensitive to charge motion under lean conditions as the engine approaches its stability limit. While these trends grow in prominence with enleanment, they remain relatively consistent across the full λ range. The following analysis highlights peak performance of the variants within the context of the λ sweeps at this condition. These peak values can be considered as starting points for engine calibration, both demonstrating approximately how an engine with each specific level and type of charge motion might be operated, and indicating the relative robustness of an engine calibration that would be based on these results.

In determining optimal conditions for a combustion system capable of significant dilution tolerance, it is important to consider the λ at which the engine would operate. For the purposes of this study, a λ corresponding to peak BTE within the λ sweep was assumed to be a primary input for determining nominal λ in an engine calibration. Figure 9 shows the peak BTE for each of the four charge motion variants at this part load condition. Note that for the swirl variant, peak BTE occurred in a λ region where COV exceeded the 3% limit. Consistent with the data presented previously, tumble and baseline variants exhibited superior BTE to the swirl and swumble variants. Figure 9 also shows the λ at which the peak BTE values occur. The baseline and tumble variants have peak BTE λ values of 1.7. Peak BTE for the swumble variant occurs at a much richer λ of 1.4. Again, the swirl results are unstable.

Leaner peak BTE λ values can be beneficial from an emissions control perspective. Engine-out $NO_x$ concentration decreases significantly at λ values leaner than approximately 1.2. A peak BTE occurring 0.1 λ leaner can mean a reduction of several hundred ppm of $NO_x$. Figure 9 shows engine-out $NO_x$ levels at the peak BTE λ values for each of the four charge motion variants. The swumble variant, with a peak BTE λ occurring 0.3 λ richer than the tumble and baseline variants, has a $NO_x$ level approximately double those of the other variants. Despite the higher exhaust temperature associated with the richer peak BTE λ, the higher engine-out $NO_x$ level likely puts a strain on the lean $NO_x$ storage catalyst.

Finally, Figure 9 shows the engine-out $NO_x$ levels at the lean stability limits. The baseline and tumble variants both have $NO_x$ levels below 100 ppm, an approximately 85% reduction from even the levels at the peak BTE λ values for the respective variants. Also notable is the wide gulf in $NO_x$ levels between the swirl and swumble variants, with lean stability limits only 0.15 λ apart. This 1000 ppm gulf illustrates the extreme sensitivity of $NO_x$ to λ in this region, slightly lean of the near-lean region. From this analysis it is evident that charge motion variants with wide dilution tolerance that have both peak BTE and lean stability limit occur at λ values well into the ultra-lean region offer significant advantages in terms of engine performance and emissions but also degree of calibration flexibility. Such combustion systems allow robust high efficiency operation with headroom to allow for precise targeting of emissions profiles, or flexibility to accept a certain amount of λ uncertainty during transient operation while still maintaining acceptable engine-out emissions. The tumble and baseline charge motion variants have the highest BTE at the leanest λ values with superior lean limit extension to the swirl and swumble variants.

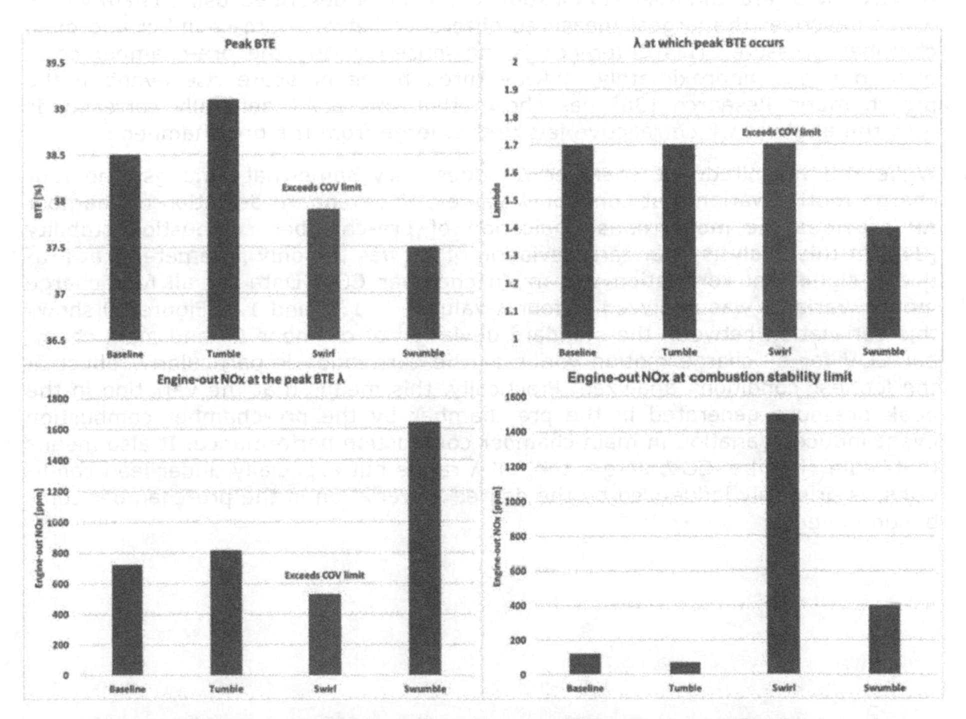


**Figure 9. BTE, peak BTE λ, and engine-out $NO_x$ at the peak BTE λ and combustion stability limit within a λ sweep from 1 to the lean limit.**

## 4.2 Pre-chamber combustion

The late burning and combustion efficiency results indicate that the differing levels and types of charge motion investigated in this study have an influence on main chamber combustion, especially in the late burning period. However, these results may be both a cause and a symptom of the differing combustion stability behavior amongst the variants. Charge motion plays a role not just in late main chamber burning processes but also in front-end processes such as pre-chamber combustion. The analysis presented in this section will seek to explain main chamber combustion stability differences as functions of pre-chamber combustion stability differences. Demonstrating a correlation between pre-chamber and main chamber combustion stability would confirm that 1) charge motion induced in the intake and developed in the main chamber impacts in-pre-chamber processes, and 2) that main chamber combustion performance is determined in large part by the pre-chamber combustion event, and main chamber COV is dictated by the stability of pre-chamber combustion. Pre-chamber combustion stability can be manipulated via active pre-chamber fuel injection strategy, which could potentially be used to correct for negative charge motion influence in the pre-chamber mixing or combustion processes.

Analysis was performed on the 300-cycle average pre-chamber high speed pressure trace. Examination of pre-chamber behavior is confined to the portion of the pre-chamber pressure trace where pre-chamber pressure rises measurably higher than main chamber pressure. This portion of the trace corresponds with the pre-chamber combustion event and expulsion of combustion products as reactive jets. Pre-chamber combustion behavior is described using chamber $\Delta P$ which describes the largest measured difference between pre-chamber and main chamber pressure. This difference is maximized during the pre-chamber combustion event, approximately midway through the pressure rise event in the pre-chamber. Research [30] has shown that this point generally corresponds with the angle at which reactive jets first emerge from the pre-chamber.

While the magnitude of chamber $\Delta P$ does vary somewhat amongst the four charge motion variants at common $\lambda$ values, the standard deviation of chamber $\Delta P$ provides the most robust indication of pre-chamber combustion stability [31]. In this analysis, standard deviation of $\Delta P$ was the only parameter that produced significant correlation with main chamber COV. Data for all four charge motion variants was analyzed at four $\lambda$ values: 1, 1.4, and 1.8. Figure 10 shows this correlation between the standard deviation of chamber $\Delta P$ and main chamber COV for all charge motion variants. The correlation is particularly robust at the leanest conditions analyzed. Practically, this means that the variation in the peak pressure generated in the pre-chamber by the pre-chamber combustion event induces variation in main chamber combustion performance. It also means that main chamber COV, across the full $\lambda$ range but especially under lean conditions, is primarily influenced by the degree of variation in the pre-chamber combustion event.

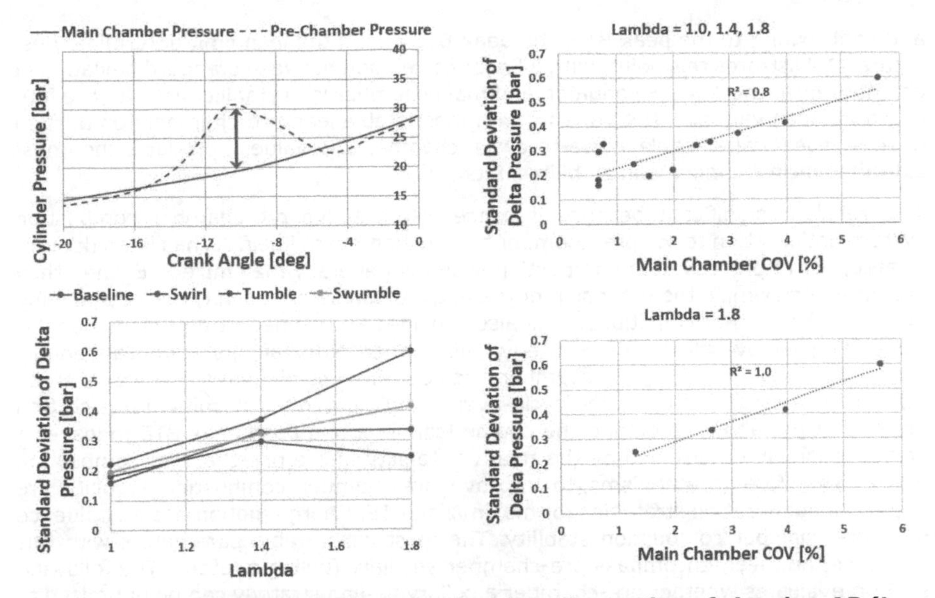


**Figure 10. Pre-chamber combustion metrics: illustration of chamber ΔP (top left); standard deviation of ΔP vs. λ (bottom left); standard deviation of ΔP vs. main chamber COV for λ = 1.0-1.8 (top right); standard deviation of ΔP vs. main chamber COV for λ = 1.8 (bottom right).**

Figure 10 shows the difference in standard deviation of chamber ΔP amongst the four charge motion variants. Notably, these results mirror both the main chamber COV and combustion efficiency trends discussed previously, with parity at the λ = 1 condition and an ever increasing disparity as the engine is enleaned. At the leanest condition considered in this dataset, λ = 1.8, the tumble variant shows the least variation in chamber ΔP and therefore the lowest main chamber COV, followed closely by the baseline variant. The swumble and swirl variants exhibited the highest degree of variation in chamber delta pressure.

The tumble variant actually displays more pre-chamber combustion stability (reduced standard deviation of ΔP) at the λ = 1.8 condition versus the λ = 1.4 condition. This is likely due to differences in pre-chamber λ at the two conditions. At the λ = 1.4 condition, auxiliary fuel is first introduced into the pre-chamber due to engine COV requirements. Due to the minimum pulsewidth limitation of the pre-chamber fuel injectors, more fuel than desired is injected into the pre-chamber, creating a richer than desired pre-chamber. Further enleanment brings a greater degree of controllability over pre-chamber λ, as background λ becomes leaner and the auxiliary fuel injection pulsewidth must increase above its minimum in order to compensate. This means that pre-chamber λ at the 1.8 condition is more optimized (and more optimizable) than it is at the 1.4 condition, helping to explain why pre-chamber stability increases for the tumble variant. This effect is equally present in the other charge motion variants, so the worsening instability in the other charge motion variants, particularly swirl and swumble, must be driven by other factors specific to the charge motion types and levels introduced, such as in-pre-chamber mixing dynamics.

The relative stability of the pre-chamber combustion event influences the main chamber COV to a high degree, thereby impacting combustion efficiency at lean conditions and contributing to the peak BTE, the peak BTE λ, and the lean limit determinations. Figure 11 illustrates this point, with a linear correlation between standard deviation of chamber ΔP at the λ = 1.8 condition and main chamber lean stability limit for the four charge motion variants. The variants with most stable lean pre-chamber combustion events, specifically stable cycle-to-cycle chamber ΔP values, produce the most extended main chamber lean stability limits.

This result is significant because it concentrates active pre-chamber combustion system optimization to the pre-chamber combustion event itself. As has been demonstrated, active pre-chamber combustion systems generally maximize BTE when they are able to maximize the extension of the engine lean stability limit, thereby pushing peak BTE λ to leaner conditions. This also provides an engine-out emissions benefit. Depending on overall engine strategy, this leaner nominal operation can enable a higher CR than in active pre-chamber engines with less dilution tolerance, thereby further increasing both peak and cycle-average BTE potential. Results in this section prove that this ability to extend the engine lean limit and push peak BTE to leaner λ values is largely determined by the relative stability of the pre-chamber combustion event. Therefore, mechanisms to improve pre-chamber combustion stability are impactful optimization strategies for maximizing BTE. Charge motion has an influence over pre-chamber combustion stability. The most calibratable parameters with this influence, however, encompass pre-chamber auxiliary fueling strategy. The following section evaluates whether pre-chamber auxiliary fueling strategy can be optimized to compensate for negative influence some of the charge motion variants have on pre-chamber combustion stability.

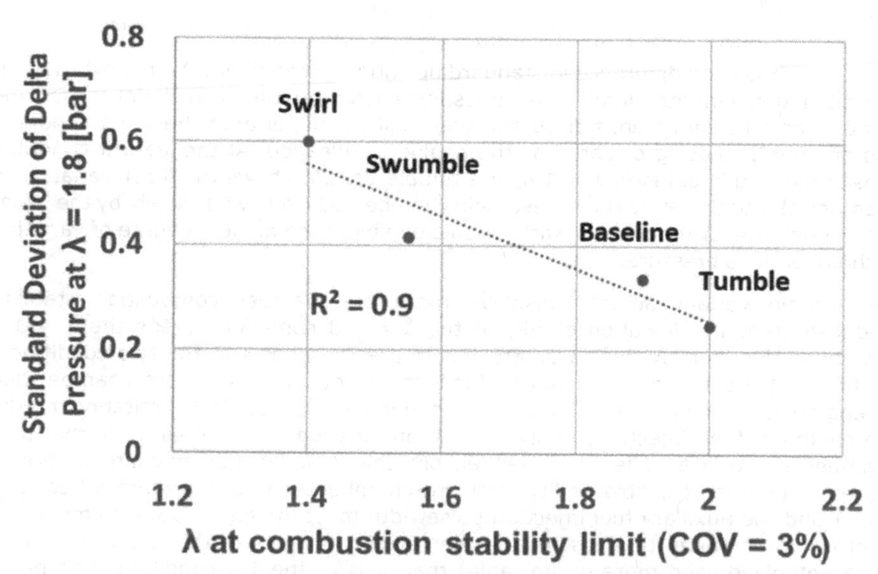


**Figure 11. Main chamber lean stability limit as a function of standard deviation of chamber ΔP at λ = 1.8.**

### 4.3 Pre-chamber fueling parameters

The DI fuel injector used in the MJI pre-chamber enables scalable fuel injection with precise control over fuel quantity. Auxiliary fueling quantity increases proportionally with main chamber enleanment across a λ sweep. Similarly, pre-chamber fueling quantity is a primary lever that is commonly used to reduce main chamber COV under lean conditions. Figure 12 shows main chamber COV response to changes in pre-chamber fuel injection quantity and start of injection angle for the four charge motion variants at the λ = 1.7 condition. At retarded start of injection angles in close proximity to spark timing, combustion becomes highly unstable for most of the charge motion variants. At these angles the pre-chamber injection event ends after the spark has occurred, leading to a high probability of insufficient fuel in proximity to the spark plug at time of spark. Other than at these extreme late injection angles, however, start of injection timing does not have a prominent influence over main chamber COV.

Auxiliary fuel quantity has a stronger influence over main chamber COV than does injection timing. All charge motion variants display lower COVs as pre-chamber fuel mass flow increases, as is expected. However, for both the swirl and swumble variants, the enhanced stability induced by increased pre-chamber fuel quantity is still above the COV limit of 3%. In fact, these variants display slightly lower sensitivity to auxiliary fuel quantity than do the tumble and baseline variants.

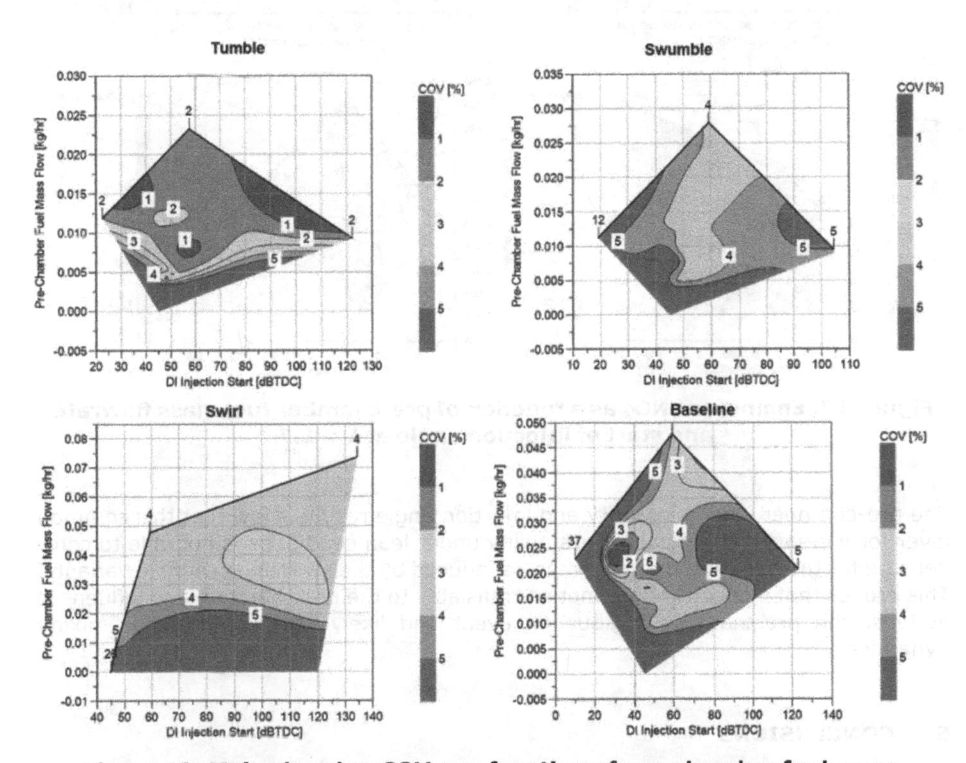


**Figure 12. Main chamber COV as a function of pre-chamber fuel mass flowrate and start of injection angle at λ = 1.7.**

Figure 13 shows the change in engine-out $NO_x$ emissions across the auxiliary fuel quantity and injection timing sweeps. Across the relatively narrow injection quantity band considered in this analysis, engine-out $NO_x$ changes significantly, by at least 500 ppm in each variant. This indicates that changes in pre-chamber λ are occurring across the sweep, with increased quantity producing richer pre-chambers, resulting in less engine-out $NO_x$. The relative magnitude of the $NO_x$ values and their sensitivity to fueling quantity also strongly indicates that the majority of $NO_x$ is being formed in the pre-chamber by the pre-chamber combustion event, a common result for active pre-chambers under ultra-lean conditions.

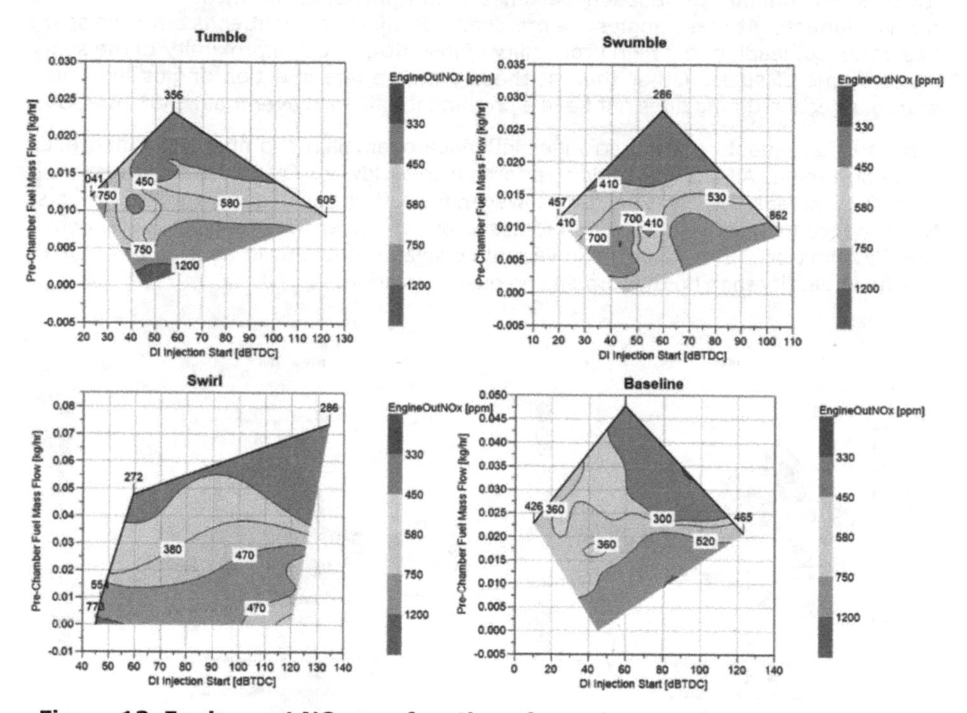


**Figure 13. Engine-out $NO_x$ as a function of pre-chamber fuel mass flowrate and start of injection angle at $\lambda = 1.7$.**

The pre-chamber fueling quantity and injection angle results show that this common lever for increasing main chamber stability under lean conditions is not able to compensate for the negative stability pressure induced by the swirl and swumble variants. This proves that induced charge motion translates to the pre-chamber, and influences at least the pre-chamber combustion event and likely also pre-chamber mixing dynamics.

## 5 CONCLUSIONS

The influence of charge motion becomes increasingly prominent under low ignitability conditions such as heavy dilution. The added tumble motion of the tumble variant used in this study produced superior lean limit extension, faster burn durations, and higher combustion efficiency in the ultra-lean region than the remaining charge motion variants. The ability to extend the lean limit also shifted the $\lambda$ corresponding to the peak BTE leaner. In most cases this meant that the peak BTE was higher than with the charge motion variants whose peaks were at less lean $\lambda$ values. A moderate increase in engine tumble produced a 0.5 percentage point increase in BTE at common CR. Shifting nominal operating $\lambda$ leaner enables the use of a higher CR in the engine, depending on operating strategy. Fully exploiting this benefit is beyond the scope of this study, but it does provide an opportunity to further increase BTE. Shifting the nominal operating λ leaner also reduced engine-out $NO_x$, providing a potential opportunity to reduce the cost and scope of the lean aftertreatment solution. While the addition of charge motion produced measurably increased in-cylinder heat losses, these were overcompensated by the reduced incomplete combustion losses with the tumble variant, proving that charge motion addition can have both a positive and negative influence over main chamber combustion.

Lean limit extension in the main chamber was correlated to a metric for combustion stability in the pre-chamber during the pre-chamber combustion event. The stark differences in this stability amongst the charge motion variants proved that charge motion not only influences main chamber combustion directly, but also indirectly via influence over the pre-chamber combustion event. This charge motion influence is prominent enough that the variants that incorporated swirl could not overcome the deleterious effect on combustion stability using increased auxiliary fueling quantity.

The results of this study demonstrate a complex and comprehensive influence of charge motion on ultra-lean jet ignition engine operation. This influence can be leveraged, as it was partially in this study, to maximize the thermal and combustion efficiencies of the engine under ultra-lean conditions and to increase the robustness of the engine calibration in this region. Future work includes coupling the experimental results with simulation to understand the influence of charge motion over in-pre-chamber mixture dynamics in more detail.

## REFERENCES


[1] Bunce, M. and Blaxill, H., "Sub-200 g/kWh BSFC on a Light Duty Gasoline Engine," SAE Technical Paper 2016-01-0709, 2016.

[2] Bunce, M., Blaxill, H., Kulatilaka, W., and Jiang, N., "The Effects of Turbulent Jet Characteristics on Engine Performance Using a Pre-Chamber Combustor," SAE Technical Paper 2014-01-1195, 2014.

[3] Quader, A. A., "Lean Combustion and the Misfire Limit," SAE Technical Paper 741055, 1974.

[4] Husted, H., Piock, W., Ramsay, G., "Fuel Efficiency Improvements from Lean Stratified Combustion with a Solenoid Injector," SAE Technical Paper 2009-01-1485, 2009.

[5] Germane, G., Wood, C., Hess, C., "Lean Combustion in Spark-Ignited Internal Combustion Engines – A Review," SAE Technical Paper 831694, 1983.

[6] Heywood, J., Internal Combustion Engine Fundamentals, McGraw-Hill, 1988.

[7] Yamamoto, H., "Investigation on Relationship Between Thermal Efficiency and $NO_x$ Formation in Ultra-Lean Combustion," SAE Journal Paper JSAE 9938083, 1999.

[8] Dober, G. G., Watson, H. C., "Quasi-Dimensional and CFD Modelling of Turbulent and Chemical Flame Enhancement in an Ultra Lean Burn S.I. Engine," Modeling of SI Engines SP-1511, 2000.

[9] Ward, M., "High-Energy Spark-Flow Coupling in an IC Engine for Ultra-Lean and High EGR Mixtures," SAE Technical Paper 2001-01-0548, 2001.

[10] Qiao, A., Wu, X., "Research on the New Ignition Control System of Lean- and Fast-Burn SI Engines," SAE Technical Paper 2008-01-1721, 2008.

[11] Ricardo, H. R., Recent Work on the Internal Combustion Engine, SAE Transactions, Vol 17, May 1922.

[12] Gussak, L. A., Karpov, V. P., Tikhonov, Y. Y., "The Application of the Lag-Process in Pre-chamber Engines," SAE Technical Paper 790692, 1979, doi:10.4271/790692.

[13] Robinet, C., Higelin, P., Moreau, B., Pajot, O., Andrzejewski, J., "A New Firing Concept for Internal Combustion Engines: "l'APIR"," SAE Technical Paper 1999-01-0621, 1999.
[14] Murase, E., Ono, S., Hanada, K., Oppenheim, A., "Pulsed Combustion Jet Ignition in Lean Mixtures," SAE Technical Paper 943048, 1994.
[15] Toulson, E., Schock, H., Attard, W., "A Review of Pre-Chamber Initiated Jet Ignition Combustion Systems," SAE Technical Paper 2010-01-2263, 2010, doi:10.4271/2010-01-2263.
[16] Attard, W., Toulson, E., Fraser, E., Parsons, P., "A Turbulent Jet Ignition Pre-Chamber Combustion System for Large Fuel Economy Improvements in a Modern Vehicle Powertrain," SAE Technical Paper 2010-01-1457, 2010, doi:10.4271/2010-01-1457.
[17] Cao, Y., Li, L. "A novel closed loop control based on ionization current in combustion cycle at cold start in a gdi engine," SAE Technical Paper 2012-01-1339, 2012. doi:10.4271/2012-01-1339.
[18] Sens, M., Binder, E., Reinicke, P.-B., Riess, M., Stappenbeck, T., Woebke, M., "Pre-Chamber Ignition and Promising Complementary Technologies," 27th Aachen Colloquium Automobile and Engine Technology, 2018.
[19] Attard, W., Kohn, J., Parsons, P., "Ignition Energy Development for a Spark Initiated Combustion System Capable of High Load, High Efficiency and Near Zero $NO_x$ Emissions," SAE Journal Paper JSAE 20109088, 2010.
[20] Qi, Y., Ge, X., and Dong, L., "Numerical Simulation and Experimental Verification of Gasoline Intake Port Design," SAE Technical Paper 2015-01-0379, 2015, doi:10.4271/2015-01-0379.
[21] Bozza, F., De Bellis, V., Berni, F., D'Adamo, A., Maresca, L., "Refinement of a 0D Turbulence Model to Predict Tumble and Turbulent Intensity in SI Engines. Part I: 3D Analyses," SAE Technical Paper 2018-01-0850, 2018, doi:10.4271/2018-01-0850.
[22] Ruhland, H., Lorenz, T., Dunstheimer, J., Breuer, A., Khosravi, M., "A Study on Charge Motion Requirements for a Class-Leading GTDI Engine," SAE Technical Paper 2017-24-0065, 2017, doi:10.4271/2017-24-0065.
[23] Loeper, P., Ra, Y., Foster, D., Ghandhi, J., "Experimental and Computational Assessment of Inlet Swirl Effects on a Gasoline Compression Ignition (GCI) Light-Duty Diesel Engine," SAE Technical Paper 2014-01-1299, 2014.
[24] Patrie, M., Martin, J., Engman, T., "Inlet Port Geometry and Flame Position, Flame Stability, and Emissions in an SI Homogeneous Charge Engine," SAE Technical Paper 982056, 1998.
[25] Hill, P., Zhang, D., "The Effects of Swirl and Tumble on Combustion in Spark-Ignition Engines," Prog. Energy Combust. Sci. Vol 20, 1994, Pgs. 373–429.
[26] Urushihara, T., Nakada, T., Kakuhou, A., Takagi, Y., "Effects of Swirl/Tumble Motion on In-Cylinder Mixture Formation in a Lean-Burn Engine," SAE Technical Paper 961994, 1996.
[27] Solomon, A., Szekely, G., "Combustion Characteristics of a Reverse-Tumble Wall-Controlled Direct-Injection Stratified-Charge Engine," SAE Technical Paper 2003-01-0543, 2003.
[28] Bassett, M., Hall, J., Cains, T., Underwood, M. et al., "Dynamic Downsizing Gasoline Demonstrator," SAE Int. J. Engines 10(3):2017.
[29] Bassett, M., Hall, J., Hibberd, B., Borman, S. et al., "Heavily Downsized Gasoline Demonstrator," SAE Int. J. Engines 9(2):729–738, 2016.
[30] Bunce, M. and Blaxill, H., "Methodology for Combustion Analysis of a Spark Ignition Engine Incorporating a Pre-Chamber Combustor," SAE Technical Paper 2014-01-2603, 2014, doi: 10.4271/2014-01-2603.
[31] Bunce, M., Peters, N., Subramanyam, S. K. P., Blaxill, H., "Assessing the Low Load Challenge for Jet Ignition Engine Operation," *Proceedings of the Institute of Mechanical Engineers Internal Combustion Engines Conference*, 2019.

*International Conference on Powertrain Systems for Net-Zero Transport*
*Institution of Mechanical Engineers, ISBN 978-1-032-11281-7*

# Experimental study of intelligent valve actuation for high efficiency spark ignition engines

**A. Minasyan, H. Zhao**

Centre for Advanced Powertrain and Fuels, College of Engineering, Design and Physical Sciences, Brunel University London, UK

**ABSTRACT**

Engine downsizing has been shown as an effective means to reduce the vehicle's fuel consumption but the full potential of engine downsizing is limited by the knocking combustion at boosted operations and the presence of pumping loss at part load conditions. In this work, an electro-mechanical valvetrain system named iVT (intelligent Valve Technology) by Camcon was used to investigate how the independently controlled variable valve timing and duration can be applied to minimise the knocking combustion by altering the effective compression ratio (ECR) at high load via Early Intake Valve Closure (EIVC) or Late Intake Valve Closure (LIVC), as well as reducing the pumping loss at part load. In particular, the effect of different valve lifts with fixed valve timings and constant duration was studied on the pumping loss, combustion process and emissions. The results show that fuel consumption was reduced up to 2.5% using iVT system compared to the baseline valve profile at 9bar net IMEP.

## 1 INTRODUCTION

IC engines have been the main power plants for various transport on land and sea. World-wide daily use of vehicles with IC engines in the 20th century has led to the very stringent legislation on their pollutant emissions over the last few decades. Moreover European Parliament and the Council set regulation for the maximum value of manufacturer's fleet average $CO_2$ emission level, targeting to 95g/km from 2020 (1) in order to combat the global warming caused by increasing $CO_2$ concentration. If the average value exceeds the limit, the manufacturer has to pay monetary penalty for each registered car (2). Another main issue associated with the use of IC engines is increasing fossil fuel consumption, which can lead to resource depletion as the amount of vehicles increases dramatically.

To fulfil the above requirements, automotive industry have been developing new technologies to improve the efficiency of modern IC engines. Downsizing is one of the successful methods of reducing fuel consumption and $CO_2$ emissions from a Spark Ignition (SI) engines. Engine downsizing can significantly reduce fuel consumption by operating the engine closer to its minimum fuel consumption region by reducing the engine displacement and with boosting. In this way pumping losses at part-load operations are reduced. However downsized engines are more prone to knocking combustion at high boost.

Variable Valve Actuation (VVA) can be used to reduce pumping losses at part load conditions and minimise knocking combustion at high loads by means of ECR reduction with Miller cycle. VVA has been studied for more than 20 years but not yet fully

DOI: 10.1201/9781003219217-2

implemented in mass production vehicles. There are a lot of prototypes and all of them can be divided into two main types: cam-based and camless systems. Cam-based systems represent engines with modified camshaft valve train where camshaft is driven by crankshaft, whereas engines with camless systems have actuators which control valve events independently of the crankshaft. The main aim of a VVA system is to modify valve events accordingly to the changing engine load and speed. The parameters that can be controlled by VVA system are: valve opening and closing timing, duration of the valve event and valve lift. Some cam-based systems such as BMW Valvetronic, Toyota Valvematic, Honda VTEC and Fiat Multi-air have been implemented on some production vehicles (3-6). However none of those systems were able to provide continuous and fully flexible variation of lift and valve timings for an individual valve.

Camcon developed an electro-mechanical valvetrain system (iVT) capable of full control of the valve events at engine speeds up to 6000rpm. This system was installed on a single cylinder research engine for both intake and exhaust valves to study valve profile effects on fuel economy and emissions. In this paper, the effect of Miller cycle with different valve profiles will be presented and compared to the baseline at various engine loads at a constant engine speed of 1500rpm.

## 2 EXPERIMENTAL SETUP

A single cylinder SI direct injection gasoline engine with 4 valves was used for this research and the specifications are given in Table 1. The engine is equipped with iVT (intelligent Valve Technology) valvetrain system for each valve as shown in Figure 1. Each valve is actuated via an independent camshaft driven by an electrical motor. Full rotation of the cam produces a full lift event whereas partial rotation allows for lift control. Varying of the motor speed controls the start and duration of the valve opening. Combination of cam rotation and motor speed allows for independent control of timings and lifts of each valve through the valve control software. The fuel was pressurised to 50bar and supplied to the DI gasoline injector, the flow rate is measured by an instantaneous fuel flow meter (Endress+Hauser Promass 83A Coriolis) before the injector. The air supplied to the engine was either at room temperature and pressure or at pre-set boost pressure from an external supercharger with closed loop control. The air mass flow rate was measured by a laminar flow meter (Hasting HFM-200) installed before the throttle. The instantaneous intake and exhaust pressures were measured by a piezo-resistive pressure transducer located just before the intake valves and another one in the exhaust port respectively. Heat release and combustion characteristics were calculated by a combustion analysis software based on the instantaneous cylinder pressure from a piezo-electric pressure transducer and crank angle from a crankshaft encoder. The emissions were measured by a Horiba 7170DEGR. The engine was coupled to an AC dynamometer and installed on the test bed with closed loop control of oil and coolant circuits. The dynamometer allowed for motored and fired operation of the engine at set speeds. The spark timing, throttle angle and AFR were controlled via an engine control software.

**Table 1. Engine specifications.**

| | |
|---|---|
| **Engine Type** | 4-stroke, single cylinder, 2 intake and 2 exhaust valves |
| **Bore x Stroke** | 81mm x 89mm |
| **Connecting Rod length** | 155.5mm |
| **Compression Ratio** | 10.8:1 |
| **Displacement Volume** | 458.6cc |
| **Intake Valves Diameter (2)** | 29mm |
| **Exhaust Valves Diameter (2)** | 26mm |
| **Fuel Injection** | Direct Injection |

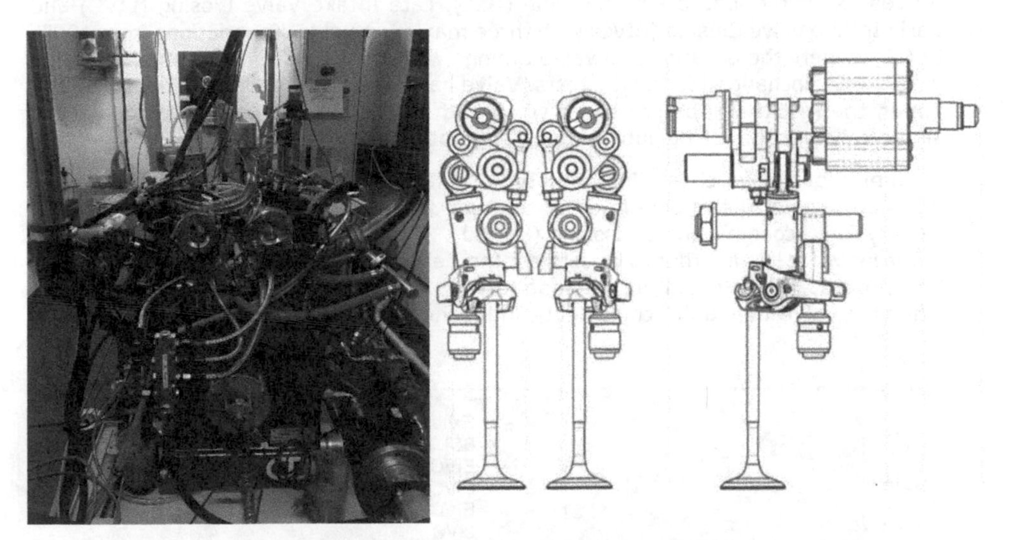

**Figure 1. Single cylinder DI gasoline engine and iVT system (7).**

## 3 TEST CONDITIONS AND MODES OF VALVE OPERATIONS

In this research, several valve profiles were applied to both intake valves (two valve mode) or one of the intake valves (single valve mode) and their effect on engine performance, combustion and emissions were investigated at 4, 6, 9 and 12.6bar net IMEP at a constant engine speed of 1500rpm. The fuel used was EU VI 95 RON Gasoline (E10) with 10% Ethanol content by volume. Fuel specifications can be found in Table 2. All tests were conducted with a relative AFR (Lambda) of 1 and the fuel injection timing was fixed at 268deg CA BTDC at an injection pressure of 50bar. The spark timing was set at MBT unless it was knock limited at higher load conditions.

**Table 2. Fuel properties.**

| **Fuel** | 95 Ron Gasoline E10 |
|---|---|
| **Density at 15 °C (kg/m³)** | 746.1 |
| **Higher calorific value (kJ/kg)** | 44220 |
| **Lower calorific value (kJ/kg)** | 41420 |
| **Stoichiometric AFR** | 13.92:1 |

The engine could be operated with both intake valves or one of them using the iVT system. In the case of the single valve mode operation, one of the intake valves was permanently closed during testing in order to induce swirl motion inside the cylinder.

Figure 2 shows the five valve profiles which were used for two valve and single valve modes for all the load cases: Baseline (BSL), Late Intake Valve Closing (LIVC) and Early Intake Valve Closing (EIVC) with three maximum valve lift variations from 100% to 64% where the duration and valve timings were kept constant. The exhaust valve profile was unchanged for all the tests. Valve parameters are shown in Table 3. In the paper, the results will be presented when both intake valves were actuated and the single valve results will be published in a separate paper.

As in previous researches, EIVC and LIVC profiles were used to reduce effective compression ratio of the engine in order to study the effects of Miller cycle on the efficiency, fuel economy and emissions (8 - 10). Additionally, EIVC was set with three different valve lifts in order to investigate their effects on pumping losses and combustion process, which could not be done in the previous studies by other researchers when the valve lift and IVC could not be independently controlled.

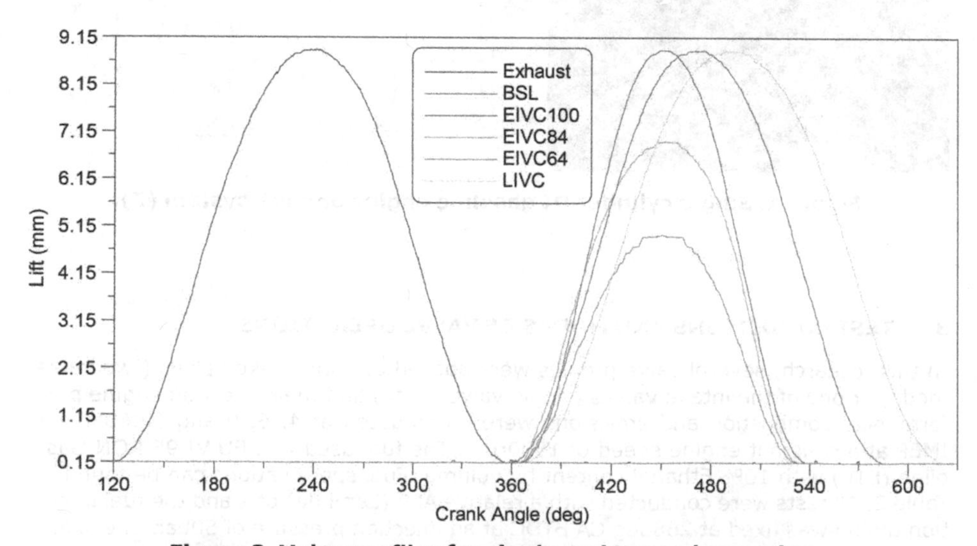


**Figure 2. Valve profiles for single and two valve modes.**

**Table 3. Valve timings and durations.**

| Valve Profile | Lift (mm) | Duration (CA deg) | IVO/EVO (CA deg) | IVC/EVC (CA deg) |
|---|---|---|---|---|
| **EIVC100** | 8.9 | 152 | 376 | 528 |
| **EIVC84** | 7 | 152 | 372 | 524 |
| **EIVC64** | 5 | 146 | 375 | 521 |
| **LIVC** | 8.9 | 226 | 382 | 608 |
| **Standard** | 8.9 | 200 | 373 | 573 |
| **Exhaust** | 8.9 | 211 | 133 | 344 |

# 4 RESULTS

## 4.1 Effect of valve profiles on the effective compression ratio

To better understand experimental results the effective compression ratio (ECR) must be evaluated for each test as IVC timing and valve lift directly affecting it. ECR was calculated based on the in-cylinder pressure for each valve profile and for each load point. The start of compression is defined at the crank angle when the polytropic compression line and intake manifold pressure are crossing on the logP-logV diagram (FIgure 3). The cylinder volume at this crank angle is then divided by TDC volume to get the ECR.

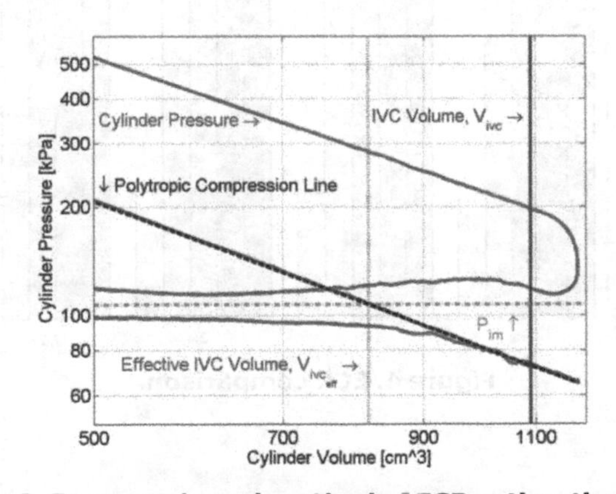


**Figure 3. Pressure based method of ECR estimation (11).**

As shown in Figure 4, the baseline profile has around 10.4 – 9.9 ECR from low to high load whereas EIVC100 has the highest ECR throughout the whole load range, between 10.8 and 10.5. The difference in ECRs between the EIVC100 and the baseline case is due to the IVC timing, which is around 10 degrees before BDC for EIVC100 and is 30 degrees after BDC for baseline profile. As a result, a small amount of fresh charge is pushed back into the intake during the compression stroke before the compression starts, resulting in a lower effective compression ratio for BSL. With increasing load more charge is being expelled into the intake manifold thus ECR is reduced further.

EIVC profiles have almost identical IVC timings, however the effective compression ratio is different for each of them. The reason for that is the variation of the valve lift. The smaller the lift the less fresh charge is sucked into the cylinder therefore lower effective compression ratio. This is clearly demonstrated by EIVC profiles across all the load points. Also with increasing load the ECR decreases respectively as more charge is needed while a certain valve lift creates a constant flow restriction throughout the whole load range.

LIVC produced the lowest effective compression ratio (8 to 8.3) as a lot of fresh charge was expelled into the intake manifold due to very late IVC. Opposite to other profiles ECR is increasing with load for LIVC. This is due to higher pressure in the intake manifold as throttle opens more. When intake pressure is close to atmospheric the pressure difference between in-cylinder and manifold pressure becomes smaller and less percentage of the fresh charge is expelled into the intake.

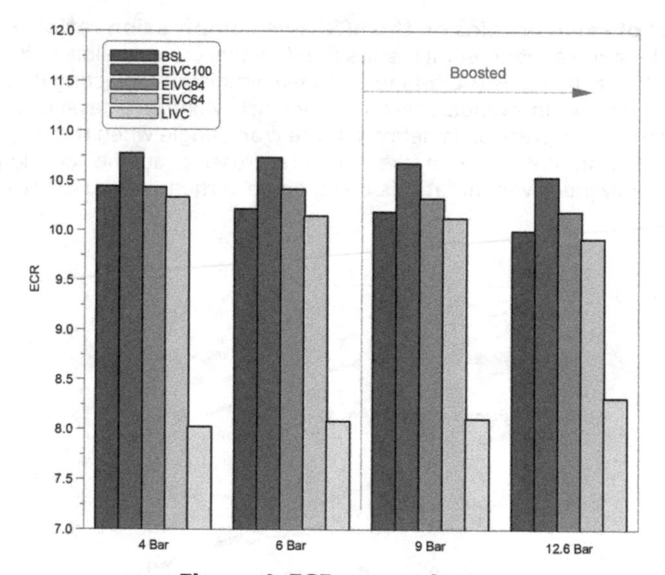


**Figure 4. ECR comparison.**

At each load in most cases the throttle was opened more for those profiles with lower ECR as higher intake pressure was needed in order to achieve the same net IMEP as shown in Table 4. However there were some cases where throttle angle was the same but ECR was different, this was due to other factors which increased the intake pressure regardless of throttle angle. This will be explained in the next section.

**Table 4. Throttle angle and intake pressure comparison.**

| Pro-file | BSL | | EIVC100 | | EIVC84 | | EIVC64 | | LIVC | |
|---|---|---|---|---|---|---|---|---|---|---|
| Net IMEP-(bar) | Throttle (%) | | | | | Intake Pressure (bar) | | | | |
| **4** | 1.83 | 0.47 | 1.80 | 0.42 | 1.85 | 0.46 | 1.85 | 0.50 | 1.96 | 0.61 |
| **6** | 2.75 | 0.64 | 2.68 | 0.59 | 2.82 | 0.65 | 2.85 | 0.69 | 3.80 | 0.83 |
| **9** | 2.18 | 0.83 | 2.15 | 0.78 | 2.24 | 0.85 | 2.24 | 0.90 | 2.44 | 1.08 |
| **12.6** | 3.75 | 1.22 | 3.40 | 1.12 | 3.65 | 1.20 | 4.65 | 1.34 | 14.00 | 1.48 |

## 4.2 Effect of valve profiles on the pumping loss

In a conventional throttle controlled SI engine the pumping loss is created due to the drop of manifold pressure below atmospheric pressure and it is one of the major causes of low engine efficiency at the low load. Comparison of Pumping Mean Effective Pressure (PMEP) for various valve profiles is shown in Figure 5.

At 4bar net IMEP, the LIVC valve profile produced the lowest negative PMEP due to larger throttle opening (0.13% more than STD) required to maintain the same IMEP with the lowest ECR. Also the charge was pushed out from the cylinder into the intake manifold during compression stroke, which increased the intake pressure for the subsequent intake strokes lowering pumping losses (Table 4).

Amongst all EIVC profiles, EIVC100 produced the highest negative PMEP, whereas EIVC64 produced the least amount of pumping loss. For EIVC100 to achieve 4bar net IMEP, the throttle was closed by 0.05% more than for the lower lifts because of higher ECR, this resulted in increased pumping loss. For EIVC84 and EIVC64 valve profiles, the throttle position was identical however intake manifold pressure was higher for the lower lift profile. A lower valve lift leads to lower flow area and hence a higher intake pressure is required to allow the same amount of air into the cylinder as the higher lift. This is usually achieved by larger throttle opening. However, in the case with EIVC64, it was noted that a bigger drop in the in-cylinder pressure at IVO caused an increase in the intake port pressure which was enough to trap the required amount of air into the cylinder without altering throttle angle. Higher intake pressure led to lower pumping losses for EIVC64. Therefore, when operating an engine at low load with a small throttle opening, the lower intake valve lift increased the intake pressure and reduced pumping loss.

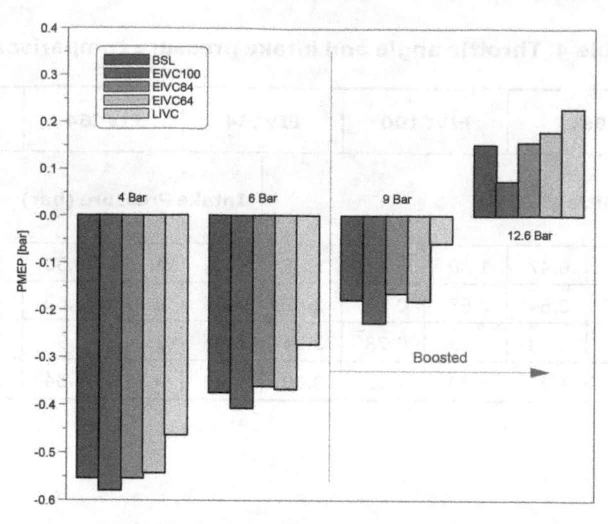


**Figure 5. PMEP comparison.**

However, an additional test with very short lift (EIVC44) produced the highest intake pressure and pumping losses larger by 0.01bar than EIVC64 (Table 5). According to J. B. Heywood this phenomenon could be explained by the relationship between valve lift and discharge coefficient (12). Depending on the valve shape, at high lifts the flow separates and discharge coefficient decreases generating higher pumping losses. At lower lifts the flow speed across the intake valves increases and the flow remains attached to the valve providing high discharge coefficient, therefore reducing pumping losses. However if the lift would be reduced further or flow speed would be increased, the flow might separate and cause an opposite effect, which was the case with EIVC44. The discharge coefficient at EIVC64 might have been higher than EIVC44 and this would reduce pumping losses, but this is purely theoretical and to quantify the results a measurement of discharge coefficient via CFD simulations is required. In the case of this experiment 64% lift was the optimum for EIVC profiles in reduction of pumping losses, therefore EIVC44 was not considered in further analysis.

**Table 5. EIVC64 and 44 comparison.**

| | **EIVC64** | **EIVC44** |
|---|---|---|
| **Intake Pressure (bar)** | 0.495 | 0.601 |
| **PMEP (bar)** | -0.54 | -0.55 |

The intake pressure for baseline profile was higher than EIVC84 due to the charge being expelled from the cylinder during compression strokes even though throttle was opened slightly less. However, negative PMEP was the same as EIVC84 due to lower effective flow area during the opening of the valve for BSL profile.

To summarise, EIVC100 got the highest pumping losses due to the lowest intake pressure. LIVC achieved major reduction in pumping losses due to increased intake pressure and the largest throttle opening.

At 6bar net IMEP a similar trend is present. LIVC had the lowest ECR and required a larger throttle angle, which increased the intake pressure, providing the lowest pumping losses among other profiles. EIVC100 had the highest pumping losses due to the lowest intake pressure followed by the baseline profile. EIVC64 had a higher intake pressure than EIVC84 but also higher negative PMEP. This indicates that EIVC84 might have higher discharge coefficient. A larger engine load requires larger volume of charge in the cylinder therefore flow velocity through the intake valves has to increase in order to accommodate for this. As discussed previously the discharge coefficient is affected by the change in flow speed, due to this the discharge coefficient could be reduced at EIVC64 causing larger pumping loss even though intake pressure was higher than at 84% lift.

For 9bar net IMEP tests, an external supercharger was used to provide the pressurised air. It was necessary to provide additional 0.6bar of boost pressure for the LIVC profile as it was not capable of achieving this load point with naturally aspirated mode, whereas other profiles didn't require boosting. Overall, the same trend was present as at 6bar. LIVC had the lowest negative PMEP due to the highest intake pressure. EIVC100 has the lowest intake pressure thus the largest pumping losses. EIVC64 and 84 had the same throttle angle but intake pressure was increased for EIVC64 due to lower lift. EIVC84 achieved lower pumping loss potentially due to higher discharge coefficient as in the case with 6bar net IMEP. However further increased flow speed could have reduced discharge coefficient even more for EIVC64 resulting in a higher negative PMEP than the baseline valve profile, even though it had much lower intake pressure.

At 12.6bar net IMEP a positive pumping work was present due to additional boost. LIVC had the highest positive PMEP due to significantly higher intake pressure compared to the rest of valve profiles, followed by EIVC64. EIVC100 had the lowest positive PMEP due to lowest intake pressure which corresponds to the lowest throttle angle. EIVC84 had slightly higher positive pumping work compared to the baseline even though the intake pressure was slightly lower, this was caused by larger effective flow area during valve opening of EIVC84.

From the above results it is evident that intake port pressure has strong influence on the pumping losses. The intake port pressure can be increased via the throttle or valve lift. It is also possible that discharge coefficient can affect pumping losses. The discharge coefficient is influenced by the flow speed which can be altered either by intake pressure or valve lift or both. Where the difference between intake pressures is not significant the discharge coefficient is more dominant in reduction of pumping losses. Across all load cases LIVC reduced pumping loses significantly compared to the baseline profile due to significant increase in intake pressure. EIVC profiles with shorter lifts also were successful in reducing pumping losses due to higher intake pressure and possibly increased discharge coefficient.

### 4.3 Effect of valve profiles on the combustion process

Changes of the intake valve profile can also influence combustion duration, which is another factor affecting fuel economy. Comparison of combustion durations and spark timings can be seen in Figure 6.

At 4bar net IMEP the baseline valve profile produced the shortest combustion duration (10-90% burn), whereas EIVC84 and 64 led to the longest duration indicating that flame speed was lower due to reduced tumble flow and hence turbulence intensity. EIVC100 and LIVC profiles had similar combustion durations which were slightly longer (less than 2 degCA) than that of the baseline valve profile. The MBT spark timing was advanced accordingly to compensate for the longer combustion duration, however the flame development angle (crank angle degrees between spark timing and 10% burn) for short lift profiles was more than 30% longer than the baseline (Table 6). LIVC also had slightly longer flame development angle indicating weaker in-cylinder turbulence than the baseline.

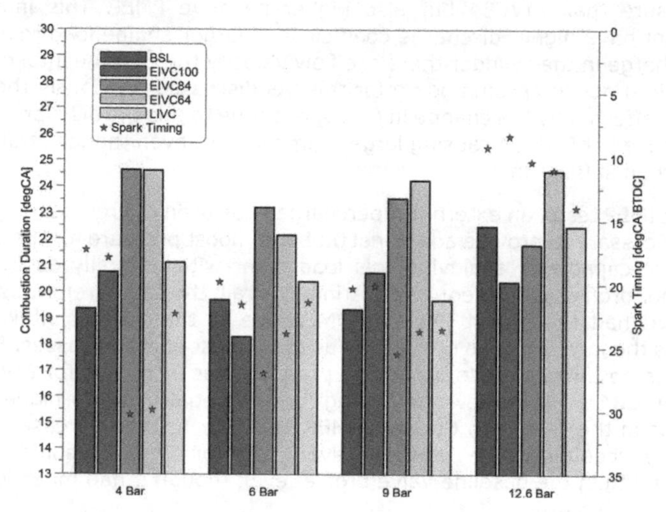


**Figure 6. Combustion duration and spark timing comparison.**

At 6bar net IMEP combustion duration was reduced for all profiles except for the BSL profile with shortest being EIVC100 and longest EIVC84. The MBT spark timing was advanced respectively for profiles with longer flame development angle, longest being EIVC with shorter lifts followed by LIVC. EIVC100 had slightly longer flame development angle than the baseline. Therefore the spark timing was also slightly advanced.

**Table 6. Flame development angle and in-cylinder lambda comparison.**

| **Profile** | **BSL** | | **EIVC100** | | **EIVC84** | | **EIVC64** | | **LIVC** | |
|---|---|---|---|---|---|---|---|---|---|---|
| **Net IMEP (bar)** | **Flame Development Angle (degCA)** | | | | | **Lambda Cylinder** | | | | |
| **4** | 19.5 | 0.98 | 19.7 | 0.98 | 26.8 | 0.97 | 26.1 | 0.98 | 22.4 | 0.96 |
| **6** | 19.5 | 0.98 | 20.3 | 0.98 | 24.3 | 0.98 | 23.9 | 0.99 | 21.1 | 0.97 |
| **9** | 18.3 | 0.98 | 20.2 | 0.99 | 22.4 | 0.99 | 23.1 | 0.99 | 19.8 | 0.98 |
| **12.6** | 16.7 | 0.98 | 16.5 | 0.98 | 18.2 | 0.98 | 19.5 | 0.96 | 19.0 | 0.98 |

At 9bar net IMEP, the combustion duration was similar to 4bar but the advance in spark timing was limited by knocking combustion. LIVC spark timing was advanced by a few degrees compared to baseline profile due to lower in-cylinder temperature at the end of compression stroke thanks to lower effective compression ratio. This resulted in reduced knocking tendency and allowed to advance spark timing. EIVC84 and 64 spark timings were also more advanced than EIVC100 due to lower ECR. The longest flame development angle was found with EIVC64 and 84, followed by EIVC100 and LIVC.

At 12.6bar net IMEP knocking had more noticeable effect on spark timing. EIVC100 has the shortest combustion duration and the most retarded spark timing due to highest ECR. Whereas LIVC has the same combustion duration as the baseline but more advance spark timing due to lower knocking tendency. This was achieved because of reduced in-cylinder temperature as effective compression ratio was lowered. The same effect can be observed with EIVC84 and 64 spark timing compared to EIVC100. The longest flame development angle was produced by EIVC64 followed by LIVC and EIVC84. EIVC100 achieved shorter flame development angle and duration than BSL.

From the discussion above it can be concluded that lower ECR causes weaker in-cylinder flow motion leading to longer combustion duration and flame development angle. This effect is more prominent at lower load and very high load with knock limited combustion. On the other hand, a lower effective compression ratio allows to advance knock limited spark timing further. Shorter valve lifts lead to slower combustion except at the highest load case. This is evidenced by longer flame development angle and combustion duration of EIVC84 and 64. In almost all the load cases shorter lift had more noticeable effect on in-cylinder turbulence than the effect of lower ECR.

## 4.4 Effect of valve profiles on the indicated fuel consumption (ISFC)

Figure 7 demonstrates how valve profiles influence ISFC at various load points. At low load (4bar IMEP) LIVC valve profile reduced fuel consumption by 3.4g/kWh compared to the baseline due to reduced pumping losses. EIVC64 was also successful in reduction of ISFC however due to very poor in-cylinder charge mixing and slow combustion it was less effective than LIVC. EIVC100 and 84 lead to increased fuel consumption due to larger pumping losses and longer combustion duration.

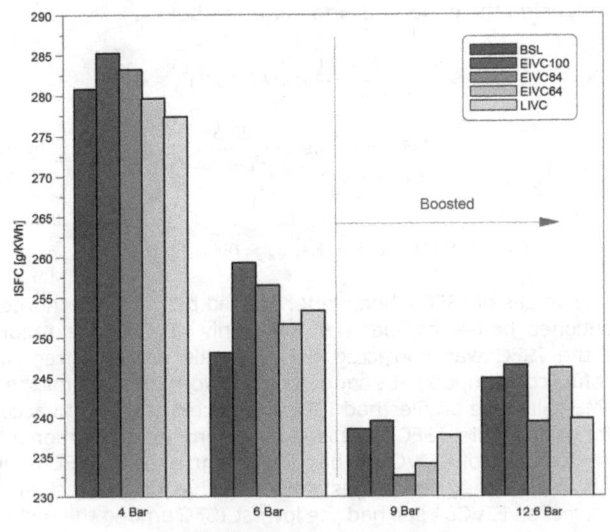


**Figure 7. ISFC comparison.**

At 6bar net IMEP the lowest fuel consumption was achieved by the standard profile due to shorter combustion duration and medium pumping losses. EIVC64 was the second best followed by LIVC profile. The reason why LIVC achieved higher ISFC value than EIVC64 even though it has shorter combustion duration and lower pumping losses, is richer local in-cylinder charge (Table 5). This could be due to some of the fuel being pushed out into the intake manifold and in the next cycle this fuel could be concentrated in one spot instead of homogeneously mixing. Whereas EIVC64 had in-cylinder AFR closest to stoichiometric. The worst fuel economy was achieved by EIVC100 due to the largest pumping loss.

A similar result for LIVC was achieved at 9bar net IMEP. Despite a shorter combustion duration and the lowest pumping loss LIVC achieved higher fuel consumption than EIVC84 and 64 due to richer in-cylinder mixture. Baseline profile also had higher fuel consumption due to lower in-cylinder lambda. EIVC100 had the largest ISFC due to larger pumping losses than the rest of the profiles and EIVC84 had the lowest fuel consumption due to stoichiometric in-cylinder AFR and lower pumping losses and shorter combustion duration than EIVC64.

At 12.6bar net IMEP in-cylinder lambda was slightly richer for EIVC84 than for LIVC, but EIVC84 had shorter combustion duration which led to lower ISFC. EIVC100 had the highest fuel consumption due to the smallest positive pumping work among other profiles. EIVC64 was the second worst due to longest combustion duration and richest in-cylinder mixture.

From the above results it is evident that valve profiles influence pumping losses and combustion duration which in turn affects fuel consumption. Moreover, in-cylinder AFR is affected by valve profiles influencing ISFC. Overall Miller cylce was succefull in reducing fuel consumption at low loads thanks to lower pumping loss and at high loads where spark timing was knock limited.

At 9 and 12.6bar net IMEP loads the net fuel consumption should also take into account of the compression work required where additional boost pressure was used. An external mechanical supercharger was used to provide the boost and was powered by electrical supply, in automotive application the supercharger would be driven by the engine crankshaft which would cause a power loss affecting ISFC. Due to this the compressor work was calculated and ISFC was then corrected taking into account supercharger work using the equations and variables below.

$$W_c = \frac{\dot{m}_{air} \times C_p \times T1}{\eta_C \times \eta_m} \times \left[ (P2 \div P1)^{\frac{\gamma-1}{\gamma}} - 1 \right]$$

$$ISFC\ corrected = \frac{\dot{m}_{fuel}}{Pi - Wc}$$

Where,

$$C_p = 1.012\ J/gK,\ \gamma = 1.4,\ \eta_c = 60\%,\ \eta_m = 90\%$$

Figure 8 shows the values of ISFC when work required by the supercharger is taken into account. As mentioned before at 9bar net IMEP only LIVC profile required additional boost therefore the ISFC was corrected only for this profile. After correction LIVC achieved highest fuel consumption at 9bar due to the work required for the supercharger. At 12.6bar net IMEP all valve profiles had ISFCs corrected taking into account additional boost required for each profile. ISFC increased for all profiles proportional to the amount of required boost, for example EIVC100 had lowest increase in ISFC as it required the least boost whereas LIVC had the largest increase due to greatest amount of boost required. After correction EIVC84 still had the lowest ISFC among the rest of the profiles.

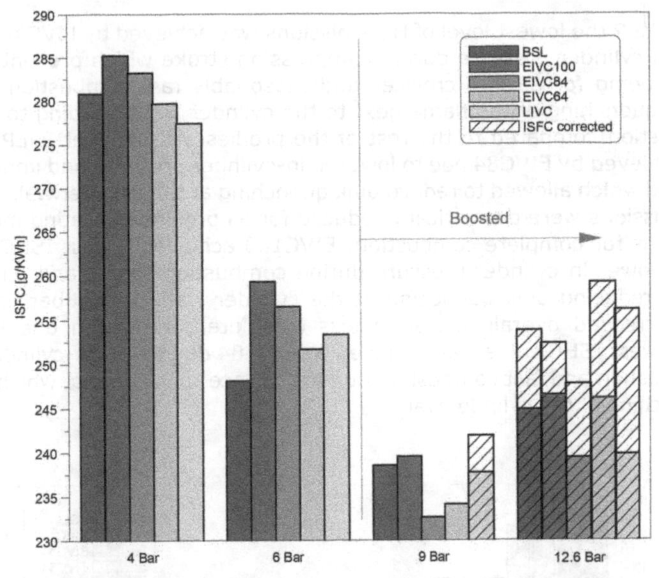


**Figure 8. Comparison of ISFC vs corrected ISFC.**

### 4.5 Effect of valve profiles on the exhaust emissions

Emissions results are shown in Figure 9a, b and c. Throughout all load cases higher concentration of ISCO was present with lower in-cylinder lambda i.e. combustion of fuel rich mixture. The lowest ISCO values were achieved when in-cylinder lambda was closer to 1 (stoichiometric combustion) with lowest value of 16 g/KWh at 9bar net IMEP with EIVC64 profile.

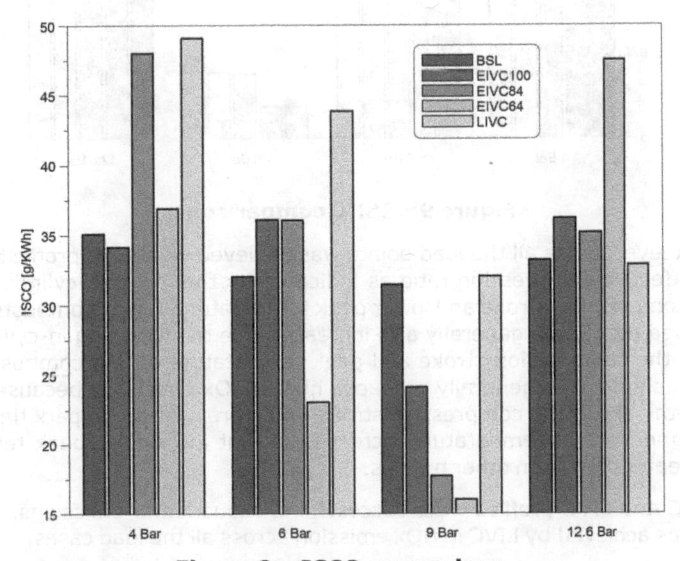


**Figure 9a. ISCO comparison.**

At 4bar net IMEP the lowest level of HC emissions was achieved by LIVC profile due to the lowest in-cylinder pressure during compression stroke which prevented some of the mixture being forced into crevices and reasonably fast combustion which prevented bulk quenching of the flame next to the cylinder walls, leading to more complete combustion compared to the rest of the profiles. At 6bar net IMEP the lowest ISHC was achieved by EIVC84 due to lowered in-cylinder pressure and improved combustion speed which allowed to reduce bulk quenching at the cylinder wall. At 9bar net IMEP HC emissions were dramatically reduced for all profiles indicating more favourable conditions for complete combustion. EIVC100 achieved lowest ISHC of 5.38 g/KWh due to lower in-cylinder pressure during combustion stroke and relatively fast combustion, reducing bulk quenching at the cylinder wall. At 12.6bar net IMEP HC emissions increased overall indicating less complete combustion due to stronger knocking. Lowest ISHC value was achieved by EIVC84 due to low in-cylinder pressure during combustion and fast combustion due to advance spark timing, which allowed to reduce quenching at the cylinder wall.

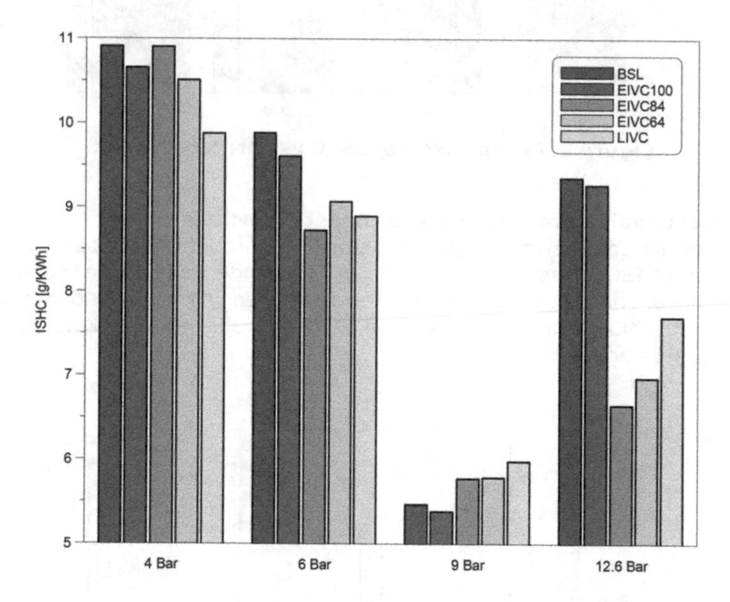


**Figure 9b. ISHC comparison.**

Lowest NOx level across all the load points was achieved with LIVC profile because of the lower effective compression ratio as indicated by the lower in-cylinder pressure during the compression stroke and lower peak temperature during combustion. As the load increased the ISNOx generally also increased due to increasing in-cylinder pressure during the compression stroke and peak temperature during combustion. EIVC profiles with shorter lifts generally produced higher NOx emissions because of higher pressure at the end of the compression stroke and more advanced spark timing which provides higher rate of temperature increase, so that in-cylinder peak temperature would be greater than from other profiles.

Overall EIVC and LIVC profiles were successful in reduction of emissions. The major reduction was achieved by LIVC in NOx emission across all the load cases.

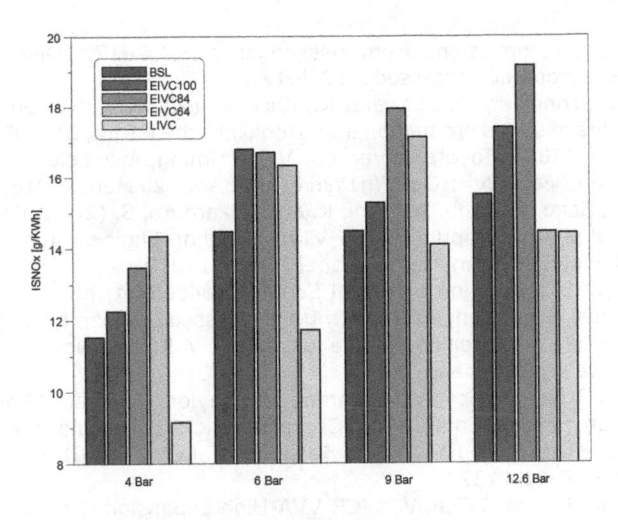


**Figure 9c. ISNOx comparison.**

## 5 CONCLUSIONS

In this study, the iVT system was employed to investigate the effect of valve lift profiles on the performance, combustion and emissions of a single cylinder direct injection spark ignition engine at different engine loads at a constant engine speed. Five intake valve profiles were tested at 1500rpm. The results are presented and analysed in this paper with main findings stated below:

- Lower valve lift increases intake manifold pressure when throttle is used
- The discharge coefficient may influence PMEP when the difference in intake pressure is not significant, not more than 0.05bar
- Lower ECR allows to advance spark timing when combustion is knock limited
- Lower valve lift reduces flame speed more than low ECR, thus combustion duration is longer (by 15% on average across the load cases between EIVC84, EIVC64 and LIVC)
- Valve profiles affect in-cylinder lambda even though exhaust lambda is kept at 1 (mixture in the cylinder is not always uniformly mixed)
- EIVC with short lifts and LIVC profiles were successful in reduction of fuel consumption up to 2.5% compared to the baseline profile at 9bar net IMEP
- LIVC reduced NOx emissions at all load cases but increased CO emissions
- Combination of EIVC profiles lowered CO and HC emissions but increased NOx almost at all load cases

In addition to the above results, studies were also carried out on the impact of operating one intake valve with different valve lift profiles and results will be published in separate papers.

## REFERENCES

[1] E. Publications, "Regulation (EU) No 333/2014 of the European Parliament and of the Council of 11 March 2014 amending Regulation (EC) No 443/2009 to define the modalities for reaching the 2020 target to reduce CO 2 emissions from new passenger cars," 2014. [Online]. Available: https://publications.europa.eu [Accessed 2 12 2017].

[2] "Reducing CO2 emissions from passenger cars," 2017. [Online]. Available: https://ec.europa.eu [Accessed 2 12 2017].
[3] Unger, H., Schneider, C., Schwarz, K., Koch, F., (2008), Valvetronic: Experience from 7 years of series production and a look into the future, BMW Group.
[4] ADPTraining (2013) Toyota Valvematic Valve Timing. Available at: https://www.youtube.com/watch?v=HOqGWg1YJh4 (Accessed: 20 March 2018)
[5] Akima, K., Seko, K., Taga, W., Torii, K. and Nakamura, S. (2006) Development of New Low Fuel Consumption 1.8L i-VTEC Gasoline Engine with Delayed Intake Valve Closing, SAE International, paper 2006-01–0192
[6] Wan, M. (2011) Autozine Technical School Available at: http://www.autozine.org/technical_school/engine/Index.html (Accessed: 18 March 2018)
[7] R. Stone, D. Kelly, "Intelligent Valve Actuation – A Radical New Electro-Magnetic Poppet Valve Arrangement" 2017.
[8] Garcia, E., Triantopoulos, V., Boehman, A., Taylor, M. et al., "Impact of Miller Cycle Strategies on Combustion Characteristics, Emissions and Efficiency in Heavy-Duty Diesel Engines," SAE Technical Paper 2020-01-1127, 2020, doi: 10.4271/2020-01-1127.
[9] Constensou, C. and Collee, V., "VCR-VVA-High Expansion Ratio, a Very Effective Way to Miller-Atkinson Cycle," SAE Technical Paper 2016-01-0681, 2016, doi: 10.4271/2016-01-0681.
[10] Osborne, R., Downes, T., O'Brien, S., Pendlebury, K. et al., "A Miller Cycle Engine without Compromise - The Magma Concept," SAE Int. J. Engines 10(3):2017, doi: 10.4271/2017-01-0642.
[11] L. Kocher, E. Koeberlein, D. G. Van Alstine, K. Stricker, and G. Shaver, "Physically-based volumetric efficiency model for diesel engines utilizing variable intake valve actuation," International Journal of Engine Research, published online, November 2011.
[12] J. B. Heywood. Internal Combustion Engine Fundamentals, Second Edition (McGraw-Hill Education: New York, Chicago, San Francisco, Athens, London, Madrid, Mexico City, Milan, New Delhi, Singapore, Sydney, Toronto, 2018).

## *Session 2: IC Engines for heavy-duty and off-highway*

# The effects of cylinder deactivation in future hybridised biogas dual fuel trucks

**A. Hegab[1], T. Yuwono[1], A. Cairns[1], J. Hall[2], R. Cracknell[3]**
[1]Powertrain Research Centre, University of Nottingham, UK
[2]MAHLE Powertrain Ltd., Northampton, UK
[3]Shell Global Solutions, London, UK

## ABSTRACT

An experimentally-correlated simulation study was carried out for commercial heavy-duty truck engine working on dual fuel using upgraded biogas as a primary fuel and diesel pilot as an ignition source. The aim was to investigate the potential benefits of cylinder deactivation (CDA) technique combined with partial restriction of the intake air as a way of improving the light load operation of dual fuel engines to make them suitable for truck propulsion in urban settings. Various air restriction techniques were examined, including intake throttling, the use of exhaust gas recirculation (EGR) and employing Miller cycle with late intake valve closing (LIVC).

## 1 INTRODUCTION

The transportation sector accounts for almost one-quarter of global anthropogenic $CO_2$ emissions. In 2015, greenhouse gas (GHG) emissions in the global transportation sector were equivalent to 10.9 billion metric tons (Gt) of carbon dioxide-equivalent emissions ($CO_2$-e) (1). As shown in Figure 1, more than 75% of global transportation $CO_2$ emissions are generated from on-road vehicles, with the heavy-duty transport sector accounts for almost half of this quantity (1). In recent years, efforts towards decarbonisation of transport has notably gained increased attention, with different strategies identified to achieve that goal (2). Despite the recent increase in electric car population worldwide, most of the increase has come from the passenger vehicles while the electrification of heavy-duty transport remains challenging due to issues over driving range, charging infrastructure and battery cost and size (3). Until these issues are addressed, the transition to the use of low-carbon fuels and hybridization in heavy duty transport is considered medium-term strategy for a more sustainable transport (4). One potent option is the use of what is referred to as "gas-diesel dual-fuel engine" as a prime mover for heavy-duty trucks. In this IC engine configuration, most of the diesel fuel is replaced with the clean, low-carbon gaseous fuel. Amongst others, natural gas (NG) remains an attractive option to be used as the prime source of energy in dual fuel engines, owing to several reasons related to energy content, properties, chemical composition, clean nature of combustion, global availability and relatively mature infrastructure (5). Whereas more decarbonisation is desirable, an upgraded biogas may also be used as the main fuel for these engines, as it could contain up to 95% methane concentration (6).

In conventional dual fuel combustion (CDFC), the diesel pilot is injected at once, late in the compression stroke, to act as an ignition source for the premixed gaseous fuel-air mixture. The engine could also revert to 100% diesel fuel operation if needed (so-called diesel fallback). Although the technique is simple and easily achievable using mechanical fuel injection systems, in this combustion mode flame propagation from the pilot region might not proceed throughout the entire charge, especially at part

DOI: 10.1201/9781003219217-3

load conditions (7). This increases the CO and HC emissions, while the combustion of the centralised pilot promotes NOx formation (8). To overcome these drawbacks, premixed dual fuel combustion (PDFC) strategy was proposed (9). In this approach, the diesel pilot is split over two injections; an early part (aka pre-injection) injected early in the compression stroke to mix with the gaseous fuel-air mixture and act as an ignition improver, and a late part (aka main injection) injected late in compression stroke to act as an ignition source of the charge. The resulting enhanced combustion reduces both CO and HC emissions, where the combustion efficiency exceeds 98% (9). NOx emissions are also reduced as long as the same EGR levels are maintained.

Despite these improvements, light load operation of PDFC engines remains unfeasible as the CDFC. At such load conditions with high substitution ratio of natural gas or upgraded biogas for diesel fuel, the engine runs at an ultra-lean mixture and low in-cylinder temperature. While the equivalence ratio dictates the flame propagation of methane/air combustion (10), the lean mixture causes the engine to encounter poor combustion and increased CO and HC emissions (predominantly methane).

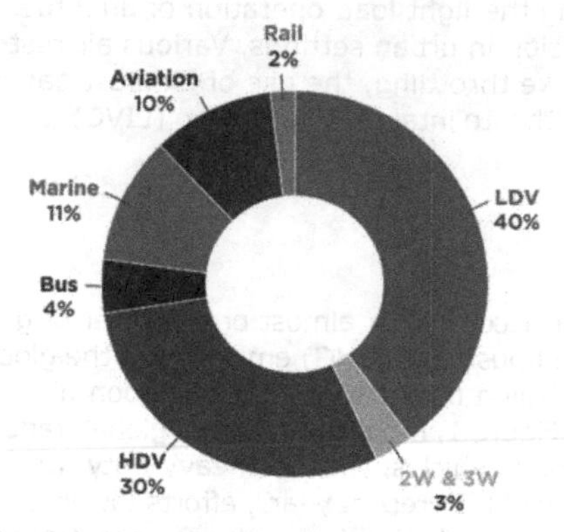


**Figure 1. Share of 2020 well-to-wheel CO2 emissions from transportation, by mode. Ref. (1).**

One proposed way to improve the light load operation of dual fuel multi-cylinder engines is the use of cylinder deactivation (CDA) (11). In this technique, only some of the engine cylinders are allowed to fire while the others are disabled (12). To produce the same brake power as if all the cylinders were firing, the firing cylinders operate with effectively richer gaseous fuel-air mixture since the total fuel amount is distributed over less cylinders. Increased mixture strength results in enhanced flame propagation hence improved engine combustion and emissions (11). Dissimilar to the cylinder cut-out technique where only the fuel supply in disabled, in CDA both the fuel injection and the valve actuation of the deactivated cylinders are disabled hence pumping loss is reduced so as the fuel consumption (13). In addition, CDA results in airflow reductions consequently higher exhaust gas temperatures and lower exhaust flow rates, which is beneficial to the aftertreatment system (13).

Along with the use of CDA to increase the specific load of the active cylinders, partial restriction of the air component of the charge while maintaining the same amount of gaseous fuel would also produce an effectively richer mixture that would benefit from the improved flame propagation (14). This could be implemented either by applying a physical restriction to the intake air flow (e.g. the use of throttling), by reducing the effective mass of air charge through using a large amount of external EGR, or by the adoption of Miller cycle with late intake vale closing (LIVC) by which the charge is partially expelled back out through the still-open intake valve hence reducing the air charge mass. Regardless of the method used, partial air restriction is considered an effective way to control the air-fuel mixture equivalence ratio ($\lambda$) in dual fuel engines and improve their light load operation (15).

The aim of the present work has been to assess and evaluate the potential benefits of employing cylinder deactivation strategy, along with different methods of partial air restriction, to improve light load operation of gas-diesel dual fuel engines. The purpose was to extend the practical operating limit of these engines so they become practical to use for truck propulsion in urban settings. The experimentally-correlated simulation study was carried out for a modern, commercial heavy-duty truck engine, at 3 bar BMEP and 1200 rpm, with 90% gaseous fuel substitution for the diesel fuel. Comparative results for cylinder deactivation (CDA) versus all-cylinder active (ACA) operation with different partial air restriction strategies (throttling, EGR and Miller-LIVC) were steered for engine combustion and performance parameters.

## 2 SIMULATION APPROACH

### 2.1 Model overview

The dual fuel one-dimensional combustion model was built for and using the data of a Volvo Penta TAD873VE 235kW/2200rpm in-line 6-cylinder 8-litre high-pressure common-rail (HPRC) production diesel engine, retrofitted for dual-fuel operation with gaseous-fuel port injection and equipped with industry standard measurements of performance, fuel consumption and exhaust gas emissions. The 1D simulation was carried out using the commercial package GT-Power. This software is an object-based engine simulation tool, in which the models are built using an adaptable, dynamic graphical user interface. Each component of the system is represented on the project map by a 'template' that is filled by the user to identify its attributes. The software architecture is based on different empirical correlations, sub-models and solvers to energy, momentum and continuity equations at each step, with calculations for heat transfer, friction etc. The complete model setup is illustrated by Figure 2.

### 2.2 Main specifics of the model

#### 2.2.1 *Diesel and gaseous fuel injection*

With diesel injection profile necessary for simulating the relevant processes to diesel fuel combustion, a detailed injection rate map was imported from GT-Power library. The map has been developed from a combined experimental and computational study for a 2000 bar HPCR injector (16). As the injector used in the experimental work archives the same pressure level with similar characteristics, this approximation was deemed acceptable since the engine simulation was run for a high gaseous fuel substitution ratio of 90%, so the diesel fuel quantity is relatively small hence any error in quantification becomes less significant.

In conventional dual fuel combustion (CDFC), pilot injection timing was iterated such that the 50% mass fraction burnt (MFB50) is attained at 8° CA ATDC, for optimum combustion. This location is also known as CA50 (17). For premixed dual fuel combustion (PDFC), the pre-injection part of the pilot was injected 15° before the main pilot injection, with the mass ratios of the two fixed at 30/70.

For gaseous fuel injection, the model was constructed to simulate the experimental setup with six gas injectors to inject the gaseous fuel into the intake runner of each cylinder. The gaseous fuel flowrate was set as 15 mg/ms, the end of injection was set as 270 °CA BTDCF (or 90 °CA before intake BDC), while the start of injection (SOI) was adjusted in keeping with the specific operating condition. The gaseous fuel used in the simulation was set as Methane/Ethane mixture at 95/5% concentration. This has been considered a good representation for natural gas and upgraded biogas.

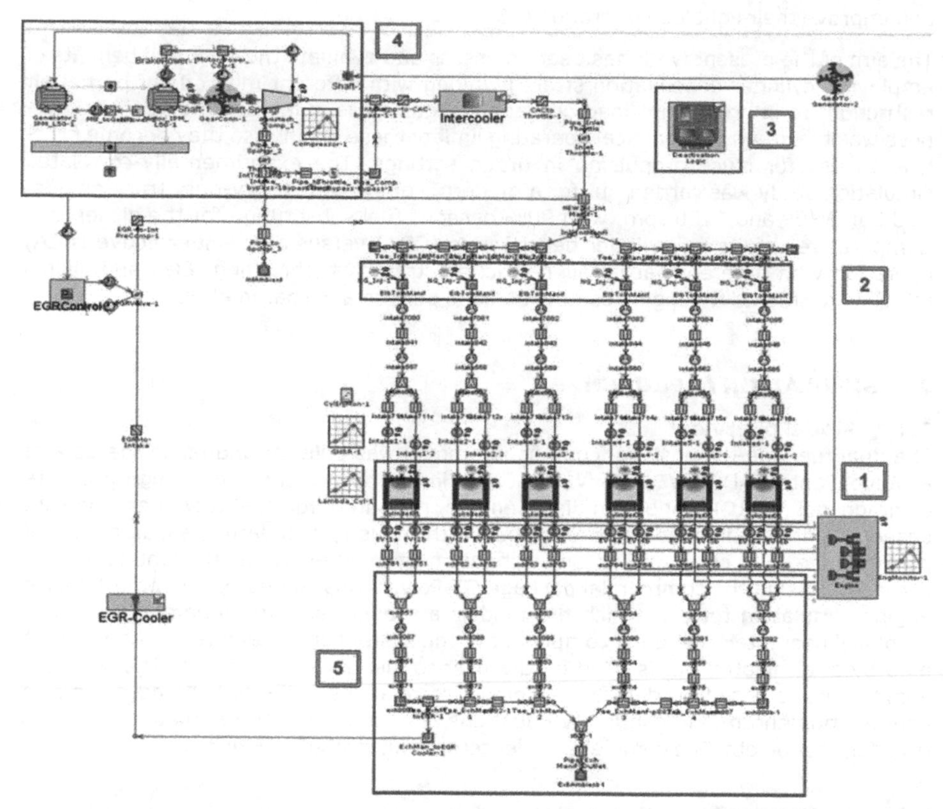


**Figure 2. The 6-cylinder heavy duty engine model setup, showing: (1) direct injection of diesel fuel, (2) port injection of gaseous fuel, (3) cylinder deactivation logic, (4) electric supercharging setup, and (5) exhaust gas ducting.**

### 2.2.2 *Intake and exhaust valves*

For engine operating modes modelled with standard valve actuation, the model used the actual valve lift data as measured on the Volvo engine. For the simulation of Miller cycle with Late Intake Valve Closing (LIVC), the intake valve was held 3.1 mm lift for a longer time (to achieve the targeted lambda), then follows the traditional lift profile for what remains. The typical lift profiles for the exhaust valve, intake value at standard timing and a late intake valve closing of 100 °CA before firing TDC (as an example of LIVC) are illustrated by Figure 3. Flow coefficients through intake and exhaust valves as well as the intake port swirl flow were measured on flow bench; results are shown in Figure 4. The data was fed into the valvetrain model to produce an accurate prediction of the air intake and exhaust gas flow processes during the simulation.

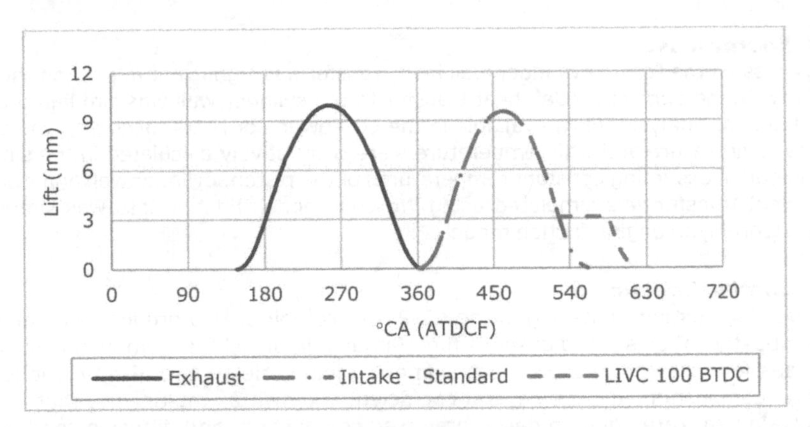


**Figure 3. Lift profiles for the exhaust valve, intake value at standard timing and LIVC of 100 °CA before firing TDC.**

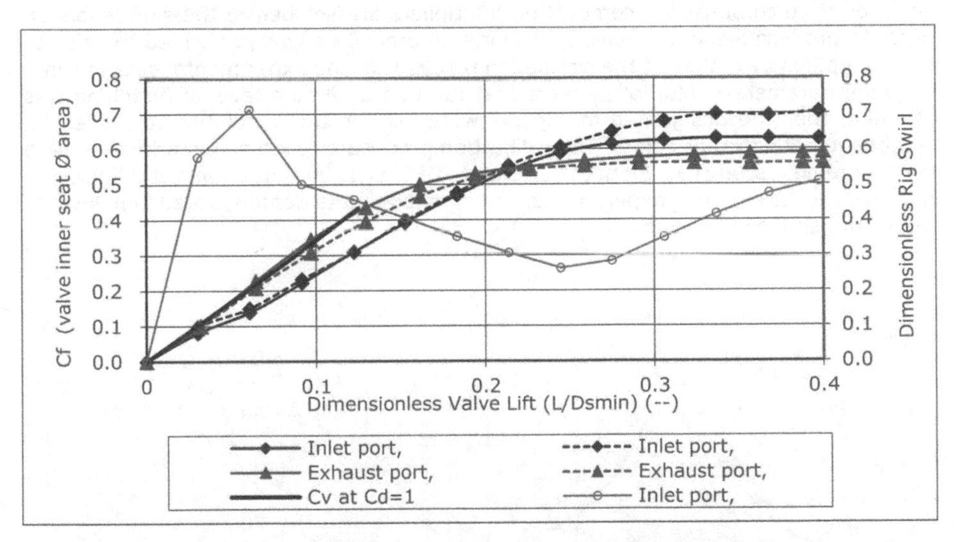


**Figure 4. Measured flow coefficient (Cf) vs dimensionless of intake and exhaust valve, and dimensionless rig swirl of the Volvo Engine.**

### 2.2.3 *Cylinder deactivation (CDA) logic*

For CDA modelling, a comprehensive logic control within the software package was used to deactivate the fuel injection as well as valves actuation of cylinder 4, 5, and 6 while keeping cylinders 1, 2 and 3 in full operation. The logic was built such that CDA becomes active after four cycles of normal operation. This was deemed necessary to ensure that the targeted BMEP was met and remained stable. The logic sequence of during the last cycle before the deactivation was: gas injection – IVC – diesel fuel injection – expansion – EVO – EVC. After the last EVC, the trapped mass undergoes compression and expansion in the closed volume. With such method, trapped mass within the deactivated cylinder was minimised.

### 2.2.4 *Energy losses*

Energy loss on the form of cylinder wall heat transfer has significant impact on the fuel economy. In the current model, heat transfer to the cylinder wall was simulated using Finite Element Analysis (FEA) available in the GT-Power. For better accuracy, the cylinder gas temperature and wall temperature were correlatively calculated in the simulation instead of assuming constant temperatures of the piston, cylinder wall and cylinder head. Heat transfer was simulated using Woschni model. Friction loss was simulated using Chenn-Flynn engine friction model.

### 2.2.5 *Combustion model*

Dual fuel combustion modelling in the GT-Power combines two predictive models for the combustion of diesel and gaseous fuel. Simulating diesel fuel combustion includes processes for spray atomisation, entrainment, ignition delay, premixed combustion, diffusion combustion. It uses several coefficients, known as multipliers, that dictate the entrainment rate, ignition delay, premixed combustion, and diffusion combustion in the simulation. Gaseous fuel combustion simulation models the flame propagation within the gas/air mixture. The multipliers used to dictate the fuel burn rate are flame kernel growth, turbulence flame speed, charge dilution, and Taylor length scale (18).

For predictive combustion, combustion multipliers are set before the simulation. In order to produce accurate results, a calibration procedure was performed to validate the multipliers by matching the simulation results with the experimental data in some targeted parameters. Multipliers were first iterated until high level of matching was attained, then the designated multipliers were used in the simulation to model the combustion at different points. The calibration procedure relied on experimental data from a previous study (9), for both CDFC and PDFC, at 1200 rpm/6 bar Net IMEP. Comparative results between experimental and simulated data demonstrated high level of matching, as shown in Figure 5.

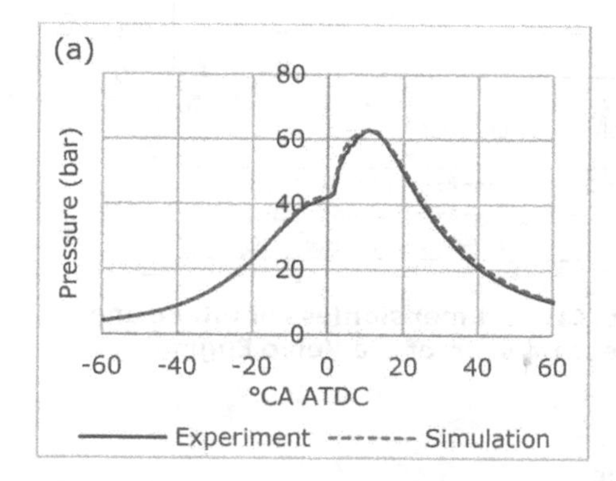


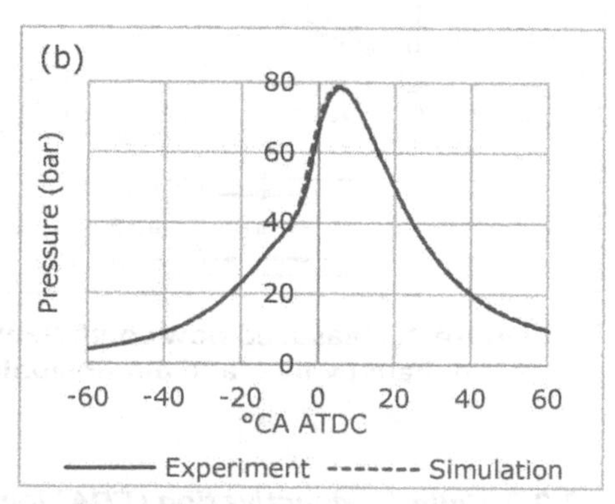


**Figure 5. Comparison between the experimental and simulation results for (a) CDFC, and (b) PDFC. Data at 1200 rpm/6 bar Net IMEP.**

## 2.3 Method

The simulation was run for two load points of 3 and 6 bar BMEP at 1200 rpm, selected to represent the low load operating conditions of heavy-duty truck engine considering their frequency of occurrence over the World Harmonised Transient Cycle (WHTC) for heavy-duty engines. At the specified operating condition, cylinder deactivation (CDA) simulation was carried out with only cylinders 1, 2 and 3 active and the remaining engine cylinders deactivated. This was then compared to the simulation of the engine operation at the same load/speed condition with all the 6 cylinders active (ACA).

With each technique, different air restriction strategies (throttling, EGR and Miller-LIVC) were examined, where dual fueling approach (i.e. CDFC or PDFC) was selected as appropriate for each strategy. For air restriction using throttling and LIVC, CDFC was chosen for more simplicity of the model. With EGR, on the other hand, PDFC was deemed more suitable as it has the capacity to improve the flame propagation that could be significantly deteriorated with large EGR quantities (19).

The simulation was run considering a 90% gas substitution ratio of the gaseous fuel for diesel fuel (energy basis), where this ratio is identified as:

$$\%\text{ESR} = \frac{\dot{m}_{GF} \times LHV_{GF}}{\dot{m}_D \times LHV_D + \dot{m}_{GF} \times LHV_{GF}} \tag{1}$$

where ($\dot{m}_{GF}$), ($\dot{m}_D$), ($LHV_{GF}$) and ($LHV_D$) are the mass flow rates and the lower heating values of the gaseous and diesel fuel, respectively (20). The values of the latter two parameters were considered as 43 and 49.5 MJ/kg fuel, respectively.

Whereas air restriction strategies were used to control the mixture strength, the latter was expressed in terms of the global air-fuel equivalence ratio, identified as:

$$\lambda_{global} = \frac{\dot{m}_a}{AFR_{GF}^{stoic} \times \dot{m}_{GF} + AFR_D^{stoic} \times \dot{m}_D} \tag{2}$$

where ($AFR_{CH4}^{stoic}$) and ( $AFR_D^{stoic}$) are the stoichiometric air-fuel ratios of the gaseous fuel and diesel fuel, respectively, and ($\dot{m}_a$) is the air mass flow rate (21).

For CDA simulation (also referred as De-Ac) at 3 bar BMEP, different strategies for air restriction were employed to examine different mixture strengths, ranging from λ=1.4 to 1.0 (stochiometric).

Throttling simulation was carried out using a butterfly valve, with the opening angle iterated to attain the targeted (λ) value. For EGR control, exhaust back pressure was adjusted to be slightly higher than intake manifold pressure to allow the EGR flow at the required quantity for the desired (λ) value. In Miller-LIVC strategy, the late valve timing determined the trapped air mass inside the cylinder hence setting (λ) value.

De-Ac simulation was also run without any air restriction (no throttling, zero EGR and standard valve timing) to reveal the leanest operational point (highest (λ) value).

For 6-cylinder operation at 3 bar BMEP, simulation was only run for λ=1.8 as this was found to be the richest mixture (lowest (λ) value) possible at the operating condition. Running the engine below this threshold would require either a significant throttling that would result in a very low manifold absolute pressure (MAP) below 0.67 bar, a considerably high EGR ratio that exceeds 34%, or a combination of both strategies corresponding to 0.8 bar MAP and 16% EGR; all lead to instable engine running. The combined strategy (throttling + EGR) with 6-cylinder was simulated for both CDFC and PDFC (given that throttling originally runs with CDFC while EGR runs with PDFC).

It was also not possible to run the 6-cylinder simulation at 3 bar BMEP for Miller-LIVC technique, since that would have meant operating at very low compression ratio that cannot sustain pilot diesel combustion.

A summary of the different configurations of the simulated conditions is provided in Table (1). Due to space limitation, only simulation results for 3 bar BMEP and 1200 rpm are presented in this work.

**Table 1. Summary of simulated conditions in the present work.**

| Air restriction method ⬇ | Technique ➡ | De-Ac | | | | 6-Cyl |
|---|---|---|---|---|---|---|
| | Lambda value ➡ | 1.0 | 1.2 | 1.4 | 1.5*** | 1.8 |
| | DF Approach ** ⬇ | | | | | |
| THRL * | C | Y | Y | Y | Y | Y |
| | P | | | | | |
| EGR | C | | | | | |
| | P | Y | Y | Y | Y | Y |
| LIVC | C | Y | Y | Y | | |
| | P | | | | | |
| THRL * + EGR | C | | | | | Y |
| | P | | | | | Y |

* THRL = Throttling.

** DF Approach: C = conventional dual fuel combustion (CDFC); P = premixed dual fuel combustion (PDFC).

*** $\lambda$=1.5 is attained without any air restriction; i.e. no throttling, zero EGR, and only with standard valve timing.

## 3 RESULTS AND DISCUSSION

### 3.1 Net IMEP and pumping loss

The impacts of cylinder deactivation on net indicated mean effective pressure (IMEP) and engine pumping loss are shown in Figure 6. It can be seen that with deactivating half of the engine cylinders at the same requested brake load, the specific load of the firing cylinders (represented by IMEPn of cylinder 1) is double its corresponding value when the engine runs on all cylinders, as shown in Figure 6(a). The increased cylinder pressure and temperature improves the combustion process. Also, with the stopped intake and exhaust processes within the deactivated cylinders and the minimised trapped mass inside these, pumping loss is reduced relative to all-cylinder operation (22), as demonstrated by Figure 6(b). This remains valid with different air restriction strategies, although pumping loss increases notably with the increased throttling (23). Still, pumping loss with De-Ac at stochiometric conditions ($\lambda$=1) remains less than that with 6-cylinder with leaner mixture ($\lambda$=1.8), despite the latter observing less throttling. LIVC offers the lowest pumping loss since the mixture flowback with this strategy reduces the vacuum degree in the intake manifold (24). With 6-cylinder operation, pumping loss remains high when throttling is employed regardless of the setting; EGR with PDFC is the only strategy that shows moderate levels of pumping loss caused by the increased exhaust back pressure required for EGR flow.

### 3.2 Brake thermal efficiency (BTE) and fuel economy

Brake thermal efficiency (BTE) trends for the different simulated cases are shown in Figure 7(a). It can be seen that all De-Ac strategies produced higher BTE values than the 6-Cyl operation, at all (λ) values. This was attributed to the combination effects of increased cylinder specific load, improved combustion in the firing cylinders and reduced pumping loss with De-Ac (25). Up to 4% improvement in BTE was attained with De-Ac with no air restriction (full throttle; λ=1.5) relative to the corresponding strategy with 6-Cyl operation (with throttling to attain richest mixture possible; λ= 1.8), where both was running on CDFC. With De-Ac, air restriction with EGR produced higher BTE than LIVC and THRL. This was due to reduced pumping loss on the one hand, and improved combustion with PDFC relative to CDFC on the other hand (9).

Throttled 6-Cyl operation produced the lowest BTE, owing to the high pumping losses. Air restriction using EGR produced the highest BTE within 6-Cyl strategies examined, but remained inferior to De-Ac. The improvement in engine fuel economy with De-Ac can be observed in Figure 7(b) that shows the brake specific fuel consumption (BSFC) trends in diesel fuel equivalent ($g_{diesel}$/kWh) for different simulated conditions. It can be seen that adapting De-Ac technique in

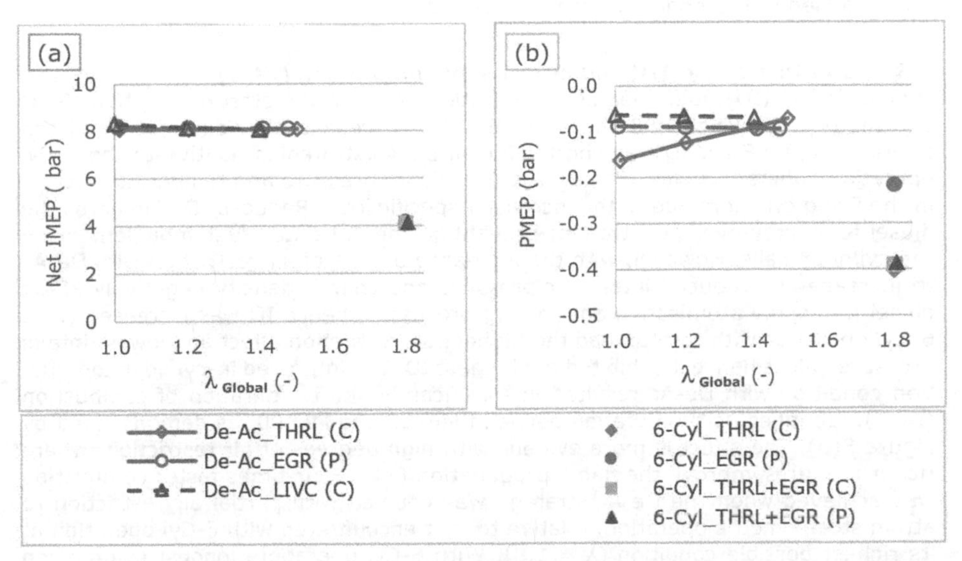


**Figure 6. (a) Net IMEP of cylinder 1, and (b) Engine PMEP. Data for 3 bar BMEP and 1200rpm.**

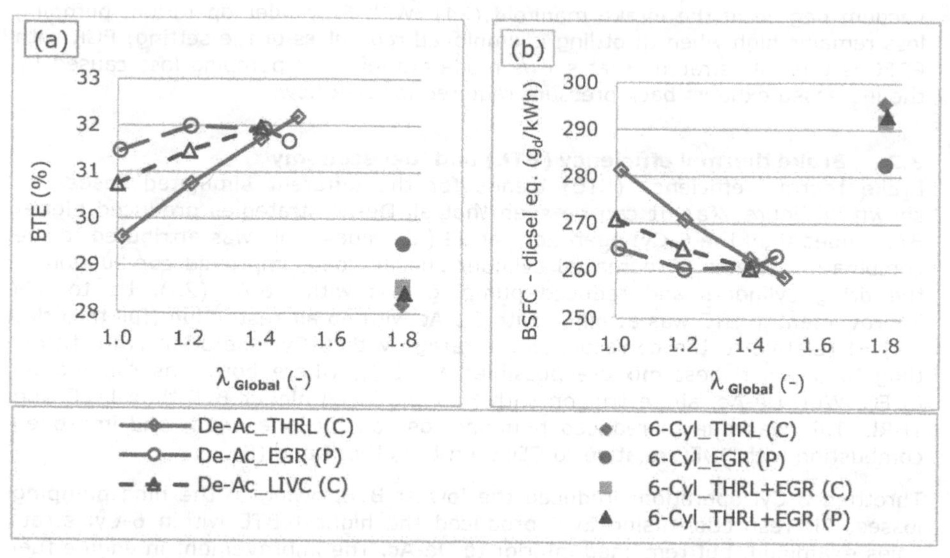


**Figure 7. (a) Brake thermal efficiency (BTE) %, and (b) BSFC in diesel equivalent ($g_{diesel}$/kWh). Data for 3 bar BMEP and 1200rpm.**

dual-fuel engines at low load operation can result in 12.5% reduction in the total fuel supplied to the engine (diesel equivalent).

### 3.3 Ignition delay (ID) and duration of combustion (DOC)

Ignition delay (ID), identified as the duration between the start of injection (SOI) and the point at which MFB equals 0.1% (26), is presented for different simulated cases in Figure 8(a). ID was shorter for all De-Ac strategies relative to the 6-Cyl operation, where this was attributed to the higher pressure and temperature levels in the firing cylinders due to the increased specific load. Reduced ID eliminates the diesel fuel spray over-penetration; preventing the unwanted fuel impingement on the cylinder walls. However, with the increased degree of air restriction with De-Ac to increase (λ), reduced intake air pressure and charge density negatively affect diesel fuel spray atomisation and mixing processes, hence ID was increased (27). 6-Cyl operation with throttle had the highest air restriction effect and lowest intake pressure values hence it exhibited the longest ID. The improved in-cylinder combustion condition with De-Ac resulted in significantly shorter duration of combustion (DOC), identified as the duration between MFB10 and MFB90, as demonstrated by Figure 8(b). The effect is more evident with high degrees of air restriction, where richer mixtures improve the flame propagation (14). Four-times faster combustion was achieved when the De-Ac strategy was coupled with partial air restriction to attain stochiometric operation, relative to that encountered with 6-Cyl operation at its richest possible condition (λ = 1.8). With 6-Cyl operation, longest combustion

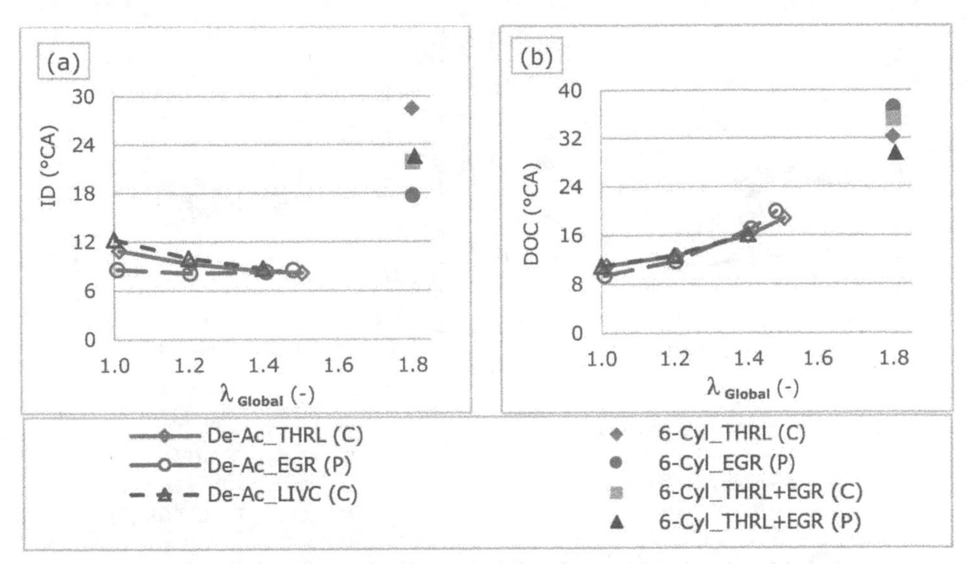


**Figure 8. (a) Ignition delay (ID) (°CA), and (b) Duration of combustion (DOC) (°CA). Data for 3 bar BMEP and 1200rpm.**

duration was observed when EGR was used to control (λ), where the high exhaust gas levels (34%) are believed to have caused considerable reduction in flame propagation (19).

### 3.4 Engine heat transfer and exhaust gas temperature

Engine cylinder wall heat transfer for different simulated cases is illustrated in Figure 9(a). It can be seen that the engine heat transfer to the cylinder walls was reduced with De-Ac technique relative to 6-cylinder operation, despite the higher specific load of the firing cylinder with De-Ac. This was attributed to the reduced surface area for heat transfer with three firing cylinders compared with that with six firing cylinders. (28). Although increasing the degree of air restriction with De-Ac technique increased the engine heat transfer, owing to the higher gas mean temperature as the mixture becomes richer, total engine heat transfer remained lower than that occurring with 6-Cyl operation using the same air restriction method and DF combustion mode. Up to 20% reduction in the engine heat transfer was attained when De-Ac was employed with no air restriction (full throttle; λ =1.5) relative to that with 6-Cyl operation (with throttling for richest mixture possible; λ = 1.8), where both was running on CDFC. In general, the use of throttling to control (λ) resulted in highest engine heat transfer, where the increased pumping loss with this strategy implied the use of larger amount of fuel

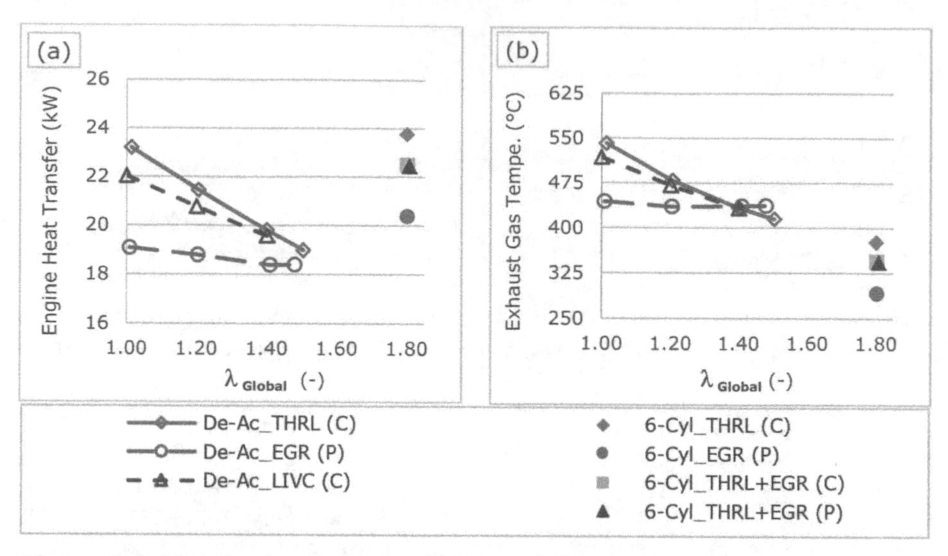


**Figure 9. (a) Engine heat transfer (kW), and (b) Mass-averaged exhaust gas temperature (°C). Data for 3 bar BMEP and 1200rpm.**

to achieve the targeted BMEP; increasing the heat release and heat transfer. The use of EGR, on the other hand, increased the mixture heat capacity and lowered the temperature inside the cylinder; reducing the heat transfer (29).

The mass-averaged exhaust gas temperature ($T_{exh}$) for different simulated cases is shown in Figure 9(b). In this work, the mass-averaged $T_{exh}$ was preferred over the time-averaged one since the former is directly related to the exhaust gas energy (as it considers the fluctuation in the exhaust gas flow rate) while the latter is not (30). It can be seen that $T_{exh}$ was always higher with De-Ac relative to 6-Cylinder operation, for all strategies and DF combustion modes. This was attributed to the higher specific load of the firing cylinders and the associated high pressure and temperature inside the cylinder. Highest $T_{exh}$ was attained with maximum air restriction at stochiometric conditions ($\lambda$=1.0), and decreased with less air restriction at leaner mixtures. Within different air restriction strategies, throttling produced the highest $T_{exh}$, owing to the increased amount of fuel used to overcome the high pumping loss. The increase in $T_{exh}$ when De-Ac is coupled with throttling at stochiometric conditions exceeds 160°C, relative to 6-Cyl dual fuel operation. The use of EGR with De-Ac did not bring about noticeable change in $T_{exh}$ despite the observed change in ($\lambda$), due to the relatively unchanged amount of fuel used. Still, $T_{exh}$ with De-Ac and any EGR ratio averaged 150°C more than the corresponding 6-Cyl operation where EGR was used to achieve the richest mixture possible. Increased $T_{exh}$ highly benefits the aftertreatment system operation (31).

## 4 CONCLUSIONS

In this work, the potential benefits of employing cylinder deactivation (CDA) strategy to improve light load operation of a gas-diesel dual fuel engine were studied. The experimentally-correlated simulation study was carried out for a modern, commercial heavy-duty truck engine at 3 bar BMEP and 1200 rpm, with 90% gas substitution for diesel fuel. The study investigated different air restriction strategies, including intake air throttling, EGR and Miller cycle with late intake valve closing (LIVC) to control the gas-air mixture strength. Results of CDA operation with 3-Cyl firing were compared to those obtained from normal engine operation with 6-Cyl active. Both conventional dual fuel combustion (CDFC) and premixed dual fuel combustion (PDFC) modes were considered. The main findings of the study are:

- CDA is an innovative technique that can be used to extend the practical operating limit of gas-diesel dual-fuel engines, so they become practical to use for truck propulsion in urban settings.
- With CDA, the specific load of the three firing cylinders increases (almost doubles) as the engine brake power is maintained. This improves the combustion process within these cylinders, while the stopped intake and exhaust processes within the deactivated cylinders reduce the pumping loss. Both effects combine to improve engine fuel economy.
- Up to 4% improvement in brake thermal efficiency (BTE) was attained with CDA with no air restriction (full throttle; $\lambda = 1.5$) relative to the corresponding strategy with 6-Cyl operation (with throttling to attain richest mixture possible; $\lambda = 1.8$).
- Employing CDA in dual-fuel engines at low load can result in 12.5% reduction in the total fuel supplied to the engine (diesel equivalent).
- Improved combustion conditions with CDA significantly shortens the duration of combustion (DOC) relative to 6-Cyl operation. Four-times faster combustion was achieved with CDA at stochiometric conditions ($\lambda$=1.0) via partial air restriction.
- Engine heat loss was reduced with CDA relative to 6-Cyl operation despite higher specific load, as a result of the reduced surface area for heat transfer. Up to 20% reduction in engine heat transfer was attained when CDA was employed with no air restriction (full throttle; $\lambda = 1.5$) relative to that with 6-Cyl operation (with throttling for richest mixture possible; $\lambda = 1.8$).
- CDA is an effective way to increase the exhaust gas temperature ($T_{exh}$), relative to 6-Cyl operation. The increase in $T_{exh}$ when CDA is coupled with air restriction to achieve stochiometric conditions ($\lambda = 1.0$) exceeds 160°C. Increased $T_{exh}$ highly benefits the aftertreatment system operation.
- Amongst different air restrictions methods studied under stoichiometric conditions ($\lambda = 1.0$) with CDA, intake air throttling endured the highest pumping loss, lowest BTH, highest heat loss yet demonstrated highest exhaust gas temperature. LIVC resulted in the least pumping loss and high $T_{exh}$, while the use of EGR resulted in the lowest engine heat loss and highest BTH.

## 5 FUTURE WORK

The work extends to investigate the simulated results for 6 bar BMEP and 1200 rpm, where an e-Supercharger will be used to provide the boosting required for simulating conditions with higher amount of intake air and higher amount of fuel at fixed mixture strength. Also, the potential benefits of employing PDFC strategy to reduce methane slip in dual-fuel engines will be evaluated, where a novel methane oxidation catalyst (MOC) will be installed in the engine aftertreatment system to provide a robust and durable methane slip control measure.

## REFERENCES

[1] The International Council on Clean Transportation (ICCT). Vision 2050: A strategy to decarbonize the global transport sector by mid-century. September 2020.

[2] European Academies' Science Advisory Council (EASAC). Decarbonisation of transport: options and challenges. EASAC policy report 37. March 2019.
[3] International Energy Agency (IEA). Global EV outlook 2019 - Scaling-up the transition to electric mobility. May 2019.
[4] Intergovernmental Panel on Climate Change (IPCC). Climate Change 2014: Mitigation of Climate Change. Cambridge, UK: Cambridge University Press, 2014.
[5] Hegab A, La Rocca A, Shayler P. Towards keeping diesel fuel supply and demand in balance: Dual-fuelling of diesel engines with natural gas. Renewable Sustainable Energy Revs 2017;70:666–97. https://doi.org/10.1016/j.rser.2016.11.249.
[6] Abbasi T, Tauseef S, Abbasi S. Biogas Energy. New York, USA: Springer; 2011.
[7] Karim GA. The dual fuel engine. In: Evans RL, editor. Automotive Engine Alternatives, New York, USA: Plenum Press; 1987, p. 83–104.
[8] Königsson F. On combustion in the CNG-diesel dual fuel engine. PhD thesis, Stockholm, Sweden: Dept. Mach. Des., Royal Institute of Technology; 2014.
[9] May I, Pedrozo V, Zhao H, Cairns A, Whelan S, Wong H, Bennicke P. Characterization and potential of premixed dual-fuel combustion in a heavy-duty natural gas/diesel engine. SAE tech pap 2016-01-0790; 2016. https://doi.org/10.4271/2016-01-0790.
[10] Dong Y, Vagelopoulos CM, Spedding GR, Egolfopoulos FN, Measurement of laminar flame speeds through digital particle image velocimetry: Mixtures of methane and ethane with hydrogen, oxygen, nitrogen, and helium. Proc Combust Inst 2002;29:1419–1426. https://doi.org/10.1016/S1540-7489(02)80174-2.
[11] Hegab AHI. Combustion characteristics of methane-diesel dual-fuel engines at light load conditions with different injection strategies. PhD thesis, Nottingham, UK: University of Nottingham; 2017.
[12] Mo H, Huang Y, Mao X, Zhuo B. The effect of cylinder deactivation on the performance of a diesel engine. Proc Instn Mech Engrs, Part D: J Automob Eng 2014;228(2):199–205. https://doi.org/10.1177/0954407013503627.
[13] Ramesh AK, Gosala DB, Allen C, Joshi M, McCarthy J Jr, Farrell L et al. Cylinder deactivation for increased engine efficiency and aftertreatment thermal management in diesel engines. SAE tech pap 2018-01-0384; 2018. https://doi.org/10.4271/2018-01-0384.
[14] Karim GA. Dual-Fuel Diesel Engines. Boca Raton, USA: CRC Press; 2015.
[15] Di Blasio G, Belgiorno G, Beatrice C. Effects on performances, emissions and particle size distributions of a dual fuel (methane-diesel) light-duty engine varying the compression ratio. Appl Energy 2017;204:726–740. https://doi.org/10.1016/j.apenergy.2017.07.103.
[16] Payri R, Salvador FJ, Martí-Aldaraví P, Martínez-López J. Using one-dimensional modeling to analyse the influence of the use of biodiesels on the dynamic behavior of solenoid-operated injectors in common rail systems: Detailed injection system model. Energy Convers Manage 2012;54:90–99. https://doi.org/10.1016/j.enconman.2011.10.004.
[17] de O Carvalho L, de Melo TCC, de Azevedo Cruz Neto RM. Investigation on the fuel and engine parameters that affect the half mass fraction burned (CA50) optimum crank angle. SAE Tech Pap 2012-36-0498; 2012. https://doi.org/10.4271/2012-36-0498.
[18] Gamma Technologies. GT-SUITE: Engine Performance Application Manual. Version 2019.
[19] Chen Y, Zhu Z, Chen Y, Huang H, Zhu Z, Lv D et al. Study of injection pressure couple with EGR on combustion performance and emissions of natural gas-diesel dual-fuel engine. Fuel 2020;261:116409. https://doi.org/10.1016/j.fuel.2019.116409.
[20] Papagiannakis RG, Krishnan SR, Rakopoulos DC, Srinivasan KK, Rakopoulos CD. A combined experimental and theoretical study of diesel fuel injection timing and

gaseous fuel/diesel mass ratio effects on the performance and emissions of natural gas-diesel HDDI engine operating at various loads. Fuel 2017;202:675–687. https://doi.org/10.1016/j.fuel.2017.05.012.

[21] Abdelaal MM, Hegab AH. Combustion and emission characteristics of a natural gas-fueled diesel engine with EGR. Energy Convers Manage 2012;64:301–312. https://doi.org/10.1016/j.jma.2017.02.004.

[22] McGhee M, Wang Z, Bech B, Shayler PJ, Witt D. The effects of cylinder deactivation on the thermal behaviour and fuel economy of a three-cylinder direct injection spark ignition gasoline engine, Proc Inst Mech Eng Part D: J Automob Eng. 2019;233:2838–2849. https://doi.org/10.1177/0954407018806744.

[23] Heywood JB. Internal combustion engine fundamentals. New York, USA: McGraw-Hill; 1988.

[24] Zhao J, Xi Q, Wang S, Wang S. Improving the partial-load fuel economy of 4 cylinder SI engines by combining variable valve timing and cylinder-deactivation through double intake manifolds. Appl Therm Eng 2018;141:245–256. https://doi.org/10.1016/j.applthermaleng.2018.05.087.

[25] J. Konrad, T. Lauer, M. Moser, E. Lockner, J. Zhu, Engine efficiency optimization under consideration of NOx- and knock-limits for medium speed dual fuel engines in cylinder cut-out operation, in: WCX World Congr Exp., SAE Int, 2018. https://doi.org/10.4271/2018-01-1151.

[26] Shingne PS, Middleton RJ, Assanis DN, Borgnakke C, Martz JB. A thermodynamic model for homogeneous charge compression ignition combustion with recompression valve events and direct injection: Part I—Adiabatic core ignition model. Int J Engine Res 2017;18(7):657–676. https://doi.org/10.1177/1468087416664635.

[27] Robert Bosch GmbH. Diesel-Engine Management. 4th ed. West Sussex, UK: Wiley; 2006.

[28] Pillai S, LoRusso J, Van Benschoten M. Analytical and experimental evaluation of cylinder deactivation on a diesel engine. SAE Tech Pap 2015-01-2809; 2015. https://doi.org/10.4271/2015-01-2809.

[29] Alger T, Gingrich J, Roberts C, Mangold B. Cooled exhaust-gas recirculation for fuel economy and emissions improvement in gasoline engines. Int J Engine Res 2017;12(3):252–264. https://doi.org/10.1177/1468087411402442.

[30] Canton JA. Comparisons of thermocouple, time-averaged and mass-averaged exhaust gas temperatures for a spark-ignited engine. SAE Tech Pap 820050; 1982. https://doi.org/10.4271/820050.

[31] Ding C, Roberts L, Fain DJ, Ramesh AK, Shaver GM, McCarthy J Jr et al. Fuel efficient exhaust thermal management for compression ignition engines during idle via cylinder deactivation and flexible valve actuation. Int J Engine Res 2016;17 (6):619–630. https://doi.org/10.1177/1468087415597413.

# Using intake port water injection to reduce NOx emissions for heavy-duty multi-cylinder diesel engine applications

**Linpeng Li, Zunqing Zheng, Hu Wang, Zongyu Yue, Mingfa Yao**
State Key Lab of Engine, Tianjin University, Tianjin, China

## ABSTRACT

In this paper, low pressure EGR and water addition were considered as candidate solutions for NOx reduction in the high BTE region, where HP EGR is insufficient to reduce NOx emission. To overcome possible wet wall and lubricant dilution issues caused by water addition, a reliable intake port water injection strategy is proposed in this study, which ensures the water entering the cylinder in the form of vapour. This water injection strategy has been studied and compared with the LP EGR strategy at high load operating conditions with respect to combustion, emission and performance. What is more, the potential of combined water injection and HP EGR strategy for further NOx emission reduction was also tested, with the aim of meeting future near-zero emission regulation.

## 1 INTRODUCTION

As the aspirations of society and government of reducing harmful emissions and carbon dioxide emission from internal combustion engine become increasingly stronger, the harmful emissions regulations get more and more stringent in the past decades [1]. Major regions or countries across the world had begun to set fuel consumption regulation or carbon emission regulation for heavy-duty vehicles in order to mitigate global warming. Considering the huge number of internal combustion engine stock, continuous effort should be made to improve the internal combustion engine [2].

Recently, many research groups and engine original equipment manufacturers have been pursuing engine brake thermal efficiency (BTE) to achieve fuel economy improvement and carbon emissions reduction [3,4], with a special interest on low speed and high load operating points due to its high BTE as depicted in Figure 1. Usually, increased compression ratio to around 23 or even higher is used, with steel pistons is adopted instead of aluminum to reduce heat transfer loss and to withstand higher combustion pressures. However, the gain in BTE is usually accompanied by the penalty in harmful emissions. O.Laguitton and Mohiuddin experimentally studied the effect of engine compression ratio on diesel engine emissions, and showed that lower specific fuel consumption and higher NOx emissions above 30g/kWh were observed with higher compression ratios [5,6]. The engine NOx emission at the highest BTE operating condition is so high that it cannot be sufficiently removed by deNOx after-treatment system alone to meet EURO 6 legislation, and so an extra technique is imperative to mitigate engine NOx emission while maintaining low BSFC.

The most common way to control engine NOx emissions in modern diesel engines is to use exhaust gas recirculation (EGR), because it lowers the combustion temperature and the oxygen concentration in the combustion chamber [7]. The decrease of NOx emissions with EGR is the result of complicated effects, sometimes even opposite

DOI: 10.1201/9781003219217-4

effects [8]. Ladommatos investigated the influence of water and carbon oxides on diesel engine combustion and emissions by a series of experiments in a single cylinder DI diesel engine, quantifying the dilution effect, the thermal effect, the chemical effect of water and carbon dioxides [9–11]. He drew the conclusion that both carbon dioxide and water vapor lower the NOx emission and increase PM emissions due to dilution effect, and water vapor has higher thermal effect due to higher specific heat capacity compared to carbon dioxide. Furthermore, if a similar amount of water and carbon dioxide is applied, water evaporation could be slightly more effective in NOx mitigation but with more particulate emission compared to carbon dioxide.

In terms of pressure in the exhaust gas recirculation loop, EGR can be categorized into high pressure EGR (HP-EGR) and low pressure EGR (LP-EGR). HP EGR is driven by the pressure difference between up-stream of the turbine and downstream of the compressor in a turbocharged engine, which means EGR flowrate is not sufficient in low speed high load operation region as explained in Figure 2. Even though many measures can be taken to increase HP-EGR ratio, such as use of variable geometry turbocharger, intake throttle, all of them increase HP EGR ratio at the expense of BTE [12]. In this context, NOx emissions in the low speed high load operating region need to be solved by other measures.

LP-EGR is achieved by redirecting exhaust from the turbine outlet to compressor inlet, between which the pressure differences generally are sufficient to drive the EGR flow of a desired amount. LP EGR can save exhaust energy to elevate boost pressure compared to HP EGR. With the elevated boost pressure and increased injection pressure, around 50% NOx reduction can be obtained at the same PM emission level compared to original HP EGR loop [8]. What's more, HP EGR and LP EGR can be combined, namely DL EGR, to further reduce NOx emission [13,14], and the strategy of adjusting HP EGR and LP EGR ratio with aim of balancing pumping loss and gross indicated thermal efficiency was put forward to further improve BSFC-NOx trade off relationship in a production turbocharged engine [15]. In the high load, LP EGR with low HP-EGR proportion can reach better boosting efficiency and fuel efficiency in a turbocharged diesel engine. From this point, LP EGR seems to be an ideal solution for NOx reduction in the high BTE region as well as engine performance optimization in the entire operating space, although LP EGR suffers from fouling damage of the compressor, response delay in the transient process and increased PM flow rate through the DPF for a given dilution ratio [16].

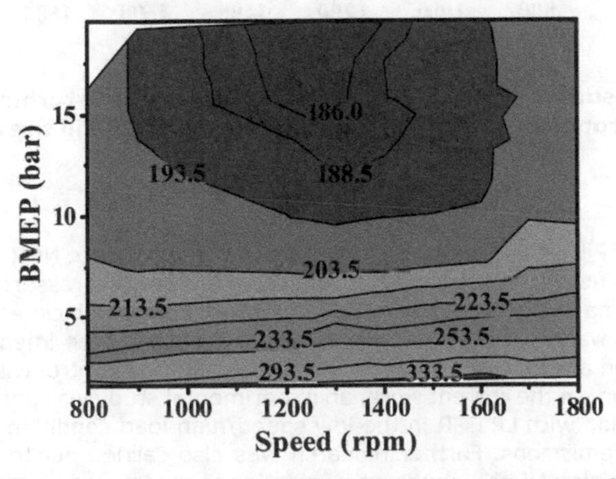


**Figure 1. BSFC universal characteristics of a high efficiency diesel engine.**

As previously mentioned in Ladommatos's research, water evaporation has more effect on NOx emissions than carbon dioxide. Indeed, addition of water to the combustion chamber has already been experimentally proved effective in reducing NOx emission [17–19]. Hountalas had comparatively evaluated EGR, intake water injection and fuel/water emulsion as NOx reduction techniques for heavy-duty diesel Engines by a using multi-zone model [20,21]. It was revealed that almost double amount of water is required for the same NOx reduction when using intake water addition compared to fuel/water emulsion. EGR and intake water addition resulted in an increase of BSFC, while water/fuel emulsion first lightly decrease BSFC, then the effect reverses with increasing load. Tauzia found that water mass of about 60-65% of the fuel mass is required to realize 50% NOx reduction in an intake port water addition method, and the NOx-PM trade-off relationship is improved in high load, and worsens in low and medium load compared to LP EGR [22,23]. While in-cylinder direct water injection succeeds in mitigating knock tendency and enhancing efficiency in GDI engine [24], direct water injection in diesel engine has also been studied. A hydraulic pump was utilized to achieve direct water injection with pressure up to 40 MPa in a common rail diesel engine, and test results showed that engine efficiency is improved by 5.35% and NOx emission is decreased by 10% by optimizing water injection timing at 1 MPa IMEP load [25], although the work consumed to inject high pressure water is not accounted for.

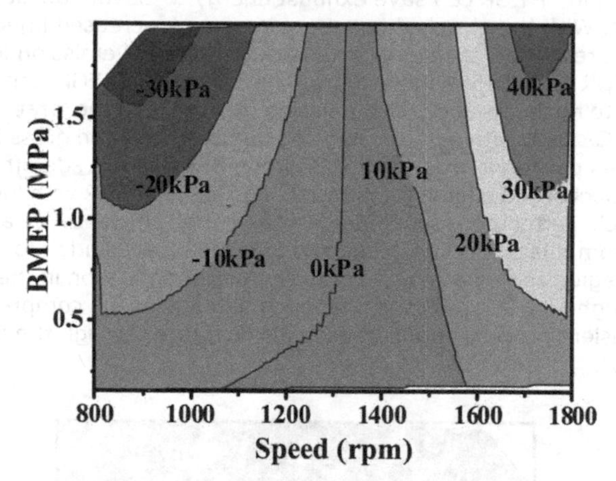


**Figure 2. Pressure differences between upstream of the turbine and downstream of the compressor of a heavy-duty diesel engine in the whole engine map [26].**

Therefore, both LP EGR and water addition in charge can reduce NOx emission. Since HP EGR is not sufficient to decrease NOx emission limited by pressure difference and LP EGR is not a mature technology to be implemented in a vehicle engine in case of response delay, water addition with electric controlled water injector into intake charge may be an alternative option, which is very easy to control without delay and inherent limitation. In the current work, an experimental study was performed to compare water addition with LP EGR in the low speed/high load condition with respect to combustion and emissions. Further research was also carried out to investigate the potential of combining HP EGR with water injection in the low load considering that HP EGR is a standard equipment in modern diesel engines.

The paper was organized as follows. Firstly, brief introduction of several water addition methods and the water injection strategy in this study will be given, Then, LP EGR and water was compared at 2.01 MPa BMEP condition with respect to combustion and emission as well as BSFC. Finally, the potential of combination of water injection and EGR on emissions and BSFC was investigated at 0.5MPa.

## 2 EXPERIMENT SETUP

### 2.1 Test bench

The experimental investigation was carried out on a four-stroke, six-cylinder, turbocharged, intercooled heavy-duty diesel engine equipped with common rail fuel injection. The detail specification of the engine is shown in Table 1 and the schematic diagram of the test bench is shown in Figure 3. More details about the test bench and data acquisition and processing can be found in reference [27].

To assess the fuel-oxy ratio more accurately, oxygen in recirculated exhaust gas is considered. The following equation is used to calculate the fuel-oxy equivalence ratio:

$$\text{Fuel} - \text{oxy equivalence ratio} = \frac{M_{fuel}*3.3}{\left[\frac{EGR*M_{air}}{100*1.293}*\frac{100}{100-EGR}*\frac{(O_2\%)\text{exhaust}}{100}+\frac{M_{air}*20.95}{1.293*100}\right]*1.429} \quad (1)$$

where EGR represents EGR ratio.

The charge mass flow is calculated by the following formula:

$$\text{Charge mass} = M_{air} + M_{water} + M_{air}*\frac{EGR}{100} \quad (2)$$

where $M_{water}$ denotes mass flow of injected water into the intake port and Mair represents fresh air mass flow.

**Table 1. Specifications of the test engine.**

| Engine Type | DI inline 6, water cooled |
|---|---|
| Engine Displacement | 8.42L |
| Rated power | 243kW@2200r/min |
| Maximum torque | 1350N·m@1200~1800r/min |
| Compression Ratio | 16.8:1 |
| Swirl ratio | 1.25 |
| Bore×Stroke | 113×140mm |
| Connecting Rod Length | 209mm |
| Number of valve | 4 |
| Combustion Chamber | Re-entrant type |
| Injection system | Common rail |
| Maximum Injection Pressure | 160MPa |
| Injector Nozzle | Φ0.163mm×8×148° |

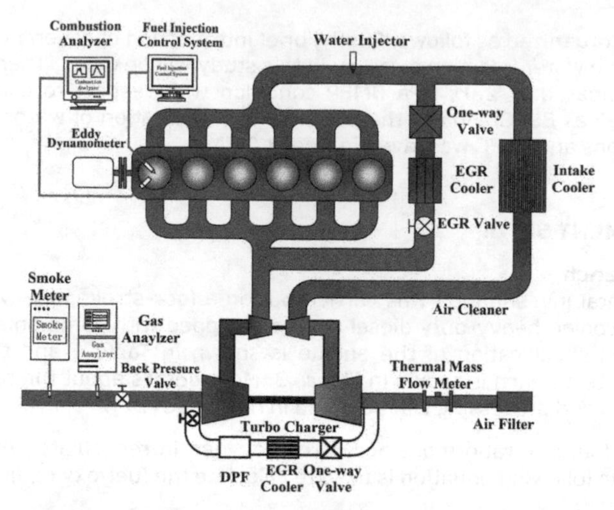


**Figure 3. Schematic diagram of the test bench.**

## 2.2 Water introduction method

Controversy always exists in engine reliability and durability when Introducing water into engine. The controversial issue is that possible wall wetting and lubricant dilution caused by water involved in combustion will lead to engine failure. In the following, the three main methods to introduce water into combustion will be concisely reviewed, and the pros and cons are concluded as follow.

### 2.2.1 *Fuel emulsion*

Emulsified diesel oil mostly exists in the form of water in oil. Since the boiling point of water is lower than diesel oil, water will evaporate earlier than diesel oil with the rapid rise of temperature, which could lead to the micro explosion/puffing phenomenon and strengthen the fuel-air mixing quality [21,28]. However, emulsified oil has changed the physical and chemical characteristics of the fuel, and there are problems such as power reduction, fuel consumption deterioration under high load, mainly due to the increase of injection duration [21]. Meanwhile, due to the immiscibility of water and oil, the problem of oil-water stratification is still a problem to be solved.

### 2.2.2 *In-cylinder water injection*

In-cylinder water injection requires a separate injector [25] or a fuel-water co-injector [17]. It has been proven that liquid water injected into combustion is the most effective way to decrease local combustion temperature and NOx emission. A special injector was invented to sequentially inject fuel-water-fuel from one injector. Unfortunately, there are still technical obstacles in the combination of high-pressure common rail and fuel-water co-injectors, since it is designed to cooperate with an in-line fuel pump.

### 2.2.3 *Port water injection*

Port water injection is the simplest and cheapest water induction method with minor modification to the engine. Compared with emulsified oil, there is no problem with oil-water stratification, but there is the possibility of lubricating oil dilution similar to in-cylinder water injection. Besides, the NOx reduction effect of port water injection is not as efficient as emulsified fuel [20].

Considering the ease of implementation and engine reliability, a special port water injection strategy was selected as the water addition approach in this study. Since water is sprayed into upstream of the intake port with length enough for the complete evaporation of water, the water enters the cylinder in the form of steam, and thus there will be no consequence of engine wear and tear. An electronic controlled gasoline injector was modified to continuously spray water into intake port with 5 bar pressure difference. During the test, the intake temperature downstream of the water injector is monitored and maintained constant with the feedback control of the intake cooler. Assuming the humidity of fresh air is 0%, that the water enters the cylinder as vapor is guaranteed by limiting the injected water mass below the moisture content of saturated wet air at the fixed temperature and pressure, which is set as maximum allowable water injection mass. The maximum allowable water injection mass is calculated by the following formula:

$$M_w = W_s * M_{air} * 10^{-3} \tag{3}$$

where $M_w$ represents the maximum allowable water injection mass flow rate in kg/h; $W_s$ expresses the saturated water content per kilogram of saturated wet air at a certain temperature and pressure in g/kg; and $M_{air}$ indicates the fresh air mass flow coming from ambient in kg/h. With known intake temperature and pressure, $W_s$ can be derived from the ASHRAE formulation [29]. The experiment was conducted at two operation points at 1137 r/min, and the boundary condition and parameter variable range for each operating condition are shown in Table 2.

**Table 2. Boundary condition and parameter variable range for each operating condition.**

| BMEP | 0.5 MPa | 2.01Mpa |
|---|---|---|
| Fuel | commercial 0# diesel | ← |
| Injection Pressure | 110Mpa | ← |
| Injection strategy | Pilot (fixed 5mg)-Main | single |
| Boost Pressure | Varied with HP EGR | 0.22~0.23MPa |
| CA50 | 4°CA ATDC | 13°CA ATDC |
| Coolant temperature | 85±2℃ | ← |
| Intake Temperature | 40℃ | 45℃ |
| EGR type | HP EGR | LP EGR |
| $W_s$ | 40g/kg | 28.4g/kg |
| $M_{air}$ | ~405 kg/h | ~650 kg/h |
| $M_w$ | 0-16.2kg/h | 0-18.4kg/h |

CA50 of combustion was controlled at 4°ATDC and 13°ATDC for the 0.5 MPa and 2.01 MPa BMEP condition respectively by adjusting the injection timing. The water used in the test was deionized water. For practical application, it may be a feasible scheme to utilize the water recycled from the exhaust.

## 3 RESULTS AND DISCUSSIONS

### 3.1 Comparison of water injection and LP EGR under high load

As depicted in Figure 4, The effect of water injection and EGR on the intake charge characteristic is shown. There is no data point for water injection on the left-hand side of the black dashed line since the injected water mass exceeds the maximum allowable value in that region. It is obvious that, constrained by the maximum water addition amount, the NOx emission can be reduced to about 9-10 g/kWh with water injection, and the water/fuel mass ratio is about 60 percent. However, EGR can reduce the NOx emission to 4 g/kwh. With the NOx emission level reduced, intake pressure increases for both the water injection case and the EGR case, resulting in increased charge mass flow. For the exhaust temperature, the EGR case makes it increase slightly, but it monotonically decreases for the water injection case, which may cause declined SCR NOx conversion efficiency in the low load because of the low exhaust temperature. Unlike the EGR replacing the fresh air with exhaust gas, water injection boosts the charge thermal capacity by adding extra water into the charge without lower oxygen concentration. As a result, the charge mass flow rate with water injection is higher than EGR, with a consequently lower fuel-oxy equivalence ratio.

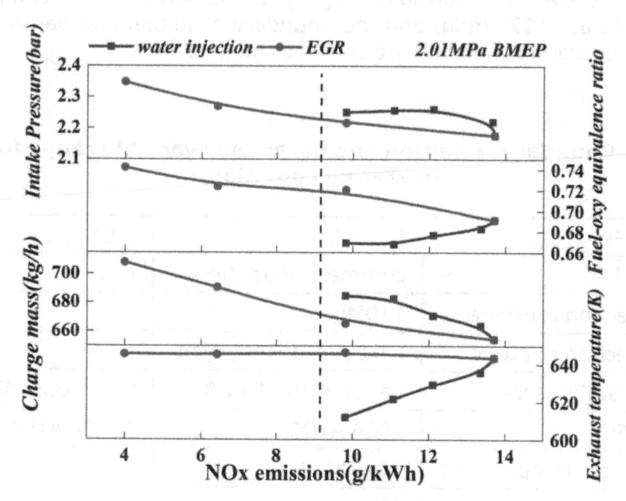


**Figure 4. The effect of water injection and EGR on intake charge characteristics at 2.01MPa BMEP.**

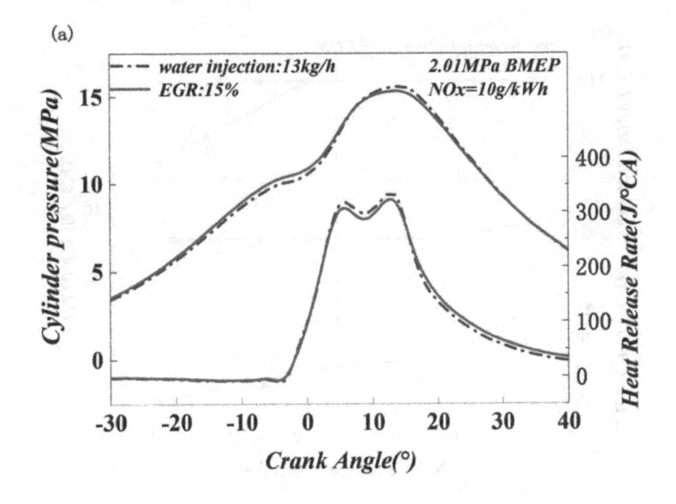


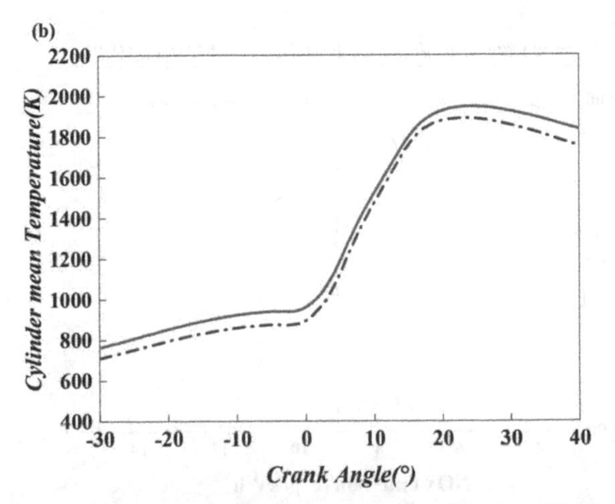


**Figure 5. Comparison of water injection and EGR on cylinder pressure HRR (a), and cylinder mean temperature (b) at 2.01MPa BMEP.**

Figure 5 reveals the effect of water injection and EGR on cylinder pressure, HRR and cylinder mean temperature at the same NOx emission level. It can be concluded from the figure that water injection can more efficiently lower the mean cylinder temperature than EGR because of the higher intake charge mass flowrate. Furthermore, the lower cylinder temperature of the water injection case has no negative effect on the main heat release compared to the EGR case, but it does slow down the reaction rate in the afterburning period of the expansion stroke.

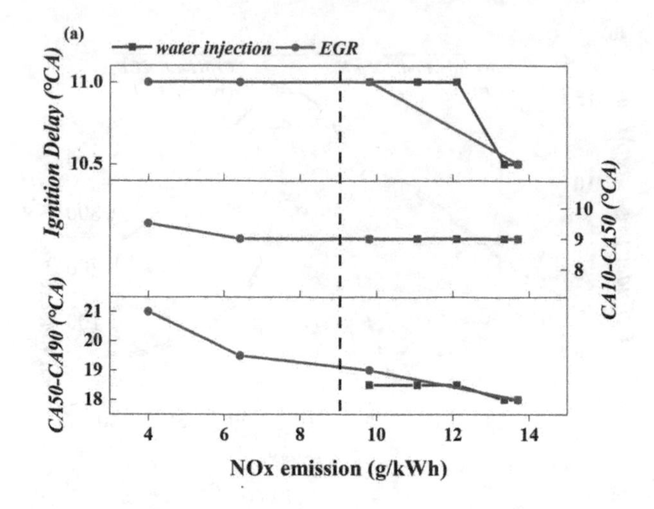


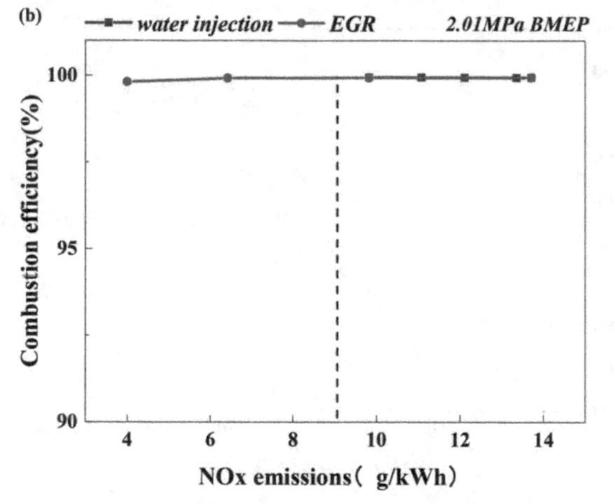


**Figure 6. Comparison of the effect of water injection and EGR on combustion characteristic (a) and combustion efficiency (b) at 2.01Mpa BMEP.**

More detailed information about the combustion characteristics is calculated and shown in Figure 6. The combustion efficiency of water injection is the same as the EGR case with lower temperatures under the high load. It is well known from the Arrhenius equation that high temperature accelerates reaction rate. Although the mean cylinder temperature of the EGR case is higher than the water injection case, the effect is likely offset by the lower fuel-oxy mean equivalence ratio, leading to longer CA50-CA90 duration for the EGR case. In addition, it is worth noting that the resolution of cylinder pressure acquisition is 0.5°CA, therefore the minor difference of the influence of EGR and water injection on ignition delay and CA10-CA50 duration cannot be identified below that resolution.

Shown in Figure 7 is the effects of water injection and EGR on emissions and BSFC at 2.01 MPa BMEP. Under high load, HC emissions of both water injection and EGR cases stay nearly constant with the NOx emission reduction. The soot and CO emissions of

EGR are inferior to the those of water injection. Therefore, the soot-NOx trade-off relationship is improved by water injection strategy. In terms of BSFC-NOx tradeoff relationship, water injection has a better performance at the high load, mainly attributable to the lower heat transfer loss of the water injection case due to lower cylinder temperature and unaffected combustion phasing.

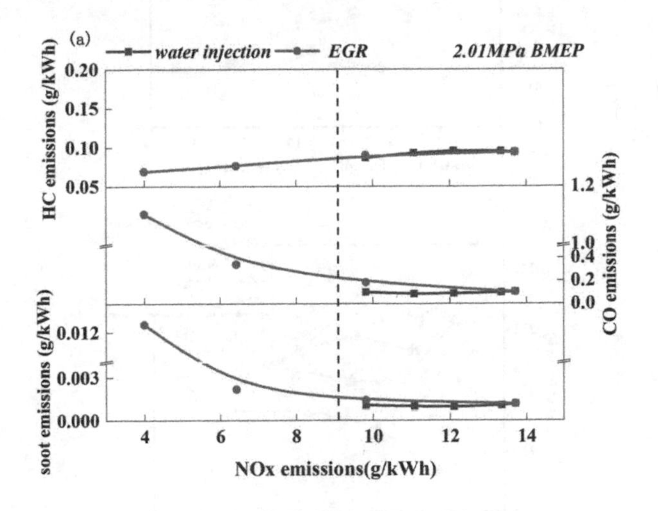


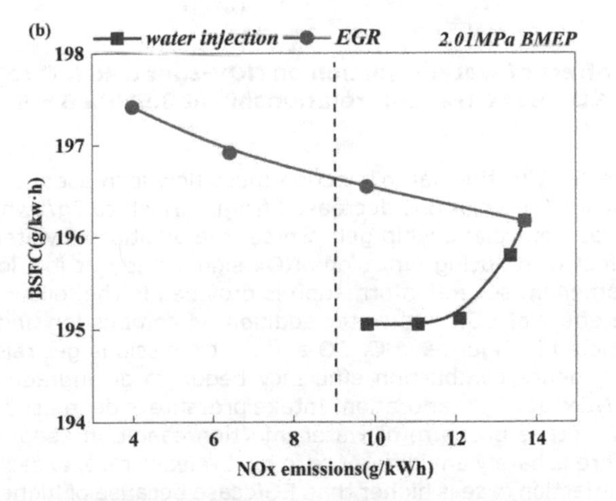


**Figure 7. The influence of water injection and EGR on emissions (a) and BSFC (b) at 2.01Mpa BMEP.**

### 3.2 Combination of water injection and HP EGR under the low load condition

It can be seen from the last section that water injection is a viable strategy to achieve NOx emissions reduction down to 9.4 g/kWh without BSFC sacrifice, which is an acceptable level for current production diesel engines. To further reduce NOx emissions at low load to meet the future emission regulation, the potential of combining water injection and HP EGR at 0.5 MPa BMEP was also tested in this study.

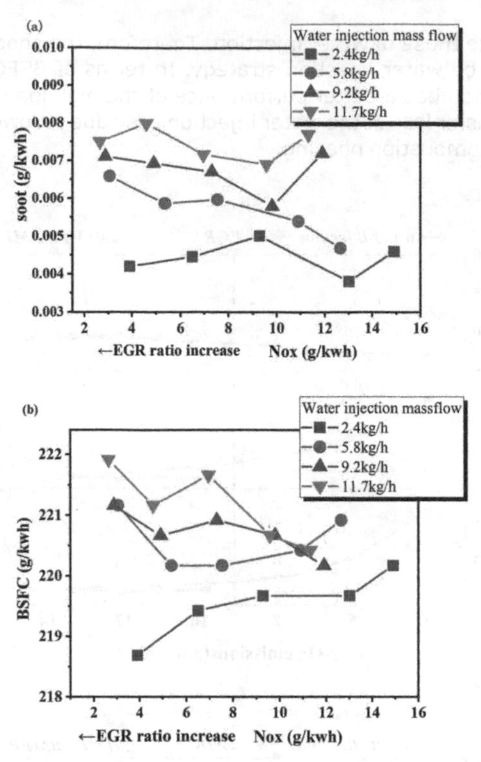


**Figure 8. The effect of water injection on NOx-soot tradeoff relationship and NOx-BSFC tradeoff relationship at 0.5MPa BMEP.**

As shown in Figure 8, With the water injection mass flow increased from 2.4kg/h to 11.7kg/h, the lowest NOx emissions decreased from 4g/kwh to 2g/kwh, at the same time, NOx-BSFC trade off relationship gets worse. the addition of water did not produce expected effect of reducing emissions NOx significantly at low load. Trying to explain this phenomenon, several information is provided in the following figures. In these figures, the effect of EGR and water addition on combustion and emissions is compared. As depicted in Figure 9, HC, CO and soot emissions get raised With NOx emissions reduced, hence combustion efficiency begun to downgrade in Figure 10, which account for NOx-BSFC deterioration. Intake pressure is decreased with increase of EGR ratio, while intake pressure of water injection condition keeps steady. Seen from Figure 12, there is barely any difference in heat release rate, except that cylinder pressure of water injection case is higher than EGR case because of higher intake pressure. Besides, the cylinder mean temperature of the water injection case is lower than the EGR case, which is accounts for the increased emissions of CO and HC.

Concerning PM (particulate matter) emissions, variations of flame temperature, global AFR and flame lift-off length all have effects on the soot production rate [30]. It is obviously that water injection has advantages in fuel-air equivalence compared to EGR. The longer the ignition delay is, the longer flame lift-off length will be [31,32]. Since the heat release rate profile has no much difference, it means that the flame lift-off length of them are almost equal. The high oxygen concentration is unfavorable for soot formation, and lower flame temperature will slow down the oxidation rate of soot, which will promote the accumulation of soot. It can be inferred that Lower flame temperature of plays dominant role in soot emissions at 0.5MPa BMEP.

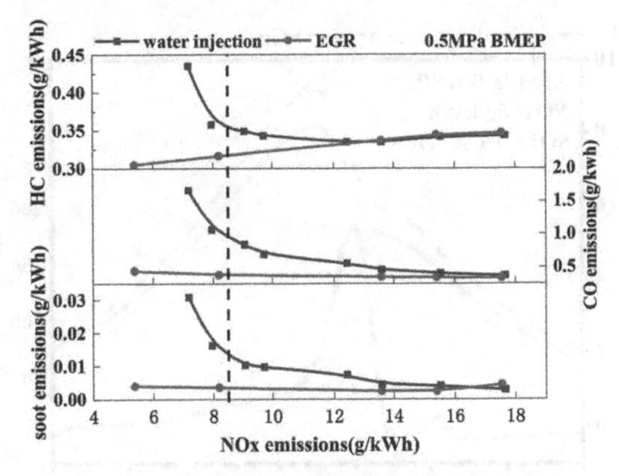


**Figure 9. CO, HC, soot Emissions relationship verse NOx emissions of water injection and EGR at 0.5MPa BMEP.**

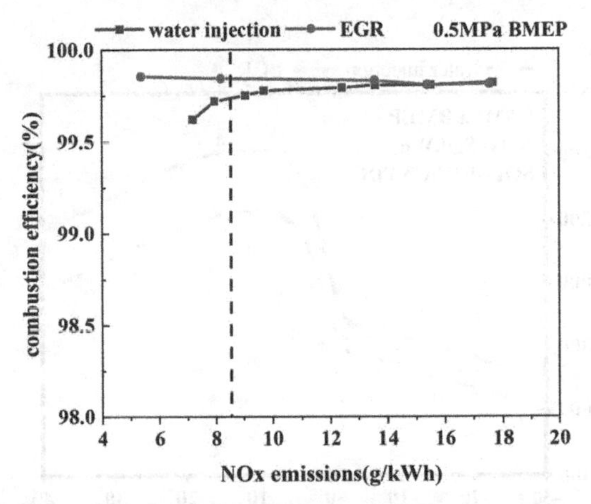


**Figure 10. The effect of water injection and EGR on combustion efficiency at 0.5MPa BMEP.**

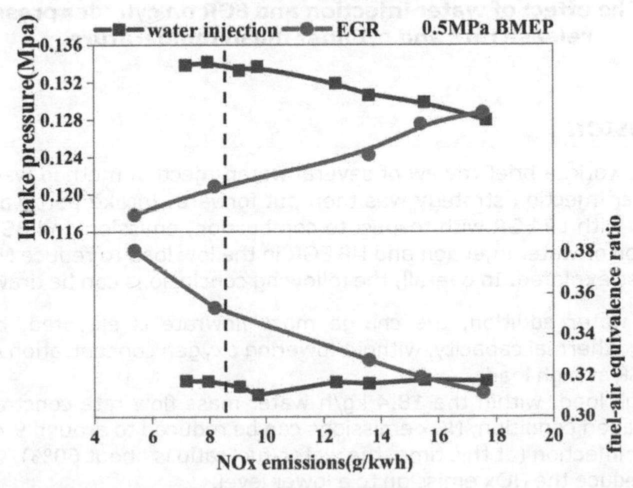


**Figure 11. The effect of water injection and EGR on intake pressure and fuel-air equivalence ratio.**

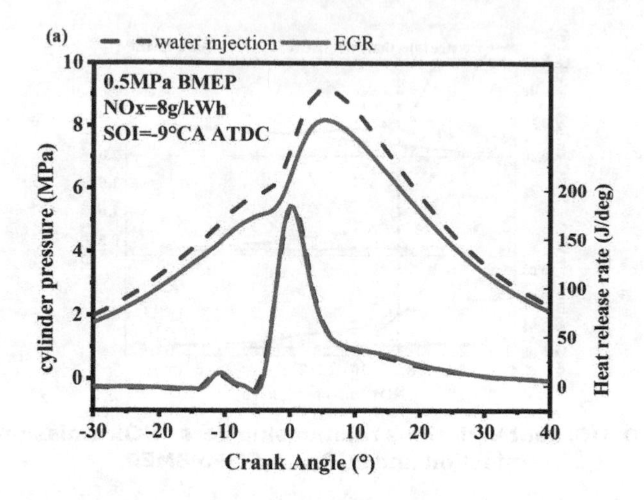


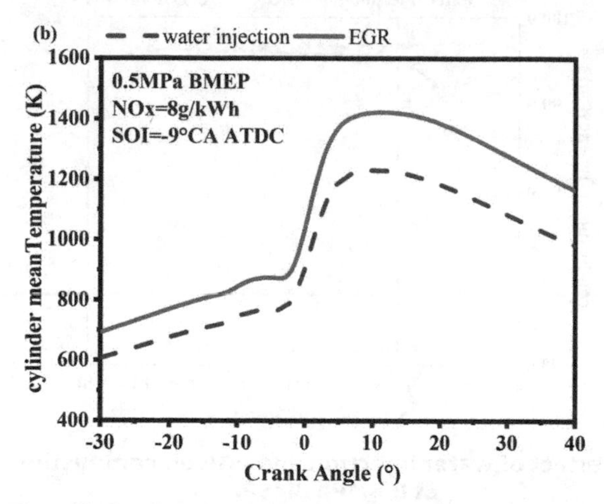


**Figure 12. The effect of water injection and EGR on cylinder pressure, heat release rate and cylinder mean temperature.**

## 4 CONCLUSION

In the present work, a brief review of several water injection method was made, and a feasible water injection strategy was then put forward. Intake port water injection was compared with LP EGR with respect to combustion, emission and BSFC. Besides, the combination of water injection and HP EGR in the low load to reduce NOx emission further was also explored. In overall, the following conclusions can be drawn:

- With water addition, the charge mass flowrate is elevated, boosting the charge thermal capacity, without lowering oxygen concentration compared to LP EGR at high load.
- At high load, within the 18.4 kg/h water mass flow rate constrained by the saturation condition, NOx emissions can be reduced to around 9.4g/kWh with water injection (at this time, the water/fuel ratio is about 60%), while LP EGR can reduce the NOx emission to a lower level.

- Although water injection can lower cylinder combustion temperature more obviously than EGR, the combustion efficiency and combustion phasing of water injection combustion is not affected.
- Water injection has better NOx-PM and NOx-BSFC relationship than the LP EGR case at high load; it is therefore better to use water injection to reduce NOx emission in high BTE regime.
- If the NOx emission at high load is expected to be controlled to a high level without BSFC sacrifice, e.g. 9g/kwh, water injection as steam strategy is a reliable and durable option.
- The lowest NOx emissions changes slightly decreased with the variation of water addition amount combined with HP EGR at 0.5MPa BMEP, and the NOx-BSFC trade-off relationship gets worse simultaneously. lower combustion efficiency, higher CO, HC emissions resulted from water injection is on account of the lower combustion temperature compared with the EGR case.
- Water injection did not take expected effect because of lower combustion temperature. In contrast to low load, the high combustion temperature of high load and air-fuel ratio advantages may be the factor that NOx-soot and NOx-BSFC get improved at 2.01MPa BMEP.
- The result of combined water injection and EGR indicate that it is better to use HP EGR solely to reduce NOx emission at low load.

**ACKNOWLEDGMENTS**

The authors would like to acknowledge the financial support provided by the National Key R&D Program of China (2017YFB0103501).

## REFERENCES

[1] Joshi A., 2021,"Review of Vehicle Engine Efficiency and Emissions," SAE International400 Commonwealth Drive, Warrendale, PA, United States.

[2] Reitz R. D., Ogawa H., Payri R., Fansler T., Kokjohn S., Moriyoshi Y., Agarwal A. K., Arcoumanis D., Assanis D., Bae C., Boulouchos K., Canakci M., Curran S., Denbratt I., Gavaises M., Guenthner M., Hasse C., Huang Z., Ishiyama T., Johansson B., Johnson T. V., Kalghatgi G., Koike M., Kong S. C., Leipertz A., Miles P., Novella R., Onorati A., Richter M., Shuai S., Siebers D., Su W., Trujillo M., Uchida N., Vaglieco B. M., Wagner R. M., and Zhao H., 2019, "IJER editorial: The future of the internal combustion engine," International Journal of Engine Research, 146808741987799.

[3] Mohr D., Shipp T., and Lu X., 2019,"The Thermodynamic Design, Analysis and Test of Cummins' Supertruck 2 50% Brake Thermal Efficiency Engine System," *WCX SAE World Congress Experience*, SAE International400 Commonwealth Drive, Warrendale, PA, United States.

[4] Vojtech R., 2018,"Advanced Combustion for Improved Thermal Efficiency in an Advanced On-Road Heavy Duty Diesel Engine," *WCX World Congress Experience*, SAE International400 Commonwealth Drive, Warrendale, PA, United States.

[5] Mohiuddin K., Kwon H., Choi M., and Park S., 2021,"Effect of engine compression ratio, injection timing, and exhaust gas recirculation on gaseous and particle number emissions in a light-duty diesel engine," Fuel, 294(5), p. 120547.

[6] Laguitton O., Crua C., Cowell T., Heikal M. R., and Gold M. R., 2007,"The effect of compression ratio on exhaust emissions from a PCCI diesel engine," Energy Conversion and Management, 48(11), pp. 2918–2924.

[7] Ming Z., Reader G. T., and Hawley J.G., 2004,"Diesel engine exhaust gas recirculation--a review on advanced and novel concepts," Energy Conversion and Management, 45(6), pp. 883–900.
[8] Maiboom A., Tauzia X., Shah S. R., and Hétet J.-F., 2009,"Experimental Study of an LP EGR System on an Automotive Diesel Engine, compared to HP EGR with respect to PM and NOx Emissions and Specific Fuel Consumption," SAE Int. J. Engines, 2(2), pp. 597–610.
[9] N. Ladommatos, S. M. Abdelhalim, H. Zhao, and Z. Hu, 1997,"The Dilution, Chemical, and Thermal Effects of Exhaust Gas Recirculation on Diesel Engine Emissions - Part 3: Effects of Water Vapour,"
[10] N. Ladommatos, S. M. Abdelhalim, H. Zhao, and Z. Hu, 1997,"The Dilution, Chemical, and Thermal Effects of Exhaust Gas Recirculation on Diesel Engine Emissions - Part 4: Effects of Carbon Dioxide and Water Vapour,"
[11] H Zhao, J Hu, and N Ladommatos, 2002,"In-cylinder studies of the effects of CO2 in exhaust gas recirculation on diesel combustion and emissions,"
[12] Liu J., Wang H., Zheng Z., Zou Z., and Yao M., 2016,"Effects of Different Turbocharging Systems on Performance in a HD Diesel Engine with Different Emission Control Technical Routes," SAE International400 Commonwealth Drive, Warrendale, PA, United States.
[13] Giorgio Z., and Massimo C., 2013,"Influence of high and low pressure EGR and VGT control on in-cylinder pressure diagrams and rate of heat release in an automotive turbocharged diesel engine," Applied Thermal Engineering, 51(1-2), pp. 586–596.
[14] Giorgio Z., and Massimo C., 2012, "Experimentalstudy on the effects of HP and LP EGR in an automotive turbocharged diesel engine," Applied Energy, 94, pp. 117–128.
[15] Youngsoo P., and Choongsik B., 2014, "Experimentalstudy on the effects of high/low pressure EGR proportion in a passenger car diesel engine," Applied Energy, 133, pp. 308–316.
[16] Millo F., Giacominetto P. F., and Bernardi M. G., 2012,"Analysis of different exhaust gas recirculation architectures for passenger car Diesel engines," Applied Energy, 98, pp. 79–91.
[17] Susumu K., Kazutoshi M., and Kenji S., 1996,"Reduction of Exhaust Emission with New Water Injection System in a Diesel Engine,"
[18] M. A. A. Nazha, H. Rajakaruna, and S. A. Wagstaff, 2001,"The Use of Emulsion, Water Induction and EGR for Controlling Diesel Engine Emissions,"
[19] R. Udayakumar, and S. Sundaram, 2003,"Reduction of NOx Emissions by Water Injection in to the Inlet Manifold of a DI Diesel Engine,"
[20] D. T. Hountalas, G. C. Mavropoulos, and T. C. Zannis, 2007,"Comparative Evaluation of EGR, Intake Water Injection and Fuel/Water Emulsion as NOx Reduction Techniques for Heavy Duty Diesel Engines,"
[21] D. T. Hountalas, G. C. Mavropoulos, T. C. Zannis, and S. D. Mamalis, 2006,"Use of Water Emulsion and Intake Water Injection as NOx Reduction Techniques for Heavy Duty Diesel Engines,"
[22] Samiur R. S., Alain M., Xavier T., and Jean-François Hétet, 2009,"Experimental Study of Inlet Manifold Water Injection on a Common Rail HSDI Automobile Diesel Engine, Compared to EGR with Respect to PM and NOx Emissions and Specific Consumption,"
[23] Tauzia X., Maiboom A., and Shah S. R., 2010,"Experimental study of inlet manifold water injection on combustion and emissions of an automotive direct injection Diesel engine," Energy, 35(9), pp. 3628–3639.
[24] Mingrui W., Thanh Sa N., Turkson R. F., Jinping L., and Guanlun G., 2017,"Water injection for higher engine performance and lower emissions," Journal of the Energy Institute, 90(2), pp. 285–299.

[25] Jiang L., Kang Z., Zhang Z., Wu Z., Deng J., Hu Z., and Li L., 2017,"Effect of Direct Water Injection Timing on Common Rail Diesel Engine Combustion Process and Efficiency Enhancement," SAE International400 Commonwealth Drive, Warrendale, PA, United States.

[26] Gui-sheng CHEN,Ying-gang SHEN, Zun-qing ZHENG, 2014,"Performance and Emission Characteristics of HD Diesel Using EGR Under Low-Speed and High-Load Conditions(in chinese)," Transactions of CSICE, 02, pp. 97–103.

[27] Liu J., Wang H., Zheng Z., Li L., Mao B., Xia M., and Yao M., 2018,"Improvement of high load performance in gasoline compression ignition engine with PODE and multiple-injection strategy," Fuel, 234(11), pp. 1459–1468.

[28] Kadota T., and Yamasaki H., 2002,"Recent advances in the combustion of water fuel emulsion," Progress in Energy and Combustion Science, 28(5), pp. 385–404.

[29] Ličina V F, Cheung T, and Zhang H, 2018,"Development of the ASHRAE Global Thermal Comfort Database II," Building and Environment, 142, pp. 502–512.

[30] Zhu S., Hu B., Akehurst S., Copeland C., Lewis A., Yuan H., Kennedy I., Bernards J., and Branney C., 2019,"A review of water injection applied on the internal combustion engine," Energy Conversion and Management, 184(1A), pp. 139–158.

[31] Zhong W., Pachiannan T., He Z., Xuan T., and Wang Q., 2019,"Experimental study of ignition, lift-off length and emission characteristics of diesel/hydrogenated catalytic biodiesel blends," Applied Energy, 235(2), pp. 641–652.

[32] Pickett L. M., Siebers D. L., and Idicheria C. A., 2005,"Relationship Between Ignition Processes and the Lift-Off Length of Diesel Fuel Jets," SAE International400 Commonwealth Drive, Warrendale, PA, United States.

*International Conference on Powertrain Systems for Net-Zero Transport*
*Institution of Mechanical Engineers, ISBN 978-1-032-11281-7*

# Heavy duty hydrogen ICE: Production realisation by 2025 and system operation efficiency assessment

**S.J. Mills CEng**
AVL Powertrain UK Ltd

## ABSTRACT

The concept of hydrogen powered internal combustion engines has seen many false dawns in recent years. But today, with the drive for zero emission powertrains and the promised development of robust fuel cell technologies, a hydrogen ICE future may well be coming to complement our future electrified world.

Issues with producing and distributing high purity hydrogen gas, and the need to realise the return on investment of existing products & capital equipment, makes a hydrogen burning ICE engine very attractive, particularly in the off highway and heavy duty sectors and in developing markets. But it is no simple conversion. To achieve an acceptable thermal efficiency, high lambda values and diesel-like peak firing pressures (PFP) are required. New combustion systems for High Pressure Direct Injection (HPDI) need to be developed, along with the associated hardware for injection and ignition strategies. Exhaust aftertreatment systems need to specified & optimised for NOx conversion, and new engine control strategies will be required.

This paper will present AVL proposals and test results in the development of H2 ICE concepts. CFD modelling results for hydrogen combustion systems will be discussed, including experience and knowledge in model parameterisation & calibration. Development status for delivery of 200+ bar PFP engines and the associated injection strategies will also be presented.

Based on extensive in-house modelling and laboratory correlation at an engine and vehicle system level, AVL will present and discuss possible H2 ICE configurations utilising parallel hybrid and waste heat recovery technologies. These will be shown to potentially offer a reduced system level H2 consumption for a number of on and off highway applications (including a typical long haul heavy duty trucks) that approach or even better fuel cell efficiency at a vehicle level.

## 1 INTRODUCTION

On highway heavy duty (HD) vehicles are responsible for around 6% of all $CO_2$ emissions in the European Union, accounting for around 25% of all $CO_2$ emissions caused by road transport. Without further technical development, it is anticipated these greenhouse gas emissions (GHG) will increase by around 10% by 2030 [1]. In order to achieve an immediate effect, the European Union has introduced stringent $CO_2$ reduction targets for HD vehicles. Starting with the 2019/20 baseline fleet emissions, each European manufacturer is required to reduce the OEM specific fleet $CO_2$ emissions by 15% and 25% by 2025 and 2030 respectively. To determine & model these $CO_2$ emissions, the Vehicle Energy Consumption Calculation Tool (VECTO) was introduced [2].

DOI: 10.1201/9781003219217-5

Significant penalty payments are being introduced for non-compliant OEM's. The emission targets can be achieved by a combination of different mechanisms. These include diesel internal combustion engine (ICE) efficiency improvements, reductions in vehicle weight & powertrain losses, as well as improvement in vehicle aerodynamic drag. However, the biggest effect is likely to be achieved by the introduction of low-emission or zero-emission HD vehicles [1].

Zero-emission HD vehicles are defined as vehicles either without an ICE, or with an internal combustion engine that emits less than 1 $gCO_2/kWh$. A hydrogen powered ICE can be classified as a Zero-emission power source; this is additionally important as it qualifies for 'super credits' in the reporting period before 2025 [1].

Hydrogen ICE technology would therefore offer a significant $CO_2$ reduction in the VECTO-based fleet calculation with conventional vehicle technology. With the possibility of early market introduction, hydrogen ICE powered trucks could therefore help to create demand for hydrogen fuelling infrastructure ahead of 2025.

A future green & renewable hydrogen economy is also seen as a critical enabler for lower GHG generation in other markets, including as Japan & the USA. The 2017 Japanese Basic Hydrogen Strategy targets the cutting of GHG's by 26% by 2030 using hydrogen as an energy carrier [3]. The U.S. Department of Energy has specifically referred to the usage of hydrogen as a fuel for ICE in their 2020 Hydrogen Program Plan [4], and China has recently defined H2-ICE as a 'clean vehicle'.

Hydrogen ICE will also offer a 'stepping-stone' to the introduction of robust Fuel Cell (FC) products to the market. Hydrogen ICE will pave the way for the introduction of the fuel storage & distribution networks, and it will enable the life of valuable current OEM manufacturing assets to be extended for maximum cost recovery.

During 2020 & 2021, AVL has undertaken significant experimentation and development work on a multi cylinder hydrogen ICE, developed from a current on-highway, heavy duty 12.8ltr engine. The thermodynamic target for the engine conversion was 350kW output with a BMEP value of up to 24 bar, and to fulfil at least Euro VI emission standards. The conversion to hydrogen combustion was undertaken internally within AVL, utilising a maximum number of carry-over components as possible. The engine has been operated in multiple configurations, including MPI & DI, and extensive & valuable results have been obtained.

In order to explore the potential of hydrogen ICE and their production realisation to 2025, thermodynamic principles, pollutant control devices, and engine architectures will be discussed in this paper, as well the potential system operational efficiency compared to FC vehicles.

## 2 HYDROGEN AS A FUEL FOR ICE

In order to identify and define the most promising concepts for hydrogen combustion ICE, an understanding of hydrogen-specific physical properties is required. Table 1 shows selected parameters and their comparison to diesel and methane fuel.

**Table 1. Properties of Hydrogen fuel compared to Diesel & Methane.**

| Property | Unit | Diesel (EU B7) | Methane ($CH_4$) | Hydrogen ($H_2$) |
|---|---|---|---|---|
| Gas density @ 1 bar | kg/m$^3$ | 0.835 | 0.716 | 0.09 |
| Lower calorific value | MJ/kg | 42.6 | 50 | 120 |
| Stoichiometric air-to-fuel ratio | kg/kg | 14.5 | 17.2 | 34.3 |
| Mixed calorific HV (MPI) | MJ/m$^3$ | – | 3.4 | 3.2 |
| Mixed calorific HV (DI) | MJ/m$^3$ | 3.8 | 3.8 | 4.5 |
| Min. ignition energy | mJ | 0.24 | 0.29 | 0.017 |
| Flammability limit – excess air ratio | – | 1.35 – 0.48 | 0.6 – 2.0 | 0.13 – 10 |
| Auto-ignition temperature | °C | ≈ 225 | 595 | 585 |
| RON/MN | – | – | 130/100 | –/0 |

Hydrogen's properties and its suitability for usage in ICE has been extensively discussed in literature [5]. The main conclusions for commercial engine application can be given as follows:

- Internal mixture formation (direct injection; DI) shows increased mixed calorific heating value in comparison to Multi-Point Injection (MPI), hence the potential for high engine power density.
- NOx generated during operation needs to be delt with.
- The minimum ignition energy is considerably lower compared to other fuels, leading to high sensitivity to undesired pre-ignition.
- Auto-ignition temperature is high and comparable to that of methane, hence achieving auto-ignition (as in a diesel combustion process) is challenging.

## 3 HYRDOGEN ICE $NO_X$ CHARACTERISTICS

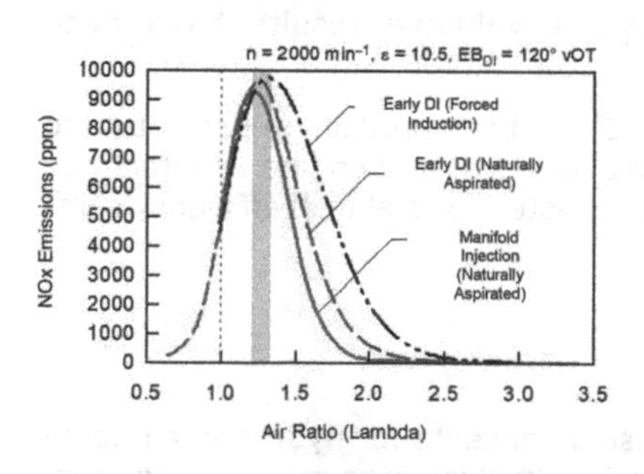


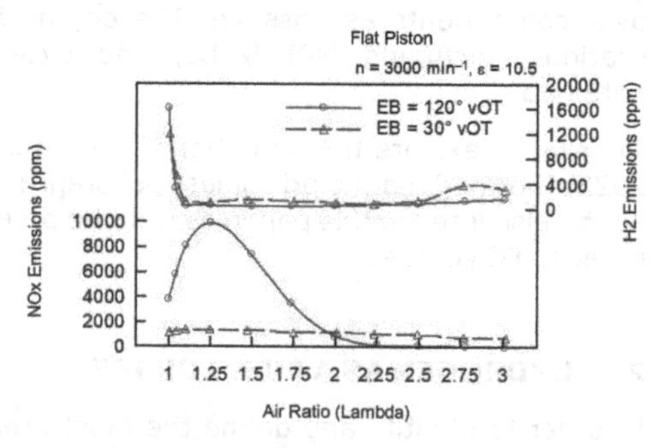


**Figure 1. Hydrogen ICE $NO_x$ characteristics [6].**

Figure 1 shows the general dependency of raw nitrogen oxide ($NO_x$) emissions and excess air ratio Lambda (λ). The maximum raw $NO_x$ emissions are reached at $\lambda = 1.2$ to 1.3. Mixture formation, engine load, combustion timing, and charging rate impact the characteristics of $NO_x$ emissions, especially in the area with high excess air ratio.

Almost $NO_x$-free combustion is reached between an excess air ratio of 2.0 – 2.8, depending on given boundaries (Figure 1, left diagram). In contrast to pre-mixed homogeneous charge, late direct injection leads to a strong stratified charge, resulting in little dependency of raw $NO_x$ emissions based on global excess air ratio (Figure 1, right diagram, red curve).

## 4 OVERVIEW OF PROMISING HYDROGEN COMBUSTION CONCEPTS

Combustion concepts can be categorised into spark ignited concepts with a homogeneous pre-mixed charge and diesel/compression ignited concepts; see Table 2.

Spark ignited hydrogen engines using pre-mixed homogeneous charge enable 100% $CO_2$ reduction potential (excluding minor emissions from burnt lube oil or SCR reagent). The achievable power density is strongly depending on the excess air ratio at full load conditions and corresponding allowable engine out raw $NO_x$ emissions. (Refer raw $NO_x$ emission behaviour as shown in Figure 1). The main engine performance related challenges for spark ignited hydrogen concepts are power density, fuel efficiency, and transient performance. For low pressure hydrogen concepts, a moderate hydrogen injection pressure is required and no additional compression system is needed [7]. The hydrogen can be directly supplied from pressure tanks (350-700 bar.)

For high pressure concepts, an injection pressure level of 250-300 bar is required. Hence an additional compression system is needed to be able to use the maximum capacity of the tank system. Liquid hydrogen ($LH_2$) could be a suitable alternative with lower on-board energy consumption. However, from a reliability standpoint the combination of high pressures and low temperatures makes the industrialization of $LH_2$-based high pressure systems questionable before 2025.

**Table 2. Commercial hydrogen combustion engine concepts.**

| | **Homogeneous Combustion/Spark Ignited** | | **Diffusion Combustion/Diesel Ignited** | |
|---|---|---|---|---|
| | **Multi-Point Injection (MPI)** | **Low Pressure DI** | **High Pressure DI (Diesel Pilot)** | **High Pressure DI** |
| **$H_2$ injection pressure** | 5 - 60 bar | | 250 - 300 bar | |
| **Mixture formation** | Swirl | Swirl/ Tumble | Swirl | |
| **Ignition** | Spark ignited | | Diesel ignited | Carbon neutral ignited |
| **Minimum excess air ratio** | 1.6 - 1.9 | | 1.3 - 1.7 | |
| **BMEP** | < 24 bar | | > 24 bar | |
| **BTE** | 42% | 42%/43% | 47% | 46% |
| **Compression ratio** | 12:1 | | 17:1 | |
| **PFP** | 180 bar | | > 220 bar | |
| **EAS** | (Oxidation catalyst) + SCR | | (Oxidation catalyst) + Particulate filter + SCR | (Oxidation catalyst) + SCR |

### 4.1 Multi Point Injection (MPI)

In order to achieve the highest engine component commonality with existing diesel or gas engines for production purposes, multi-point injection concepts are attractive. The lower mixed calorific heating value (compared to direct injection concepts) can be partly compensated for by lower excess air ratio required at full load for the same engine out $NO_x$ emissions due to better homogenisation of charge. However, the risk of undesired pre-injection and backfire, increasing over engine operational lifetime, could inhibit application in commercial vehicles with long usage requirements, such as heavy-duty long-haul applications.

### 4.2 Low Pressure Direct Injection (LP DI)

The risk of backfire can be inhibited with hydrogen direct injection. Additionally, the higher mixed calorific heating value compared to multi-point injection gives potential for increased power density and full load performance. The performance needs of typical heavy-duty long-haul trucks, as sold in Europe, can hence be realised.

The main challenge with the view to market introduction by 2025 is the availability of mature hardware for injection components and ignition systems. For multi-point injection, hydrogen variants of natural gas fuel injectors are currently available. For low pressure direct injection, functional samples of dedicated hydrogen injectors are being developed and tested by different fuel system manufacturers.

For these reasons, AVL decided to equip the AVL Hydrogen Engine with the capability to run both, LP DI as well as MPI. This setup allows for an accurate & robust comparison of the advantages and disadvantages of the two injection principles on the same base engine architecture.

Based on current injector operational performance, it would appear likely these components will reach sufficient maturity by 2025.

For spark ignited hydrogen engines, the positive ignition source is of vital importance. Cold open or housed pre-chamber spark plugs are required to avoid undesired pre-ignition events. As seen in Table 1, hydrogen requires very low ignition energy. Based on this fact, it is of major importance to ensure accurate control of the ignition and to consider ignition coils which are capable of avoiding uncontrolled coil discharge via the plug electrodes. These components are generally available right now. However, to ensure reliability for commercial vehicle operation, further improvement of the components is needed.

In order to fully exploit the potential of a spark ignited combustion concept, an optimisation of mixture formation in the cylinder is the logical next step. A dedicated cylinder head with focus on ideal combination of charge motion and optimal positioning of injector will further support mixture formation. Tumble cylinder head concepts are considered as being the most promising for spark ignited $H_2$ combustion.

### 4.3 High Pressure Direct Injection (HP DI)

In parallel with the spark ignited hydrogen ICE concepts, a diesel-like combustion of the hydrogen charge mix would also be possible.

In order to achieve the highest power density, best fuel efficiency and transient performance, a diesel-like diffusion combustion is targeted for future applications. To stably initiate the hydrogen ignition, a pilot injection of diesel would be a possible approach. The combination of a diesel pilot injection to ignite methane gas is already in production and ensures diesel substitution rates of around 95% in the WHTC cycle [8]. To fully realize the potential of a carbon free energy carrier, substitution of the diesel ignition would be the logical next step. The availability of such a non-carbon-based ignition system, in sufficient maturity, seems unlikely before 2025.

## 5 CHARGING CONCEPTS

In order to achieve full load engine performance and without exceeding maximum permitted engine out $NO_x$ emissions (<10 g/kWh), a high excess air ratio (~1.9) is required. State of the art single-stage charging systems are suitable to satisfy these demands at standard ambient conditions and steady-state operation. Figure 2 shows simulated operating points in the turbocharger compressor map for the targeted 350 kW full load curve of the single stage charged AVL Hydrogen Engine.

Although application of EGR at high engine speeds reduces air mass flow through the compressor, (the operating point in compressor map moves to the left to lower turbocharger speed), there is only a small margin to maintain an excess air ratio for high altitude operation.

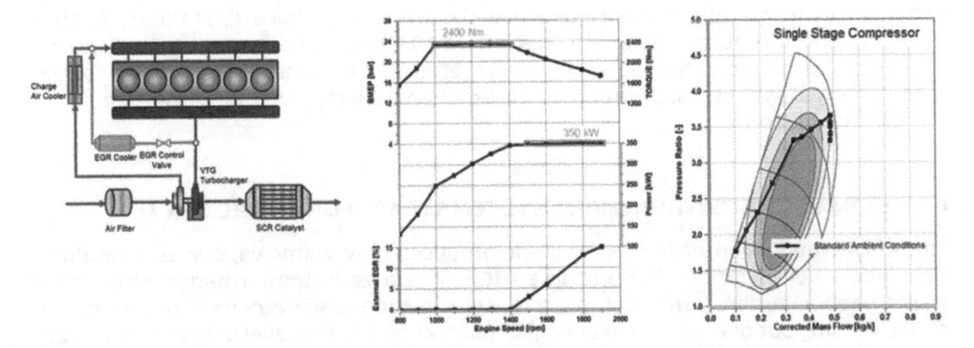


**Figure 2. Single-stage charging concept with VTG.**

An acceptable transient response of the hydrogen engine is strongly affected by un-favourable raw $NO_x$ emission behaviour at low excess air ratio. In order to ensure reasonable emission levels during transient operation, a high excess air ratio needs to be maintained. 1D transient thermodynamic simulations with AVL Boost$^{TM}$ show load response results with different excess air ratio limitations; see Figure 3.

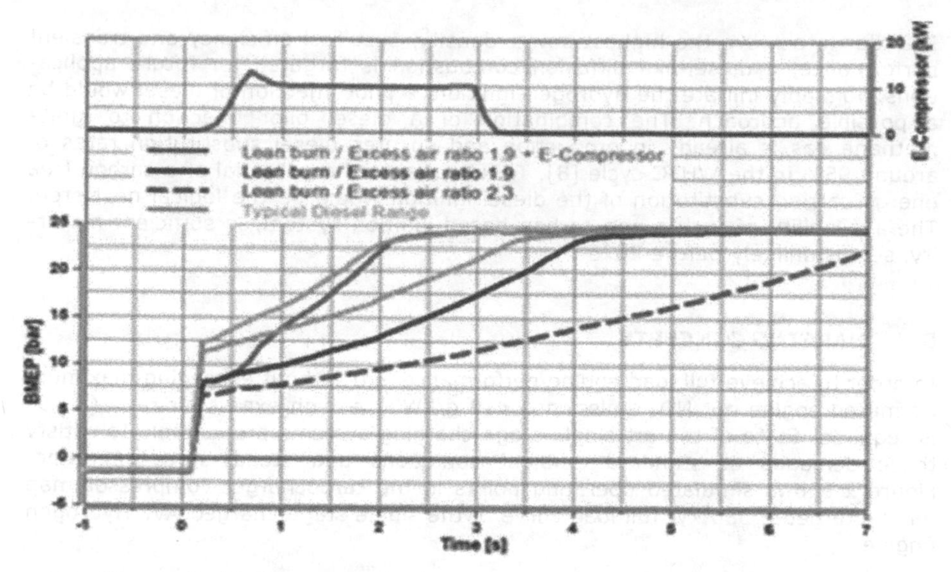


**Figure 3. 1D thermodynamic transient simulation at 1,000 rpm.**

These results show that the single-stage charging system is feasible for demonstration of the engine operation under standard conditions on a test bed, but has penalties in terms of transient response. To overcome these limitations, further development is required. Should it not prove possible to resolve the gap between achievable dynamic response and the targeted dynamic behaviour with a single turbo charger system, application of an additional electrical compressor would be a possible solution.

## 6 EXHAUST AFTERTREATMENT SYSTEMS (EAS) FOR HYDROGEN ICE

The ideal combustion of hydrogen in air produces only water vapour as a product. However, when hydrogen is burnt in an ICE, it results in the formation of not only water vapour but also $NO_X$ compounds, $H_2$ slip, and trace amounts of particulate matter arising out of the burning of engine lubrication oil. The level of these pollutants, especially the raw $NO_X$ emissions, in the $H_2$-ICE exhaust strongly depend on the engine operating conditions.

In the case of applications which operate predominantly under steady state conditions (e.g.: gensets, lighting towers, etc), where a favourable excess air ratio can be ensured ($\lambda > 2.5$), no EAS would be technically needed. In an application like the AVL Hydrogen Engine, which requires highly transient operation, an EAS is required in order to process the raw $NO_X$ emissions.

The $H_2$-ICE exhaust gas contains a much higher water vapour content compared to a conventionally fuelled ICE. Additionally, there will be some $H_2$-slip that will also be present. The EAS for the $H_2$-ICE hence has to be designed to handle these species during regular operation while also ensuring that there is no drop of performance over lifetime.

Although there are no particulate emissions arising out of the fuel combustion in the $H_2$-ICE, there will be trace amounts of particulate emissions that can be expected in the exhaust gas as a result of the burning of lubrication oil in the combustion chamber. In order to avoid combustion anomalies caused by high lube oil consumption, optimised piston bore interfaces (e.g.: ring package, honing, etc.) have to be introduced. These will reduce particulate emissions sufficiently to eliminate the need for a particulate filter.

## 6.1 Synthesis gas test results on conventional Scr Tech.

AVL has conducted synthesis gas test bed investigations to evaluate the performance of conventional SCR technologies with urea dosing using the exhaust gas composition from a $H_2$-ICE. These tests were conducted with a copper SCR and a vanadium SCR.

In Figure 4, the results of the evaluation of the $NO_X$ reduction performance of the two SCR catalysts are depicted. The two charts in the upper row show the $NO_x$ conversion efficiency of the two SCR catalyst technologies with a constant dosing ratio of 1.2 against increasing concentrations of $H_2$. This was undertaken to evaluate the performance of a conventional SCR catalyst when encountering $H_2$ slip from the engine that would invariably occur during operational use. The results indicate that there is no perceivable dip in performance with increasing $H_2$ concentrations.

The two charts in the lower row show the $NO_X$ reduction performance of the two SCR catalyst technologies versus temperature. The performance was evaluated for fresh and aged conditions, with varying $NO_2/NO_X$ ratios and water vapour content as described. The results with the copper SCR catalyst indicate that the higher water content only has a negative impact on the $NO_X$ reduction at low temperatures and at $NO_2$ /$NO_X$ ratio of zero. The impact due to the higher water content is lower at a $NO_2/NO_X$ ratio of 0.5, and the low temperature aging does not have a significant effect on the performance. The results with the vanadium SCR catalyst indicate that while higher water content has negligible impact on the $NO_x$ conversion efficiency, low temperature aging shows a considerable negative effect on $NO_X$ reduction performance when compared with the copper SCR catalyst.

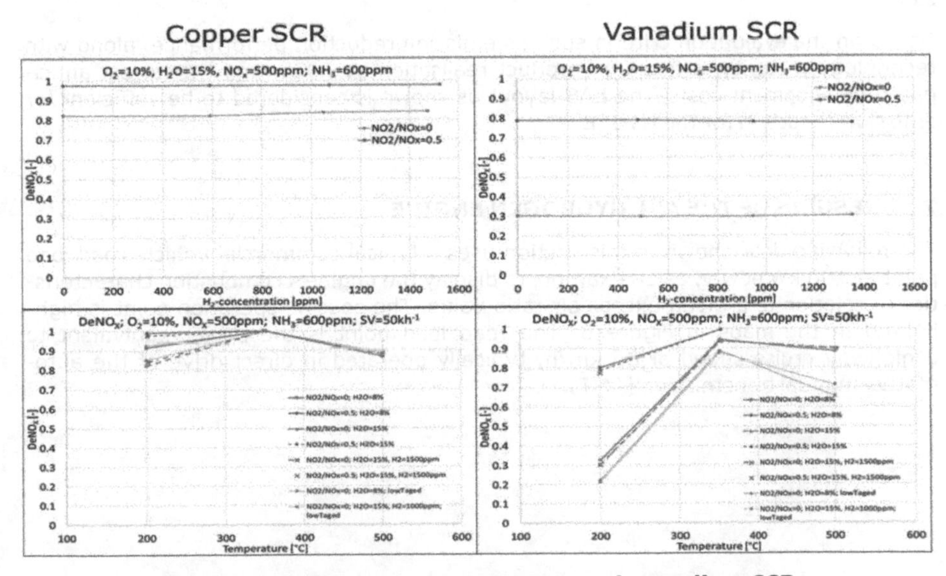


**Figure 4. Synthesis gas test: copper and vanadium SCR.**

## 6.2 Eas layout for the AVL hydrogen engine

Based on the hydrogen exhaust gas composition, which was measured on the AVL engine test bed and the synthesis gas rig results, the EAS layout depicted in Figure 5 was chosen for the AVL Hydrogen Engine.

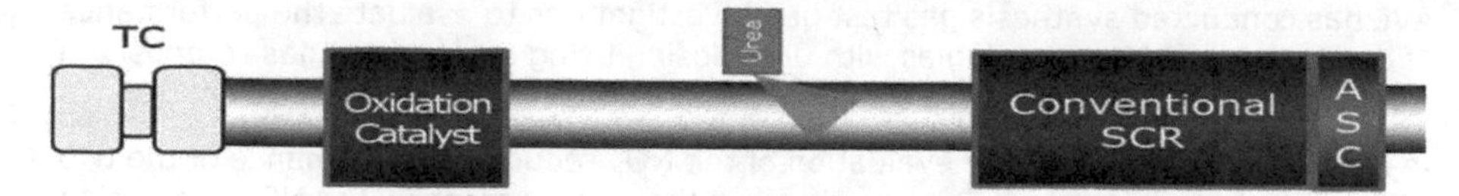


**Figure 5. EAS Layout for the AVL Hydrogen Engine.**

This consists of an oxidation catalyst followed by a conventional SCR catalyst, with urea dosing and an Ammonia Slip Catalyst (ASC). A copper SCR was chosen as the SCR technology and therefore the oxidation catalyst is necessary to promote $NO_2$ formation upstream of the SCR and ensure optimum performance.

Alternatives for the SCR technology used would be iron-copper or vanadium, depending on the application. The use of an oxidation catalyst is optional when using a vanadium SCR but recommended when using a copper SCR or an iron-copper SCR. Additionally, if the system is vulnerable to urea deposit formation or sulphur poisoning of the SCR catalyst, the use of the oxidation catalyst is recommended in order to trigger high temperatures for urea deposit removal and desulfurisation respectively. Alternatively, if direct dosing of $NH_3$ as reductant is done, completely $CO_2$ free operation can be achieved from the aftertreatment side.

Based on the evaluation criteria such as emission reduction performance, along with technology maturity, a focus on product realisation by 2025, and the lowest anticipated development costs, the EAS layout as shown is considered to be sufficient for emission reduction for the $H_2$-ICE.

## 7 RESULTS OF THE AVL HYDROGEN ENGINE

The following data shown in this section uses a typical commercial vehicle road load point at 100 kW power, as an example to display the engine's combustion characteristics in relation to a given excess air ratio value. The specific operation point is highlighted in the maps in Figure 6. The road load point is the power equivalent to a highway cruise speed of 85 km/h, typically operated in direct drive of the automated-manual transmission (AMT).

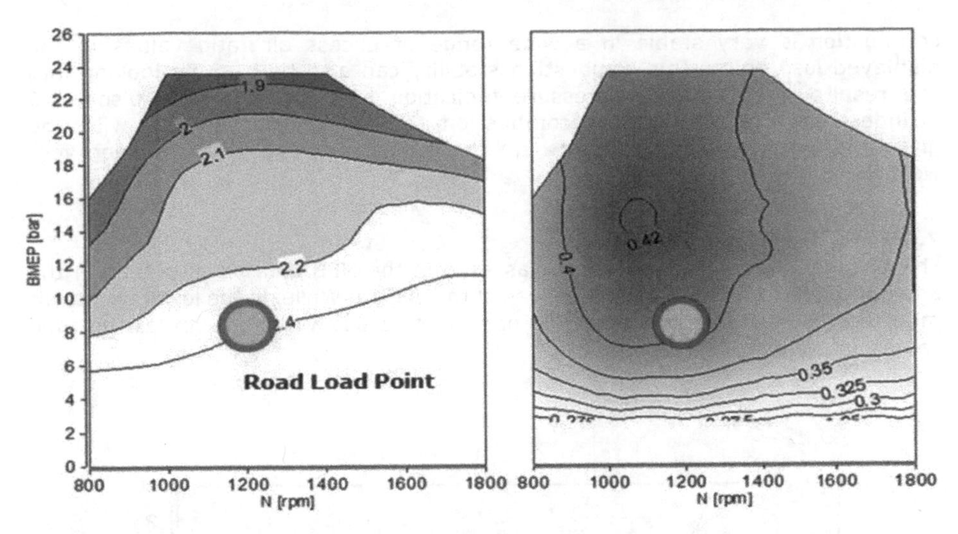


**Figure 6. Excess air ratio and BTE of the AVL Hydrogen Engine.**

In Figure 7, additional engine operational results are shown.

### 7.1 $NO_x$ Emissions (Figure 7; Upper)

The variation of excess air ratio in the road load point shows exactly the $NO_x$ emission behaviour that is expected and was described earlier. The sensitivity of the $NO_x$ emissions in relation to the given excess air ratio can be clearly seen, especially in the area between $\lambda$ of 1.6 – 2.0, where the steep gradient leads to $NO_x$ emission differences in magnitude by a factor of 5. From excess air ratio values >2.2 onwards, the engine shows the expected flat emission curve on a very low $NO_x$ level.

The $NO_x$/excess air ratio trade-off is a result of the combustion optimisation performed under stationary operation conditions and indicates good mixture homogenisation.

### 7.2 Mass Fraction Burnt (MFB) (Figure 7, Upper Middle)

The duration of the combustion event is strongly influenced by the excess air ratio. While the MFB50 is constant over the whole range of the excess air ratio variation, the MFB10 shows a shift towards earlier combustion. MFB90 moves with an even higher significance towards later combustion, resulting in an increase of combustion duration over excess air ratio.

### 7.3 Coefficient Of Variation (COV) (Figure 7, Lower Middle)

The stability of the combustion is usually validated by the COV of the Indicated Mean Effective Pressure (IMEP).

Looking at the COV from the test bed results, even though a slight increase towards higher excess air ratio values can be seen, the results indicate that the combustion is very stable in a wide range of excess air ratio values in the displayed load point. This combustion stability can also be seen by looking into the results of the cylinder pressure indication in Figure 8. (Note also all 6 cylinders show a very stable combustion for a hydrogen engine, with the given fuel injection window. Also, here the rather combustion can be seen very well.)

### 7.4 $H_2$ Slip (Figure 7, Lower)

The change of combustion duration (as seen in the MFB data) is also seen in the amount of unburnt $H_2$ in the exhaust gas of the engine. While, in the lower excess air ratio region, the $H_2$ slip is around 200 ppm, it increases with higher excess air ratio values up to around 750 ppm.

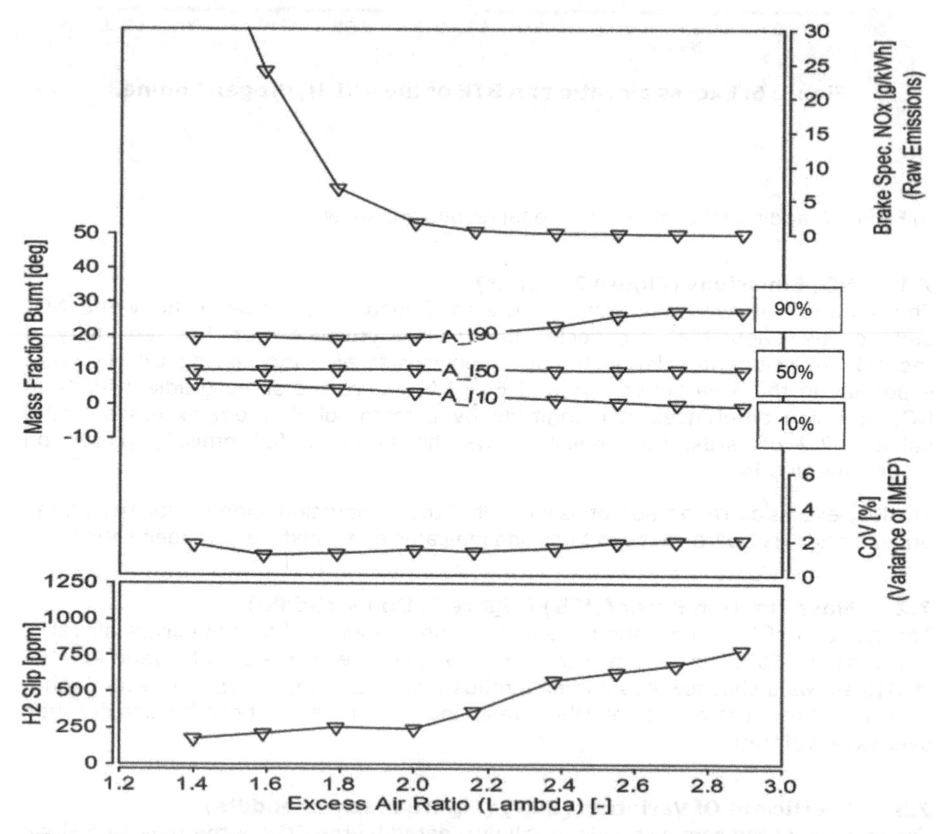


**Figure 7. $NO_x$, MFB, $H_2$ slip and IMEPC over excess air ratio at a typical road load point.**

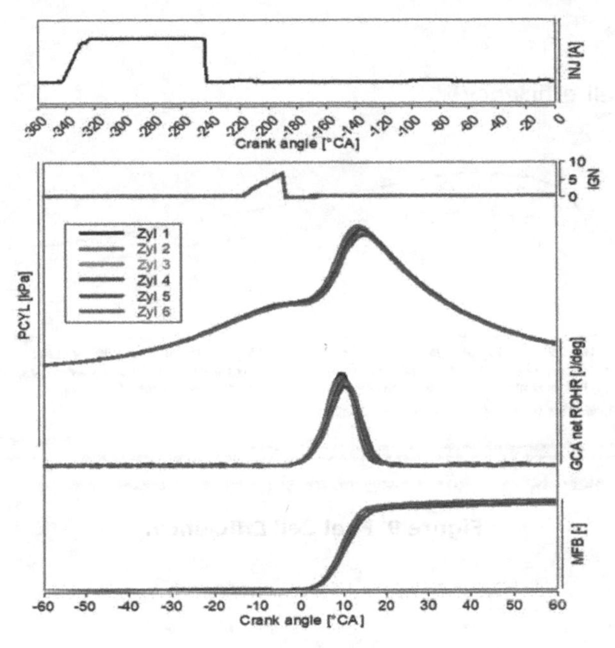


**Figure 8. Combustion stability.**

As mentioned above, this engine concept was defined to demonstrate the capability of the $H_2$ combustion principle on a full engine at the test bed.

## 8 HYDROGEN ICE: POWERTRAIN EFFICIENCY POTENTIAL

Hydrogen ICE can be viewed as a bridging technology to a future state of mass fuel cell (FC) adoption. As a stepping-stone, hydrogen ICE offers the market a number of positive advantages, including:

- Extending the life & recovery period of existing ICE manufacturing assets & equipment
- Developing strategies & methods for on machine hydrogen storage, re-fuelling and control
- Relative insensitivity to fuel contamination
- Reduced development costs & lead times compared to FC & BEV products
- Promote & facilitate the widespread acceptance and introduction of a hydrogen based economy
- Offer market re-assurance (maintenance/repair, operator experience, training, etc) as new fuels & technologies are developed
- Robust and understood BOM cost
- Accurate assessment of powertrain life expectancy based on ICE history

However, at a powertrain level, the thermal efficiency of the hydrogen ICE would always appear to be a disadvantage compared to FC.

Maximum brake thermal efficiency for a HP DI hydrogen ICE is around 47%. (Table 2.) For a pressurised fuel cell system (FCS), efficiencies of up to 60% are achievable at lower loads. (Figure 9.) Note, however, system efficiency tends to decay at loads above 30%.

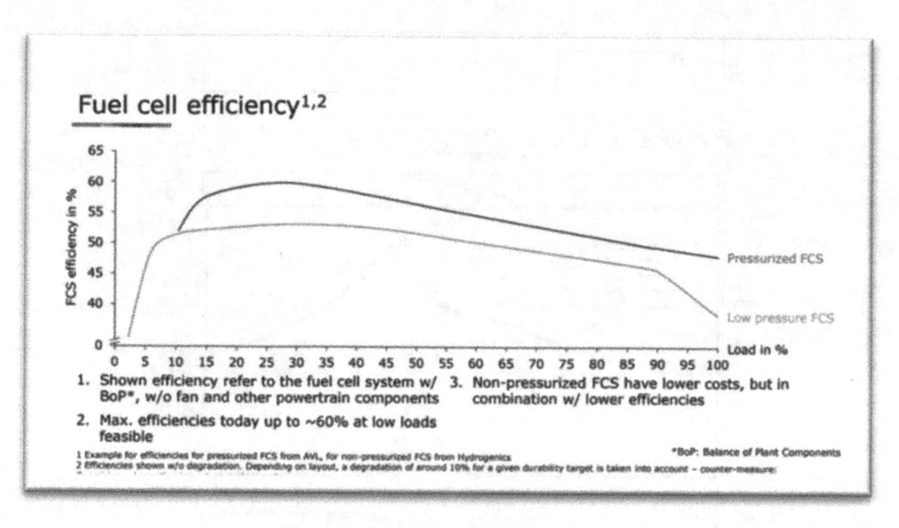


**Figure 9. Fuel Cell Efficiency.**

Using AVL CUISE-MTM modelling tools, AVL had undertaken extensive simulation of an on-highway HD vehicle. Powertrain models were completed & calibrated for both a current (conventional) system and a FC arrangement. Excluding the primary power sources (ICE & FC), system efficiencies were calculated at 97% and 85% to 90% respectively. (Figure 10.)

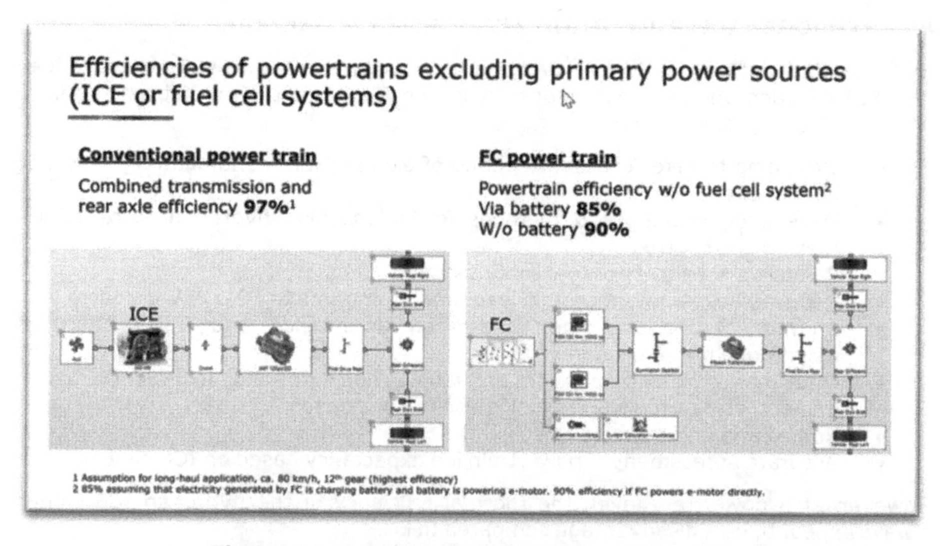


**Figure 10. Powertrain Efficiency – ICE vs FC.**

A conventional ICE power transmission system can thus be shown to have some advantage over that required for a FC vehicle.

Various duty cycles were then applied to the AVL-CRUISE-M models to simulate hydrogen consumption for both ICE and FC vehicles. For long haul operations, where the vehicle is predominantly operated at high speed and consistent load conditions, the ICE system is approximately 7% less efficient than a FC product. However, for more transient operations, where a combination of operating points is considered (city, urban & highway), the difference is more marked & the ICE system is considerably less efficient due to the lack of energy recuperation opportunities in the transient operation profile. (Figure 11.)

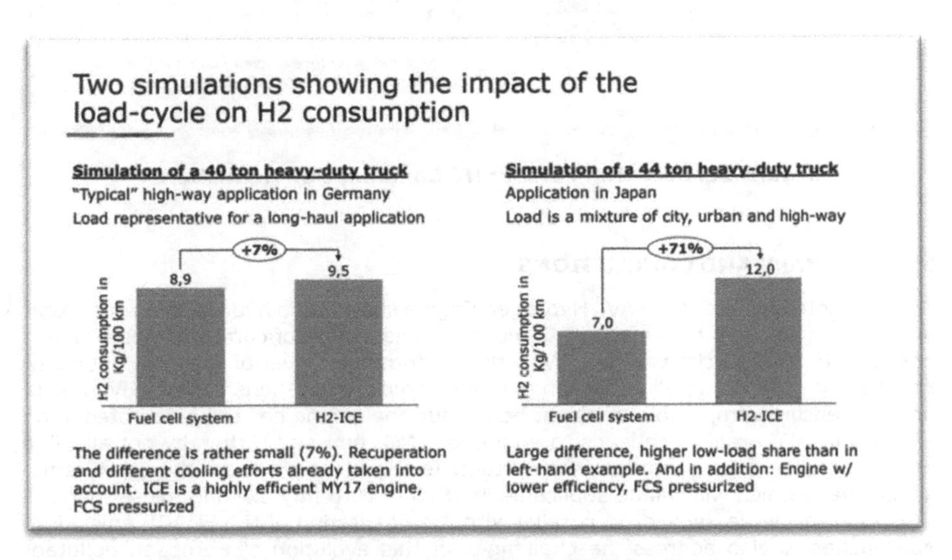


**Figure 11. H2 Load Cycle Effect on H2 Consumption.**

## 8.1 Next Steps

Whilst is accepted the hydrogen ICE vehicle is not comparable to a FC product for overall thermal efficiency, it is also clear that the gap is relatively small in specific applications & use cases. With further development of the system powertrain architecture, modelling in AVL CRUISE-M suggests a comparable or even better system level efficiency is possible.

Use of a parallel hybrid ICE could see a reduction in fuel consumption of approx.. 6% compared to the base line. Integration of a waste heat recovery (WHR) system would yield an additional 2% saving. Both technologies are available & have been demonstrated by AVL. Utilising both features would hence give an overall saving of up to 8% from the hydrogen ICE base line, and offer a fuel consumption level below that of a hydrogen FC vehicle. (Figure 12.)

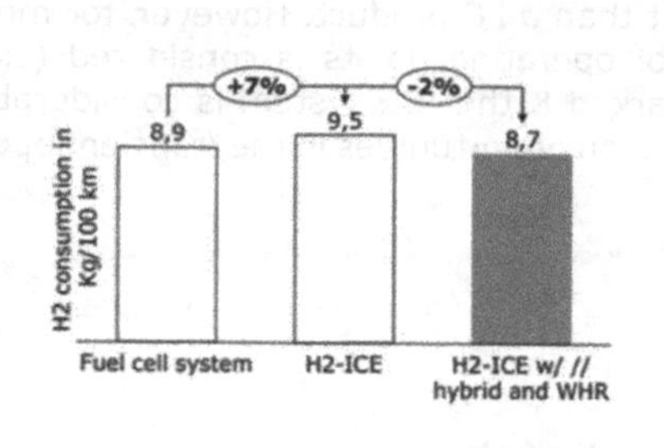

Potential to further reduce hydrogen consumption for H2-ICE based powertrains
+7%
-2%
8,9
9,5
8,7
H2 consumption in Kg/100 km
Fuel cell system
H2-ICE
H2-ICE w/ // hybrid and WHR
Building on the previous simulation and investigations done for EU's CO2 limits 2025 and 2030, AVL estimated the possible benefits of a parallel hybrid system and WHR for a long-haul application.
The parallel hybrid could contribute w/ approx. 6% fuel consumption reduction, a WHR system in the range of min. 2%.
Taking the MY17 engine efficiency as basis, this would lead to lower hydrogen consumption as the base-case fuel cell system.
A combination w/ future, even more efficient engines, would lead to a further reduction.

**Figure 12. Potential Further H2 Consumption Reductions.**

## 9 SUMMARY AND CONCLUSIONS

The presented results of the AVL Hydrogen Engine show that a hydrogen internal combustion engine could be a viable, $CO_2$ neutral replacement of currently diesel fuelled heavy duty commercial vehicles. With the performance level of the AVL Hydrogen Engine, the majority of all European long haul truck applications can be covered. In order to ensure comparable transient behaviour, the engine has to be operated in an excess air ratio area which leads to engine out $NO_x$ emissions, thereby not allowing operation without an aftertreatment system. To ensure compliance with EU VI emissions levels, which will still be applicable in 2025, a currently existing, diesel-derived SCR system can be applied. In parallel with the finalisation of the EU VII legislation, approaches to also address the challenges of this evolution of European pollutant legislation will be investigated by AVL.

The achievable dynamic response of a hydrogen internal combustion engine is lower in comparison to current best-in-class diesel engines. This current situation will improve based on the further development of turbocharger systems, which is also driven by the ongoing developments for diesel applications. In case that performance reserve is needed for certain applications, introduction of bigger displacements would be an option to also operate under high altitude conditions with reduced de-rating.

Operation of combustion engines with hydrogen fuel results in several specific challenges in terms of reliability, which have to be addressed in course of further development. Keeping lube oil consumption low, dealing with combustion irregularities and ensuring the durability of valves as well as spark plugs are already identified fields of activity in this regard. Proven systematic approaches established for diesel and natural gas engine applications and extended to hydrogen operation will ensure to achieve the expected durability and reliability targets.

With the results derived from the AVL Hydrogen Engine, the realisation of a hydrogen internal combustion engine by 2025 can be seen as feasible. Therefore, it will serve as an enabler to build up commercial vehicle hydrogen fuelling infrastructure that will also be shared with fuel cell electric trucks.

Longer term, the overall system efficiency of a hydrogen ICE vehicle can be demonstrated to approach or even exceed the thermal efficiency of a hydrogen FC machine for certain use cases and applications. Hence hydrogen ICE can provide both a 'stepping stone' to a zero carbon, hydrogen FC fleet, and also a competitive technology in its own right, given appropriate application & use case.


## ACKNOWLEDGEMENTS

The author of this paper would like to thank all of the external partners and supporters but especially the employees of the Institute of Internal Combustion Engines and Thermodynamics of Graz University of Technology, Robert Bosch GmbH, Garret Motion as well as Ventrex Automotive GmbH.

DI Anton Arnberger, DI Admir Zukancic, DI Martin Wieser, Neil Kunder MSc, Dr.-Ing. Mirko Plettenberg, Ing. DI (FH) DI Bernhard Raser, DI Rolf Dreisbach all of AVL List GmbH, Graz


## REFERENCES


[1] Commission Regulation (EU) 2019/1242 of the European Parliament and of the Council of 20 June 2019 setting CO2 emission performance standards for new heavy-duty vehicles and amending Regulations (EC) No 595/2009 and (EU) 2018/956 of the European Parliament and of the Council and Council Directive 96/53/EC

[2] Commission Regulation (EU) 2018/956 of the European Parliament and of the Council of 28 June 2018 on the monitoring and reporting of $CO_2$ emissions from and fuel consumption of new heavy-duty vehicles

[3] Japan Ministry of Economy, Trade and Industry, Basic Hydrogen Strategy. https://www.meti.go.jp/english/press/2017/1226_003.html, Download: March 3rd 2021

[4] U.S. Department of Energy, Department of Energy Hydrogen Program Plan. https://www.energy.gov/articles/energydepartment-releases-its-hydrogen-program-plan, Download: March 3rd 2021

[5] Eichlseder, H., Klell, M.; Wasserstoff in der Fahrzeugtechnik; 2. Auflage, 2010, Wiesbaden, S 152f

[6] Grabner, P.: Potentiale eines Wasserstoffmotors mit innerer Gemischbildung hinsichtlich Wirkungsgrad, Emissionen und Leistung. Dissertation, Technische Universität Graz, 2009.

[7] Arnberger, A. et al.: Der Wasserstoffmotor im Nfz: Brückentechnologie oder langfristige Lösung?; ATZ, MTZ Conference; Baden Baden, 2021.

[8] Arnberger, A. et al.: Commercial Natural Gas Vehicles: Tomorrow's Engine Technologies for most stringent $NO_x$ and $CO_2$ Targets; ATZ, MTZ Conference; Baden Baden, 2018.

[9] Velten, C., Hammer, M., Wohlfart, J., Holland, B.G. et al.: "Durability Test Suite Optimization Based on Physics of Failure," SAE Technical Paper 2018-01-0792, 2018, doi:10.4271/2018-01-0792

## *Session 3: Engines with sustainable fuels (e.g hydrogen, e-fuels, biomethane)*

*International Conference on Powertrain Systems for Net-Zero Transport*
*Institution of Mechanical Engineers, ISBN 978-1-032-11281-7*

# Future liquid & gaseous energy carriers: A key prerequisite for carbon-free mobility

**Th. Körfer**

FEV Group GmbH, Germany

## ABSTRACT

Effective and affordable protection of the climate represents one of the utmost challenges of the current time. The overall reduction of greenhouse gases is a fundamental mission for all associated stakeholders. The transport sector covers nearly a quarter, accurately 24 %, of the global CO2 emissions - therefore the potential lever for reductions and optimizations is correspondingly huge. In order to meet the ambitious targets, like the European Green Belt initiative, cross-sectoral, multi-disciplinary and holistic measures are mandatory.

An increased emphasis on sustainability throughout the society will not only drive the use of sustainable raw materials, it will also motivate and stimulate a closed loop economy for non-renewable materials, esp. for so-called critical materials, such as rare earths, whose supply is potentially threatened by political conflicts, increased use or by restrictions in trade or mining. As refining such materials from alloys is sometimes even very energy-intensive, Europe's need for energy will likely not become smaller but rather greater, despite all activities aiming at better overall energy efficiency. Consequently, it is predicted that major markets in Europe will continue to import large amounts of energy from abroad.

This paper in hand shares ongoing research activities and latest project results on sustainable, carbon-neutral mobility using green electricity to produce renewable fuels. This approach represents a complementary way to shift traffic to electrically powered propulsion systems. In view of the real-world market requirements, exist still many applications in the transport sector, where purely electric propulsion systems under all environmental conditions do not offer a suitable solution. This applies at least to the heavier CV's and to long-distance operation as well as to large non-road mobile machinery. For such applications, gaseous and liquid fuels with their high energy densities will remain the prime choice for the near future. In the mid-term future, some of these energy carriers will be produced cost-efficiently from renewable energy sources in worldwide areas with the most favourable climatic conditions. Thus, besides powertrain electrification, either H2-operated propulsion systems, incl. H2-powered ICE's but as well newly designed Power-to-X (PtX) fuels from renewable electricity and CO2 from various sources, so called e-fuels, are a highly attractive alternative to ensure mobility with a closed carbon cycle.

Synthetic e-fuels also show considerable potential for solving the classic trade-off's between $CO_2$ and pollutant emissions of ICE based powertrains. Consequently, they provide a worthwhile solution for a clean and sustainable mobility in the next decades. However, to achieve a short-term reduction of CO2 emissions of the transportation sector, these fuels must be compatible with the technology of the current vehicle fleet and the existing re-fueling infrastructure, and, ideally, they should be miscible with fossil fuels. Fischer-Tropsch (FT) products for example are very similar to petroleum-based fuels and meet the latter requirement so that they can be mixed with conventional fuels without any problems. However, despite the principal similarity to fossil Diesel fuel, major

DOI: 10.1201/9781003219217-6

differences in the combustion behaviour of FT-Fuels can arise. The presentation concludes with a summary on functional benefits and associated cost estimations.

## 1 INTRODUCTION

The transformation and development of the entire transportation sector in the EU towards sustainability and 100% environmentally friendly operation has been initiated with the 2015 Paris agreement and even more concrete with establishing the Green Belt Initiative in 2020. As the initiative follows a global approach, the minimization of transport-related $CO_2$ emissions is needed on a global scale and the United States, China, Japan, and India have issued standards for tailpipe $CO_2$ reduction [5, 6]. As the EU has set strict regulations as well to reduce the GHG emissions by 15% in 2025 and by 25% in 2030 with 2019 as baseline, alternative solutions must be developed to de-fossilize all segments of the transportation sector. In the mid- to long-term strategy, it is the absolute target to even find on solutions for Zero-$CO_2$- or $CO_2$-neutral-powertrains, as aimed for by global greenhouse gas (GHG) regulations as well. With dedicated specific needs and strict market requirements, such as reliability, versatility and, above all, total extreme cost of ownership (TCO), these ambitious $CO_2$ targets are challenging for all applications, but especially for heavy-duty vehicles, agricultural and construction machinery as we as for marine use cases and the entire air travelling sector. This goes hand in hand with increasing travel frequency, freight traffic and growing vehicles sizes as well and weight restrictions as additional constraints. Besides partial or complete electrification of the propulsion system, assuming 100% renewable and sustainable electrical energy in the same time frame, also new liquid or gaseous energy carriers, like synthetic fuels or hydrogen come more and more into the focus. As for achieving the necessary conditions for the planet to maintain the temperature increase below the important 1,5% threshold, it is crucial to consider The Well-to-Wheel (WtW) evaluation, especially to understand and identify the different sources of $CO_2$ emissions in the transport sector and to derive the most effective and appropriate steps. The WtW represents finally a result of the contribution of Well-to-Tank (WtT) emissions plus tailpipe emissions from Tank-to-Wheel, (TtW) [8]. The WtT emissions include the $CO_2$ generated during the production of the fuel, its transportation to the fuel station and dispensing to the vehicle tank. The reduction of TtW in the transportation sector is possible by improving the fuel efficiency of the powertrain, waste heat recovery measures, drag, weight, and rolling resistance reduction [9]. To reduce the WtW emissions renewable energy carriers play an important role. On one hand, due to their high specific energy, liquid energy carriers are very attractive in mobile applications, as they need to cover long distances. On the other hand, renewable fuels offer the potential to impact the $CO_2$ output rather quickly as they feature the potential to be used as drop-in fuel to conventional fuels in existing vehicles and with increasing the drop-in quantity of renewable fuels the overall CO2 emissions can be mitigated with an increasing gradient, depending on the growing availability of alternative fuels. Furthermore, different pathways are also available to produce renewable and sustainable alternative fuels, even tailored to support ultra-low pollutant emissions in the next decade to reach a zero-impact status for ICE-powered applications.

## 2 APPROACH AND METHODOLOGY

Optimized Power-to-X processes are the appropriate technological reaction for real sector coupling. For the last decades, combined heat and power generation (CHP) has been a major success factor for maximizing the efficiency of fossil energy utilization by using waste heat for the heating sector. Electromobility, the direct use of electrical energy via batteries, is currently heavily promoted in numerous countries, mainly for

LD vehicles, but also for other, heavier applications. Without doubt, this approach represents for sure a massive disruptive trend. However, electromobility is most likely not the full and comprehensive answer for all markets, all regions and all kinds of mobile applications as well, e.g. long-haul heavy transportation, marine propulsion and aviation. Here, energy-dense renewable fuels generated from green and sustainable electrical energy, called E-Fuels, should replace quickly and increasingly fossil fuels. These E-Fuels are anticipated to be fully compatible to and can be easily mixed with conventional, fossil fuels in order to reduce the carbon content in the fuel mixture over time, without the demand to change cars or the fuel logistics infrastructure. In this manner, the transition from a fossil world to a largely carbon-neutral environment can be done relatively fast and smoothly with direct $CO_2$ emissions reductions on the way, while simultaneously lowering pollutant emissions.

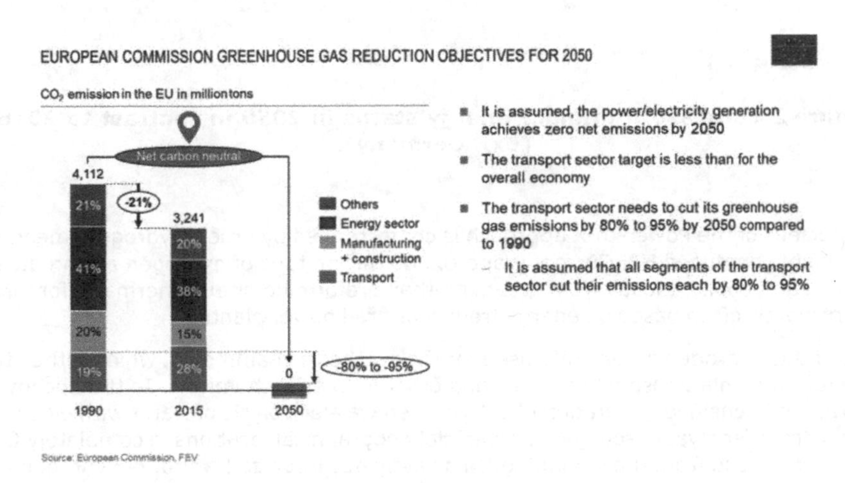


**Figure 1.**

Moving from the overall global perspective more to the concrete situation in the EU, Figure 1 displays the scenario, foreseen by the European Commission, to develop Europe within the next 30 years into a fully carbon neutral continent resp. society with a prospering and clean economy without GHG emissions.

In contrast to the reference year 1990 the GHG emissions have to be cut from more than 4 million tons of $CO_2$ to net-zero. This means for the transport sector a lowering of 80...95% versus 1990 and even more with respect to the 2015 figures.

Looking from the other side, also from the perspective of the primary energy carriers, certain hurdles and challenges are still identified on the pathway to the carbon-neutral economy in 2050.

As Germany, but also other big markets in Europe, is today strongly depending on energy import from outside, this situation will also not change in the mid-term future. Even considering an ambitious reduction of the primary energy demand of 50% in 2050, nearly half of the necessary energy will have to come from outside Germany. Up to one third of the total energy consumption might be covered by Power-to-Fuels, and nearly three quarters are probably by liquid energy carriers like PtX fuels from outside Europe. Import of synthetic fuels from renewable energy sources will make up almost 50% of today's mineral oil imports by 2050.

UTILIZATION AND IMPORT OF PRIMARY ENERGY CARRIERS IN GERMANY IN 2018 AND 2050

Primary energy carriers in PJ

13,730
-50%
14,400 in 2008
9,520
~7,200
15%
22%
6%
23%
34%
32%
32%
Imports
Imports
2018
2050

Renewables
Hard coal and lignite
Nuclear energy
Mineral oil
Natural gas
Power-to-Fuel

- Primary energy use will be shortened by approx. 50% in 2050 - compared to 2018
- Fossil fuels will provide only 5% of the total primary energy consumption in 2050
- Import of renewable electricity to Germany will increase
- Up to one third of the total energy consumption might be covered by Power-to-Fuels, thereof
  - *75% are imported from outside Europe*
  - 17% are imported from inside Europe
- Import of synthetic fuels from renewables will make up *almost 50% of today's mineral oil imports*

Source: European Commission, BMWi Energiedaten, AG Energiebilanzen, ewi gGmbH "Evolution scenario"

**Figure 2. Forecast of primary energy status in 2050 in contrast to 2018 (Ex.: Germany).**

The nucleus of the Power-to-X approach is characterized by green Hydrogen, meaning sustainable produced H2. Currently approx. 80 million tons of hydrogen are produced every year globally, mostly from steam methane reforming or autothermal reforming, unfortunately often based on energy from coal-fired power plants.

50% of this hydrogen is currently used for the synthesis of ammonia, which is then the base for ammonia phosphate or urea and other industrial chemicals. In the upcoming future, the intensified generation of e-Hydrogen via electrolysis of water with electrical energy from renewable sources in beneficial geographical locations is completely free of $CO_2$ emissions from the beginning. In a vastly decarbonized world, e-Hydrogen will support and ensure a long-term, independent and reliable power-to-power storage on a large scale. Re-usage will be either realized in H2-capable gas turbines, or with derivative products like MeOH or other e-fuels in all kinds of engines or fuel cells to safeguard the market related energy supply. A simplified description of the process scheme for H2/e-fuels is provided in Figure 3.

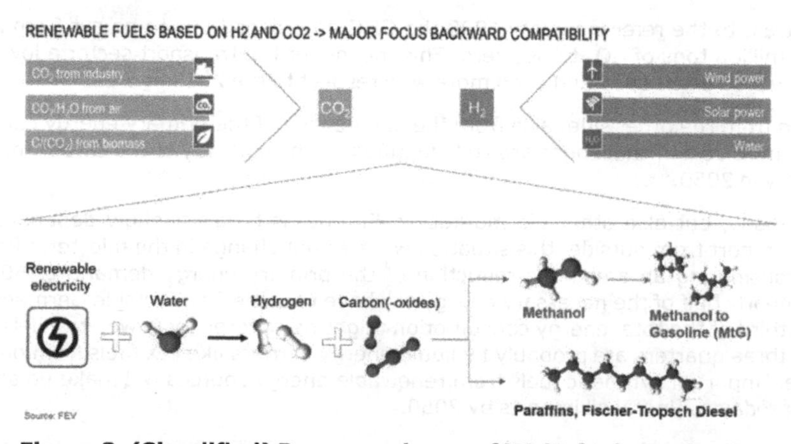


**Figure 3. (Simplified) Process scheme of H2/E-fuels production.**

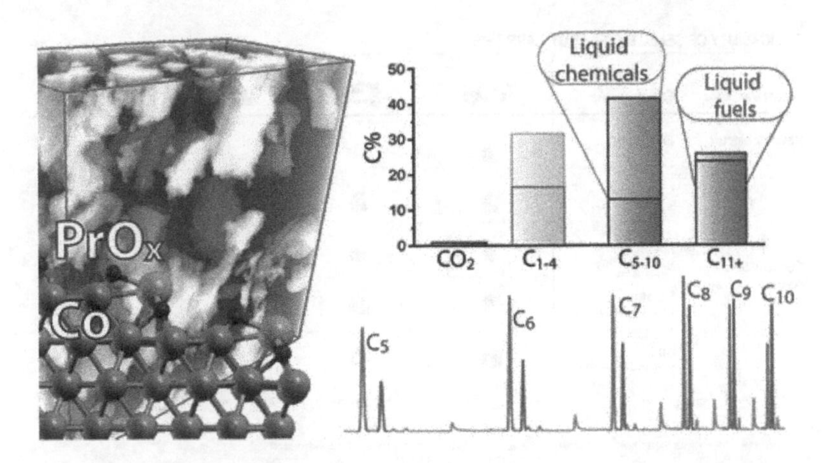


**Figure 4. Simplified depiction of a cobalt fischer–tropsch catalyst specification for the combined production of liquid fuels and olefin chemicals from hydrogen-rich syngas.**

For further steps in decision making, often the parameter of energy efficiency is considered in public discussions, but with a closer look to the variety of applications and use-cases, also considering environmental and infrastructural boundaries, the decision-making becomes more difficult and complex.

For future, sustainable hydrogen (H2) synthesis, only water and electricity from so-called renewable energy are mandatory. H2 is a reactant for all other synthetic fuels and it is not only the simplest, but also cheapest fuel to produce. However, today even green H2 from renewables is typically more expensive than any other fossil fuel. But new electrolysis technologies, such as High-temperature PEM and Solid Oxide Fuel Cells, jointly with an increasing market size, will continuously reduce the costs. Current estimations fore-see a H2 sales price at the pump of 3-7 USD/kgH2 in 2030 as realistic.

The major driver for the near-term direct use of $H_2$ as a fuel is the EU's increasingly stringent EU $CO_2$ regulations, which require a reorientation of powertrain concepts in on-road transportation. With the currently enacted Tank-to-Wheel legislation (new) carbon-free or low-carbon fuels produced from renewable sources are coming into focus. Hydrogen as a zero-carbon fuel provides enormous potential but poses as well major hurdles and technical challenges in terms of storage and short-term availability.

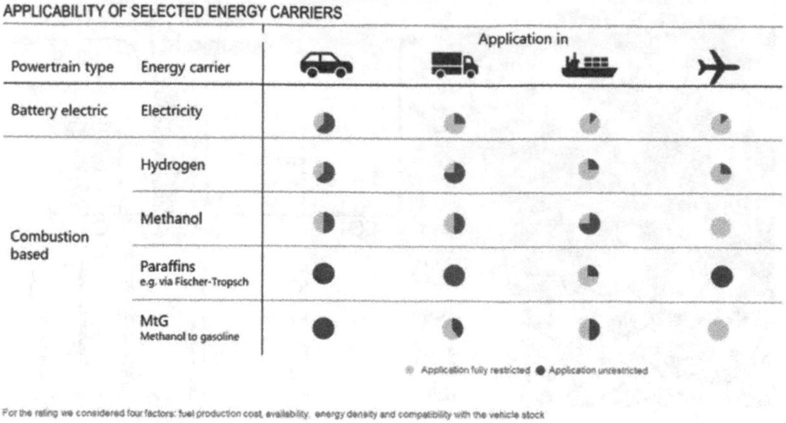

APPLICABILITY OF SELECTED ENERGY CARRIERS

Application in

Powertrain type | Energy carrier

Battery electric | Electricity

Combustion based | Hydrogen | Methanol | Paraffins e.g. via Fischer-Tropsch | MtG Methanol to gasoline

Application fully restricted ● Application unrestricted

For the rating we considered four factors: fuel production cost, availability, energy density and compatibility with the vehicle stock

**Figure 5. Ranking of energy carriers for different applications in the transportation sector.**

Furthermore, as a gaseous fuel, hydrogen is not well suited for high pressure direct injection, unless liquefied hydrogen is used. This is due to the power requirements for pressurization. Hydrogen´s key feature is definitely its extreme fast combustion leading to superior lean burn and EGR tolerance. [5] Overall, hydrogen features a wide range of possible combustion system layouts and aftertreatment concepts as depicted in Figure 6. As a direct conclusion it can be easily derived out of the below chart, which base is typically used for the conversion of ICE's towards H2 operation.

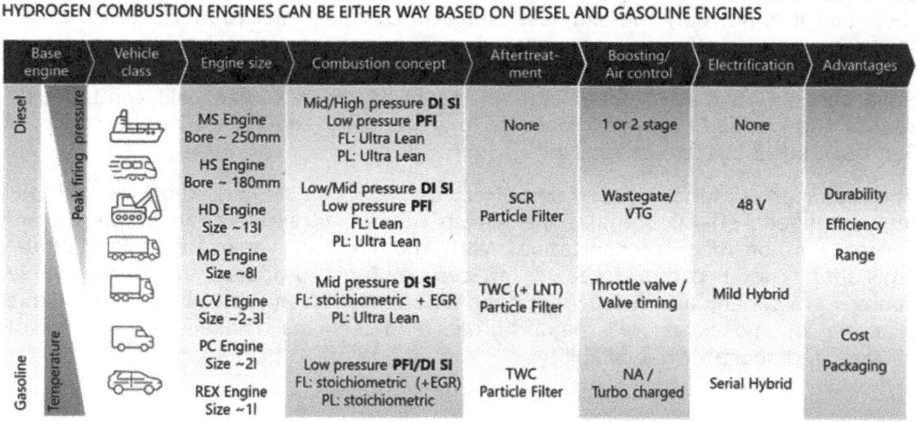

HYDROGEN COMBUSTION ENGINES CAN BE EITHER WAY BASED ON DIESEL AND GASOLINE ENGINES

| Base engine | Vehicle class | Engine size | Combustion concept | Aftertreatment | Boosting/ Air control | Electrification | Advantages |
|---|---|---|---|---|---|---|---|
| Diesel (Peak firing pressure) | | MS Engine Bore ~ 250mm<br>HS Engine Bore ~ 180mm | Mid/High pressure **DI SI**<br>Low pressure **PFI**<br>FL: Ultra Lean<br>PL: Ultra Lean | None | 1 or 2 stage | None | |
| | | HD Engine Size ~13l<br>MD Engine Size ~8l | Low/Mid pressure **DI SI**<br>Low pressure **PFI**<br>FL: Lean<br>PL: Ultra Lean | SCR<br>Particle Filter | Wastegate/ VTG | 48 V | Durability<br>Efficiency<br>Range |
| | | LCV Engine Size ~2-3l | Mid pressure **DI SI**<br>FL: stoichiometric + EGR<br>PL: Ultra Lean | TWC (+ LNT)<br>Particle Filter | Throttle valve / Valve timing | Mild Hybrid | |
| Gasoline (Temperature) | | PC Engine Size ~2l<br>REX Engine Size ~1l | Low pressure **PFI/DI SI**<br>FL: stoichiometric (+EGR)<br>PL: stoichiometric | TWC<br>Particle Filter | NA /<br>Turbo charged | Serial Hybrid | Cost<br>Packaging |

**Figure 6. Overview of H2-operation boundaries for multiple engine/ application classes.**

As with increasing weight of the H2-ICE to ensure the propulsion of the application the demand for high efficiency raises, factors like max. peak firing pressure, boosting grade for ultra-lean operation and engine robustness evolve, the tendency to select a former Diesel-fueled engine as base is extremely high. In this context, the optimization of the combustion chamber for optimal mixture formation as well as for rapid and complete combustion becomes absolutely crucial, which also drives the key question for the best location of the high-pressure injector in the cyl. head. Figure 7 provides an exemplary comparison for the mixture formation at 20° BTDC for two different cyl. head designs and two in-cylinder swirl levels.

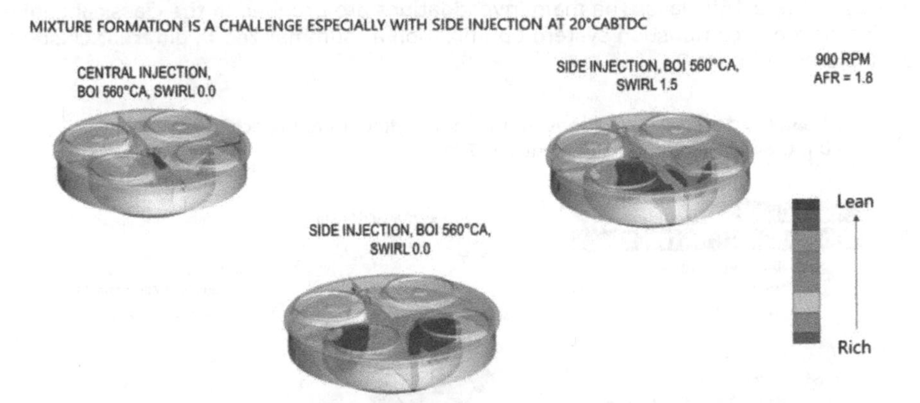


**Figure 7. Combustion characteristics for H2 combustion w/ different injector positions.**

For proper and precise layout and definition of high-efficiency H2 combustion system definitions, a detailed understanding about H2 fuel introduction and mixture formation is essential. For this purpose, fundamental research work and adjustment of high-resolution CAE tools is mandatory. Figure 8 displays key elements of a full chain of proven FEV Group development tools and methodologies.

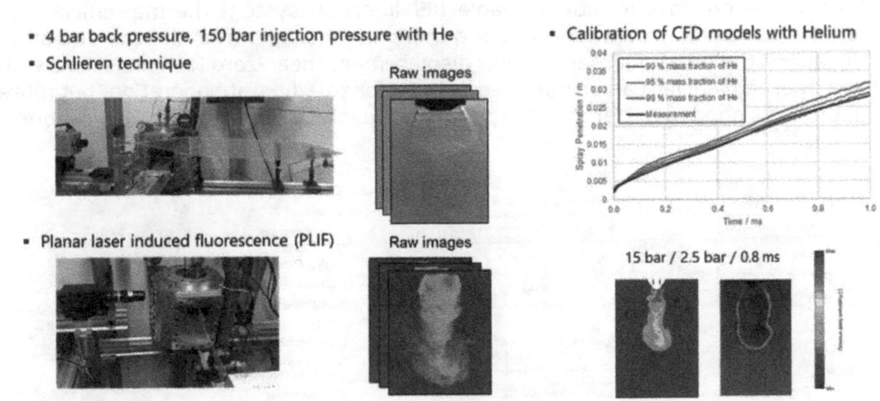


**Figure 8. Fundamental research activities for efficiency-optimization of H2-combustion.**

After definition and specification of optimized combustion system characteristics, optimization and calibration of the H2-ICE for best $h_e$-NOx trade-off represents the next step towards vehicle application, considering EATS layout and characteristics.

Based and synchronized with the basic research work to understand and judge the key challenges of an H2-ICE in terms of the dominating features and attributes, experimental investigations and studies are performed on different engine platforms and towards various application types, ranging from Light-Duty on-road application with approx. 2,0L engine displacement up to large, off-highway use cases with engine capacities in the ≥ 16L class. The main investigations are grouped in the classical engineering frame of combustion system optimization as summarized in different clusters, as depicted in Figure 9.

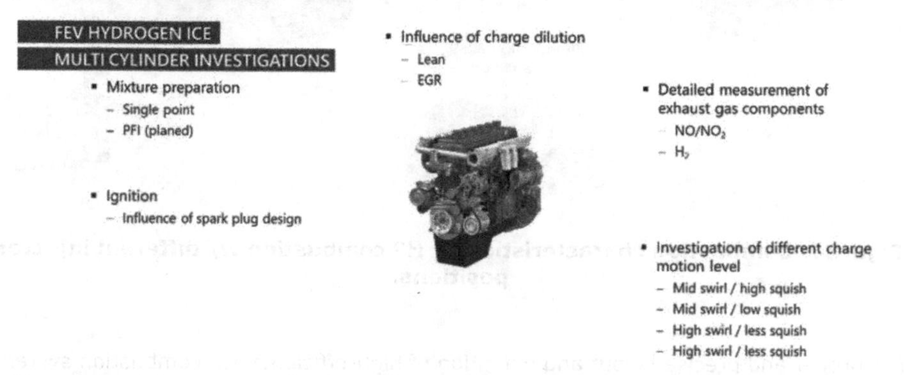


**Figure 9. Main areas of base engine development and combustion system optimization for H2-ICE programs to meet upcoming requirements.**

According to these identified development areas, the subsequent Figure 10 provides a compressed overview about the different development stages. For a MD multi-cylinder engine w/ 7,7L capacity and featuring a state-of-the-art and an optimized combustion system layout with available PFI injection system thermal efficiencies above 40% can already be achieved. For a more advanced, extremely lean operating DI-concept on the base of larger cylinder displacement, near-zero NOx emissions with high thermal efficiencies can be obtained already for steady-state operation, but more development is needed to transfer this performance into full-size engine applications.

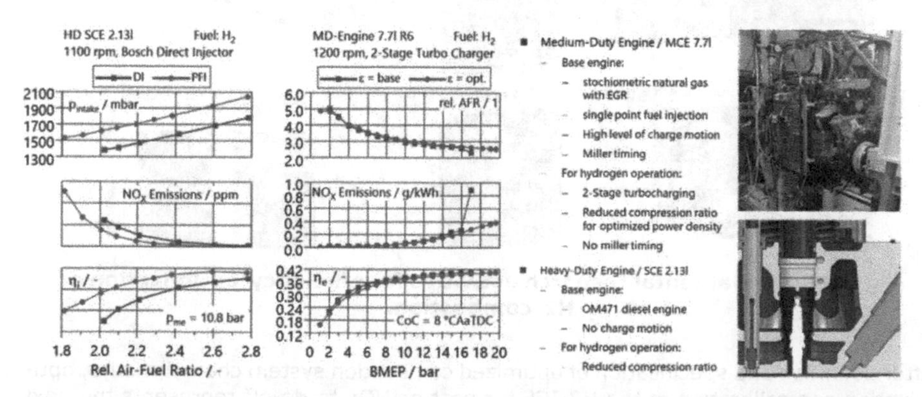


**Figure 10. Functional performance results for different engine development stages – MCE w/ PFI technology and SCE w/ DI technology.**

As an outcome of a fine-tuned full H2-ICE system, a high-efficiency, zero $CO_2$ and near-zero pollutant emission propulsion system can be achieved, as displayed in Figure 11 for a large truck engine in the given legislative certification cycle.

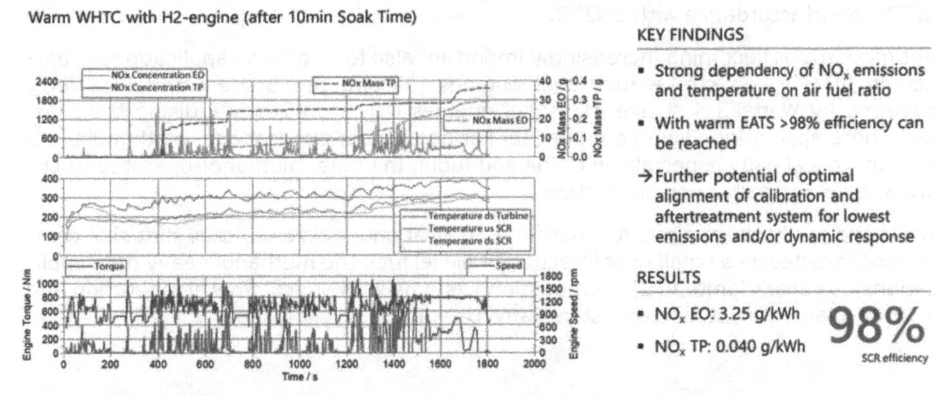


Warm WHTC with H2-engine (after 10min Soak Time)
KEY FINDINGS
Strong dependency of NOx emissions and temperature on air fuel ratio
With warm EATS >98% efficiency can be reached
→Further potential of optimal alignment of calibration and aftertreatment system for lowest emissions and/or dynamic response

NOx EO: 3.25 g/kWh
NOx TP: 0.040 g/kWh

Engine Torque / Nm
Engine Speed / rpm

**Figure 11. Representative example of converted diesel truck engine, optimized and refined for operation with H2.**

Switching over to another key group of future renewable fuels. Besides the paraffins also the alcohol group represents a hot candidate for future transportation. Pure Methanol is here a representative candidate, as it can be produced in a quite cheap manner and it features an established compound of production and building-block. However, certain issues and hurdles are seen on the fuel distribution and compatibility side, which might limit the use cases on applications, which can make use of individual fueling chains and depots.

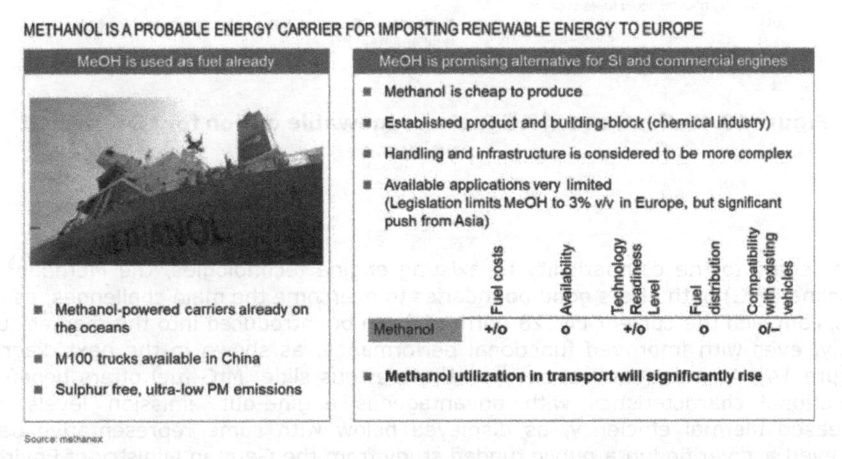

METHANOL IS A PROBABLE ENERGY CARRIER FOR IMPORTING RENEWABLE ENERGY TO EUROPE
MeOH is used as fuel already
Methanol-powered carriers already on the oceans
M100 trucks available in China
Sulphur free, ultra-low PM emissions
MeOH is promising alternative for SI and commercial engines
Methanol is cheap to produce
Established product and building-block (chemical industry)
Handling and infrastructure is considered to be more complex
Available applications very limited
(Legislation limits MeOH to 3% v/v in Europe, but significant push from Asia)
Fuel costs
Availability
Technology Readiness Level
Fuel distribution
Compatibility with existing vehicles
Methanol
+/o
o
+/o
o
o/–
Methanol utilization in transport will significantly rise
Source: methanex


**Figure 12. Assessment of MeOH as fuel for large bore engines for marine applications.**

Methanol is – as known - the simplest and easiest to produce liquid synthetic fuel that can be synthesized from CO2 and renewable H2. It is expected that renewable methanol can be imported to Europe at costs below 1 €/lDiesel$_{equivalent}$ by 2030, e.g. from the Middle East and North Africa as well as from Chile and Australia.

Today, in Europe, the use of methanol as an additive to conventional gasoline is limited to 3% v/v in accordance with EN228.

But methanol is becoming increasingly important also for maritime applications in particular. MAN´s two stroke methanol engines [7] and the Stena Germanica ferry powered by Wärtsilä 4-stroke methanol engines [8], are already cruising the seas, with more applications being announced. For heavy duty applications, methanol plays a larger role already especially in China and India. In China, methanol-operated trucks are already in small series production.

While the methanol engines for maritime applications realize typically diffusive combustion initiated by a small pilot injection of diesel fuel, the methanol heavy duty applications use spark ignited (SI) combustion systems with stoichiometric operation and a Three-Way-Catalyst for exhaust gas aftertreatment.

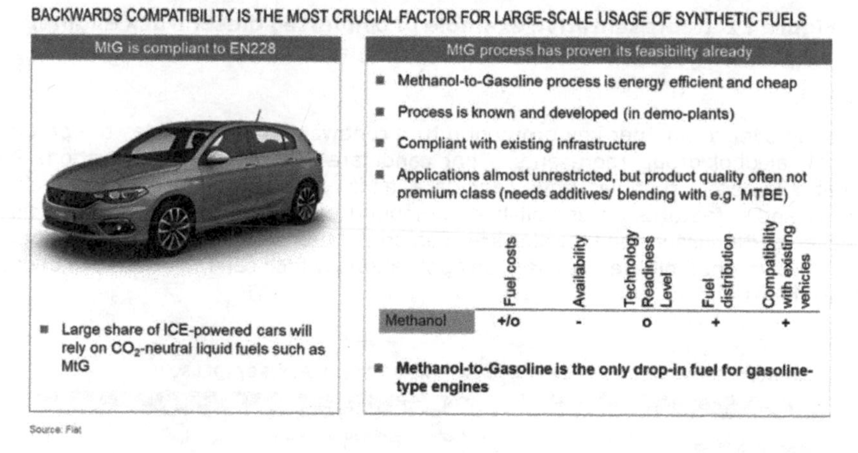


**Figure 13. Reflection of MtG fuel as renewable option for LDV with SI engines.**

With regard to the compatibility to existing engine technologies, the Methanol-to-Gasoline (MtG) path offers good boundaries to overcome the main challenges, as it is compliant with the current EN228 norm and can be introduced into the market quite easily, even with improved functional performance, as shown in the next diagram (Figure 14). As mentioned already on the previous slide, MtG fuel offers beneficial operational characteristics with advantageous engine-out emission levels and increased thermal efficiency, as displayed below with some representative data, achieved and verified in a public funded study from the German Ministry of Environment within the C3 Mobility program.

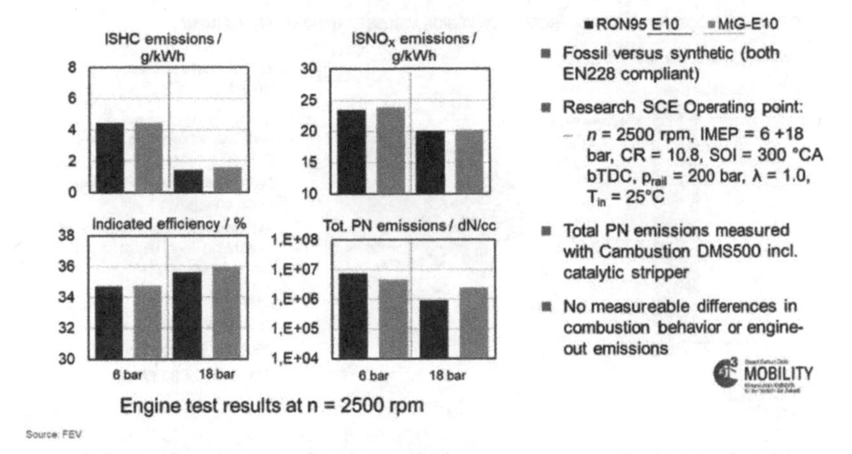


**Figure 14. Representative results of MtG operation of an in-series vehicle application.**

Especially with view on Diesel-based applications, mainly commercial vehicle installations and agricultural and constructional machinery, well-known "Fischer-Tropsch" fuel types represent a very attractive energy carrier for Diesel-based Powertrains with regard to fuel costs, availability, technology maturity, fuel distribution and retail and compatibility with in-service technology, while offering simultaneously benefits on the tailpipe emission behaviour, important for the Clean Air Initiatives.

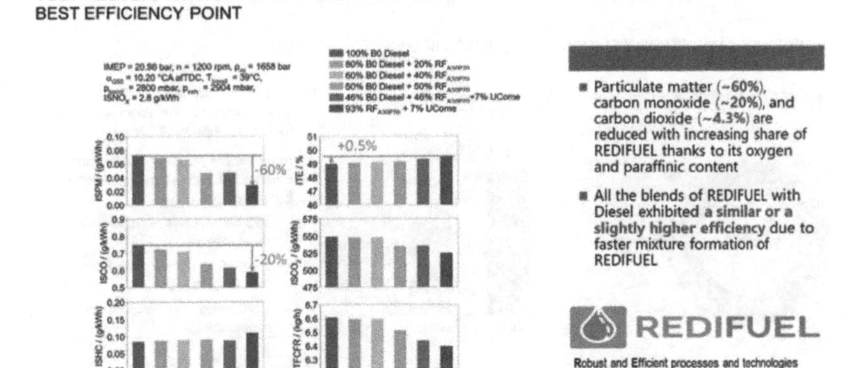


**Figure 15. Improvements of renewable drop-in fuel additions of diesel engine behaviour.**

A detailed depiction of the merits and advantages of these new fuel types, even in drop-in conditions, can be taken from this representative example. Within the referred public funded project REDIFUEL these results have been obtained on a large, truck-like Diesel Engine (Figure 15). Here in this representative mid-load point, at Iso-NOx conditions, PM-emissions have been exemplarily reduced by approx. 60%, CO by up to 20% and $CO_2$ output could be lowered by more than 4%.

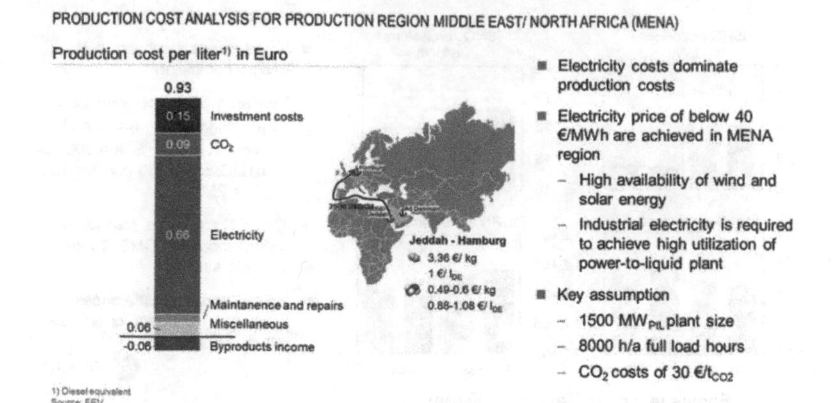


**Figure 16. Outlook on near-term price options for H2 from beneficial locations with excessive solar power.**

The before described technical parameters and the corresponding achievements lead finally to the key topic of energy costs. In the presented slide you can find an estimation of potential future costs of climate-neutral fuels under certain quite likely key assumptions. Taking this into consideration, renewable fuels could be highly interesting as well as highly attractive to be a key element on the pathway to carbon-neutral transport and mobility in the upcoming decades.

Just recently, Saudi-Arabia's $2^{nd}$ PV tender gained a world-record ultra-low price offer of 1,04 US-$cent per kWh electric power, realized in the projected solar production plants "Sakaka" and "Dumat Al-Jandal", covering together a capacity of 3600 MW.

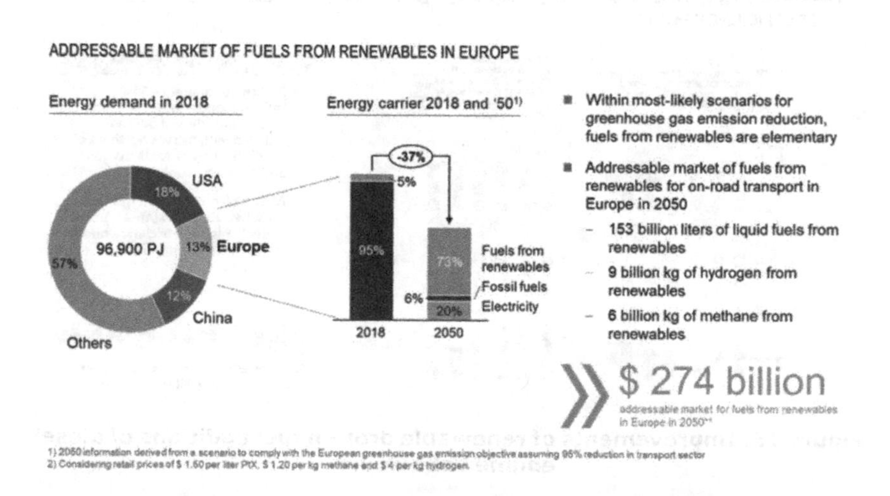


**Figure 17. Comprehensive description of business potential for H2-related activities.**

Summarizing all the before said, we can conclude, that there will be a very attractive business place for renewable fuels, which contains of liquid fuels mainly, but other types like hydrogen and methane as well. In total, the forecast of the addressable market amounts to more than 270 billion € in 2050

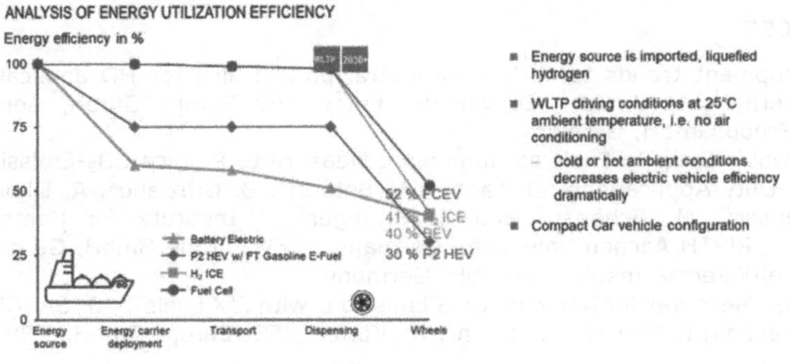


**Figure 18. Full chain depiction of efficiency chain for energy-import countries/regions.**

As said before, assuming full availability of the electric energy directly at the grid, the BEV strategy features some advantageous features in terms of primary energy usage. However, already here, it's worth to mention, that this is only a 2,4-3 times better efficiency against P2-HEV's with renewable fuels, besides all other favourable aspects for the liquid $CO_2$-neutral energy carriers. But, if we consider, that also in the future large European markets need to import energy from outside, the picture changes, as electrical power can't be transferred so easily. So, if in direct consequence, primary energy has to be transported, the efficiency chain alters strongly, as now the chemical energy has to be converted in the target market again to electrical energy, which brings all the different technical options quite closely together, so that the remaining aspects and criterions get more and more important.

## 3 SUMMARY AND CONCLUSIONS

For realization of a fast and broad reduction of the $CO_2$ emissions from the transportation sector, the application of all suitable technology options is necessary. In the short-term, these are primarily powertrain and vehicle optimizations, including electrification, for further efficiency gains. Nevertheless, it is obvious that the cost/benefit ratio varies greatly between the multiple measures and applications. Since the required $CO_2$ emission reduction, both relative and absolute, is different for each OEM, there is no one and only optimal strategy, but it is highly dependent on the powertrain line-up and the vehicle portfolio. Powertrain and fleet optimization measures have natural limitations when it comes to total energy savings.

To achieve the long-term goal of a $CO_2$-neutral transport, a switch to non-fossil energy carriers is required. These energy carriers could be synthetic fuels produced from $CO_2$ and hydrogen by using electricity from renewable sources or biofuels using various biomass feedstocks. The realization of electric mobility and the use of hydrogen, which will certainly make a major contribution to reducing $CO_2$ fleet emissions, are strongly favoured over alternative technologies by the "tank-to-wheel" accounting, and the use of e-fuels and biofuels has been left out of the current $CO_2$ legislation. A revised legislation that credits all $CO_2$ mitigation technologies, including synthetic fuels, should be a high priority to accelerate the process of addressing climate change most efficiently. There are several options available to make synthetic fuels more attractive, such as a tax on $CO_2$ or on carbon from fossil sources. Another option would be to credit carbon-neutral fuels against fleet emissions through a certificate system.

## REFERENCES

[1] "Development trends towards $CO_2$-neutral powertrains for HD applications"; M. Muether[1], L. Virnich[1], D. van der Put[2]; [1]FEV Europe GmbH, Germany; [2]FEV Group GmbH, Germany

[2] "Renewable Drop-In Fuels as Immediate Measure to Reduce $CO_2$-Emissions of Heavy-Duty Applications"; J. Yadav[1], V. Betgeri[1], B. Graziano[2], A. Dhongde[2], B. Heuser[2], M. Schönen[2] and N. Sittinger[3], [1] Institute for Combustion Engines, RWTH Aachen University, Germany; [2]FEV Europe GmbH, Germany; [3] OWI Oel-Waerme-Institute gGmbH, Germany

[3] "Tailored Measures for Net-Zero GHG Emissions with PtX Fuels"; 10th SASCI Technical Meeting; B. Heuser[1], H. Busch[2], Th. Körfer[2];[1]FEV Europe GmbH, [2]FEV Group GmbH,

[4] "Combustion system development for H2 fueled HD ICE"; 8th International Engine Congress 2021; L. Virnich, B. Lindemann, M. Müther, A. Dhongde, M. Schönen, J. Geiger; FEV Europe GmbH, Germany

[5] "Holistic approach for the development towards a $CO_2$-neutral powertrain for HD applications", ATZlive Event 2021, M. Muether[1], K. Deppenkemper[1], B. Heuser[1], L. Virnich[1], Th. Lüdiger[2] and D. van der Put[3]; [1)]FEV Europe GmbH, [2)]FEV Consulting GmbH, [3)]FEV Group GmbH

[6] "Design of Cobalt Fischer–Tropsch Catalysts for the Combined Production of Liquid Fuels and Olefin Chemicals from Hydrogen-Rich Syngas", ACS Catalyst 2021, 11, p. 4784-4798, K. Jeske, A.C. Kizilkaya, I. López-Luque, N. Pfänder, M. Bartsch, P. Concepción, G. Pietro; Max-Planck-Institut für Kohlenforschung, Mülheim/Ruhr, Germany; ITQ Instituto de Tecnología Química, Universitat Politecnica de Valencia-Consejo Superior de Investigaciones Cientificas (UPV-CSIC), 46022 Valencia, Spain

# Effect of diesel-ethanol fuel blends on engine performance and combustion behaviour of Compression Ignition (CI) engines

**F.O. Olanrewaju[1,2]*, H. Li[1], G.E. Andrews[1], H.N. Phylaktou[1]**

[1] School of Chemical and Process Engineering, Faculty of Engineering and Physical Sciences, University of Leeds, LS2 9JT, UK

[2] Department of Engineering Infrastructure, National Agency for Science and Engineering Infrastructure (NASENI), Nigeria

## ABSTRACT

Diesel-ethanol (DE) fuel blends, also called diesohols, are known to reduce NOx emissions from Compression Ignition (CI) engines. Engine researchers in the past have investigated the effects of DE fuel blends on the combustion and emissions performance of CI engines. However, conflicting results have been reported about the effect of DE fuel blends on engine-out $NO_x$, and CO emissions. Furthermore, the emission results of previous researchers in this field were largely focused on the levels of the regulated emissions ($NO_x$, CO, Total Hydrocarbons (THC), and Particulate Number (PN)). This has created an information gap in terms of the emission levels of Volatile Organic compounds (VOCs) such as benzene, and aldehydes, which are harmful to human health. The aim of this work was to extensively investigate the effect of DE blend fuels on the performance as well as the combustion behaviour of CI engines. A 5.7 kW engine output or 4.3 kW generator output, single-cylinder diesel Gen-set was used. The investigated fuel blends were 0%, 5%, 10%, and 15% ethanol in diesel while the tested conditions of generator output were idle, 2 $kW_e$ (kilowatt electric medium power), and 3 $kW_e$ (high power). The speciation of the engine exhaust was carried out by a Fourier Transform Infrared (FTIR) analyser. A dynamic electric mobility particle spectrometer (DMS500) was used to measure the particle size distribution of the exhaust from the engine. It was found that the diesohols increased the Brake Specific Fuel Consumption (BSFC) of the Gen-set, and contrary to what was reported in literature, the DE blends decreased the Brake Thermal Efficiency (BTE) of the engine compared to diesel baseline. The ethanol-blended fuels caused an increase in CO and THC emissions relative to baseline. The maximum reduction in $NO_x$ was 37% below the baseline for DE15 at 3 $kW_e$. The investigated DE fuel blends reduced the engine-out particulate emissions compared to the baseline at the high power condition but led to a significant rise in the aldehydes emissions. Ethanol and benzene emissions also increased compared to the baseline, but the levels were only significant at idle. The use of 15% green ethanol (a zero-carbon fuel) in diesel will reduce $CO_2$ emissions from transport in the UK by 9% (equivalent to 5.5 million tonnes reduction in transport $CO_2$ emissions). The results of the current work will enhance the use of DE biofuel blends in sub-saharan African countries for emission reduction and sustainable power generation. Diesel Gen-sets (without emission after-treatment) are widely used in sub-saharan African countries, where there is a good supply of feedstock for the production of ethanol from Sweet sorghum.

**Keywords:** Diesohol, combustion, emissions, FTIR, speciation.

*Corresponding author
DOI: 10.1201/9781003219217-7

## 1 INTRODUCTION

The global need to switch from fossil fuels to sustainable (renewable) energy sources as well as clean combustion technologies has inspired research into the use of diesel-ethanol (DE) fuel blends (diesohols) in diesel engines. Ethanol has the potential to reduce the combustion temperature in Internal Combustion Engines (ICEs) due its relatively high heat of vaporization compared to diesel. Combustion strategies that can reduce the combustion temperatures in diesel engines are desirable because reduction in engine-out NOx can be achieved when the flame temperature is lowered. The strategies that were proposed in literature for the utilisation of ethanol in diesel engines include fumigation (Port Fuel Injection (PFI) of ethanol), in-line mixing (for unstable DE blends), use of stable DE fuel blends (direct injection of splash-blended ethanol and diesel), and the use of straight ethanol with enhancers (lubricity and ignition enhancers) (1). Fumigation and the direct injection of splash-blended DE fuel blends are less costly strategies compared to the other approaches. Fumigation involves the injection of ethanol (the low reactivity fuel) via the air intake port to form a background premixed charge in the cylinder prior to the Direct Injection (DI) of diesel (the high reactivity fuel) near the Top Dead Center (TDC). The dual-fuel engine is operated in Reactivity Controlled Compression Ignition (RCCI) mode when ethanol is introduced by fumigation. Yu and Zheng (2) and Divekar et al. (3) investigated the effect of PFI injection of ethanol with direct injection of diesel on the combustion behaviour and performance of diesel engines. Direct injection of stable DE fuel blends (splash-blended DE fuel) is the simplest strategy to introduce ethanol into diesel engines as the approach does not require a major retrofit on the existing engine.

Engine researchers in the past have studied the effect of increasing the percentage of ethanol in DE fuel blends on the emissions and performance of DI diesel engines. Salih (4) studied the effect of ethanol fuel blends on a Petter (DI) diesel engine. The author reported 30% reduction in $NO_x$ when diesohol was used with 7.5% naphtha. Rakopoulos et al. (5) utilised a fully instrumented, 177 kW heavy duty DI Mercedes-Benz engine in their investigation. The authors reported that the Brake Specific Fuel Consumption (BSFC), the Brake Thermal Efficiency (BTE), the Ignition Delay (ID), and the Total Hydrocarbons (THC) increased above the baseline while the levels of soot and CO decreased as the percentage of ethanol increased in the blends. According to the authors, the level of $NO_x$ decreased slightly below the baseline for the investigated blend fuels (DE5 and DE10 with Betz GE emulsifier). The 30% reduction in $NO_x$ by DE blends that was reported by Salih (4) contradicts the trend that was reported by Rakopoulos et al. (5). Lapuerta et al. (6) reported that the BSFC and the THC increased above the baseline for DE fuel blends while the BTE was similar for the tested fuels (DE0, DE7.7, DE17 with 0.62% $O_2$Diesel additive) at low loads. According to the authors, the $NO_x$ emission from the tested DE blends was slightly lower than the baseline while the reduction in Particulate Matter (PM) was significant at high loads. Kass et al. (7) investigated the emissions from a 5.9 litre Cummins B series diesel engine that was run on DE fuel blends. The authors reported that the levels of $NO_x$ and aldehydes were not significantly affected by the diesohol fuel blends. According to the authors, the levels of CO and THC increased above the baseline while PM decreased below the baseline. Furthermore, the authors reported that the raw engine-out ethanol was <20 ppm for the tested DE fuel blends (DE10 and DE15). He et al. (8) studied the emissions from a 4-cylinder, 59 kW DI diesel engine when the engine was run on pure diesel, DE10, and DE30 at different Brake Mean Effective Pressure (BMEP) values. The authors found that the aldehydes emissions increased with increase in the concentration of ethanol. The authors also reported that, as the percentage of ethanol increased, the levels of the unburned ethanol increased to maxima at medium loads. The results of He et al. (8) for aldehydes and unburned ethanol emissions contradict what was reported by Kass et al. (7). Li et al. (9) investigated the effects of DE fuel blends on the emission and performance of a water-cooled, single cylinder DI diesel engine. Low, medium, and high loads were tested at 1,760 rpm and 2,200 rpm (rated speed). The tested blends were DE5, DE10, DE15, and DE20 with 1.5% emulsifier. The authors reported that the BSFC and the BTE of the engine increased above the baseline as the percentage of ethanol increased in the fuel blends. The authors also reported that, at the high and full loads (at the rated speed), the CO decreased for the DE blends as the load on the engine increased. According to the authors, DE10 and DE15 led to a drastic reduction in $NO_x$ at low and medium loads at the rated speed while the THC increased by up to 40% above the baseline.

The contradicting reports of previous researchers underscore the need for more investigations on the effect of DE fuel blends on the emissions and performance of DI diesel engines. The aim of the current work was to investigate the effect of the direct injection of stable DE fuel blends on the performance and the combustion behaviour of a single cylinder DI diesel Gen-set engine. The speciation of the engine-out exhaust gas was done by FTIR spectroscopy so that both the regulated and the unregulated harmful emissions were measured. The emission levels for benzene, ethylene, formaldehyde, acetaldehyde, and 1,3-butadiene are also reported in the current work. 1,3-butadiene is a major precursor to Polycyclic Aromatic Hydrocarbons, PAH's (10). The reduction in transport $CO_2$ emissions that is achievable by using ethanol-blended fuel in diesel engines was estimated to illustrate the significance and relevance of the current work to Net-Zero transport.

## 2 METHODOLOGY

### 2.1 Fuel properties

The fuels that were used to prepare the diesohol blends for the tests were off-road Ultra Low Sulphur Diesel (ULSD) and anhydrous ethanol. Two (2) litres of 5%, 10%, and 15% by volume of anhydrous ethanol in diesel (DE5, DE10, and DE15 respectively) were splash-blended for the test (DE0 represents pure diesel). The properties of the ULSD and the anhydrous ethanol that were used to prepare the blends are shown in Table 1. The properties of the ULSD fuel that was utilised comply with EN590.

**Table 1. Properties of pure diesel and anhydrous ethanol.**

| Property | Pure diesel | Anhydrous ethanol |
|---|---|---|
| Density at 15 °C, (kg/m$^3$) | ~840 | ~795 |
| Kinematic viscosity at 40 °C (mm$^2$/s) | ~2.7 | 1.2 |
| Cetane Number, CN | 48 | 8 |
| Lower Heating Value, LHV (MJ/kg) | 44 | 26.9 |
| Energy density (MJ/litre) | 36 | 21.4 |
| Latent heat of evaporation (kJ/kg) | 544-795 (11) | 840 (12) |
| Boiling temperature at 1 bar (°C) | 246-388 | 78 |
| Flash point (°C) | >62 | 13 |
| Aromatics (mg/kg) | 11 (upper limit) | - |
| Sulphur content (wt%) | <0.0015 | - |

### 2.2 Engine description and instrumentation

The details of the engine and the instrumentation are presented in Tables 2 and 3. The Gen-set engine that was used in the current work was a DI diesel engine with a re-entrant bowl piston (modern combustion chamber design). The load on the engine was varied through the Hillstone AC load bank (HAC240-10). The measured temperatures and fuel consumption were logged by LabView software. The Particulate Matter (PM) values were calculated from the measured Particulate Number (PN) distribution by assuming spherical particles and raising the diameter of the particles (Dp) to the power 2.65 (13).

**Table 2. Engine specification.**

| Parameter | Specification |
|---|---|
| Make | Yanmar |
| Type | 4-stroke, single cylinder |
| Rated power (kW) | 5.7 |
| Speed (rpm) | 3,000 |
| Bore (mm) | 86 |
| Stroke (mm) | 75 |
| Compression ratio | 20.9:1 |
| Displacement ($cm^3$) | 435.66 |
| Total cylinder volume ($cm^3$) | 457.55 |
| Injection pressure (MPa) | ~20 |
| Injection timing (bTDC) | 13° |

**Table 3. Instrumentation.**

| Parameter | Equipment specification |
|---|---|
| Temperature | K-type thermocouple |
| Fuel consumption | Scale (ADAM CPW plus-35), accuracy: 0.01 kg |
| Gaseous emissions | Gasmet FTIR Analyser (DX-4000) |
| Exhaust particle distribution, PN | Differential Mobility Spectrometer (DMS500) |

### 2.3 Test matrix

Table 4 presents the test matrix. Each of the investigated DE blends was tested at three conditions of kilowatt-electric ($kW_e$) load as shown in Table 4. The Gen-set engine had an alternative power loss of about 25%. Therefore, the tested conditions of power: idle, 2, and 3 $kW_e$ were equivalent to engine-out power conditions of idle, 2.7, and 4 kW respectively.

**Table 4. Engine test conditions.**

| Fuel blend | Load (Generator-out) ($kW_e$) | Engine-out power (kW) |
|---|---|---|
| DE0 (Baseline) | 0 | 0 |
| | 2 | 2.7 |
| | 3 | 4 |
| DE5 | 0 | 0 |
| | 2 | 2.7 |
| | 3 | 4 |
| DE10 | 0 | 0 |
| | 2 | 2.7 |
| | 3 | 4 |
| DE15 | 0 | 0 |
| | 2 | 2.7 |
| | 3 | 4 |

The BSFC and the BTE of the engine were estimated from Equations (1) and (2) respectively.

$$BSFC = 3600 \times \left(\dot{m}_f / P\right) \tag{1}$$

$$BTE = 1000 \times P / \left(\dot{m}_f CV\right) \tag{2}$$

In Equations (1) and (2), *P* is the power of the engine (in kW), $\dot{m}_f$ is the fuel flow rate (in kg/s) while *CV* is the Calorific Value of the fuel in MJ/kg.

## 3 RESULTS AND DISCUSSION

The set of data that was obtained at the end of the experiment (engine temperatures, fuel consumption, emissions data) was analysed to investigate the effect of ethanol-blended fuels on the performance of the engine and the engine-out emissions.

### 3.1 Effect of diesel-ethanol (DE) fuel blends on engine performance

Figure 1 shows the effect of the DE fuel blends on the engine-out exhaust temperature (measured at the exhaust manifold), the BSFC and the BTE of the engine.

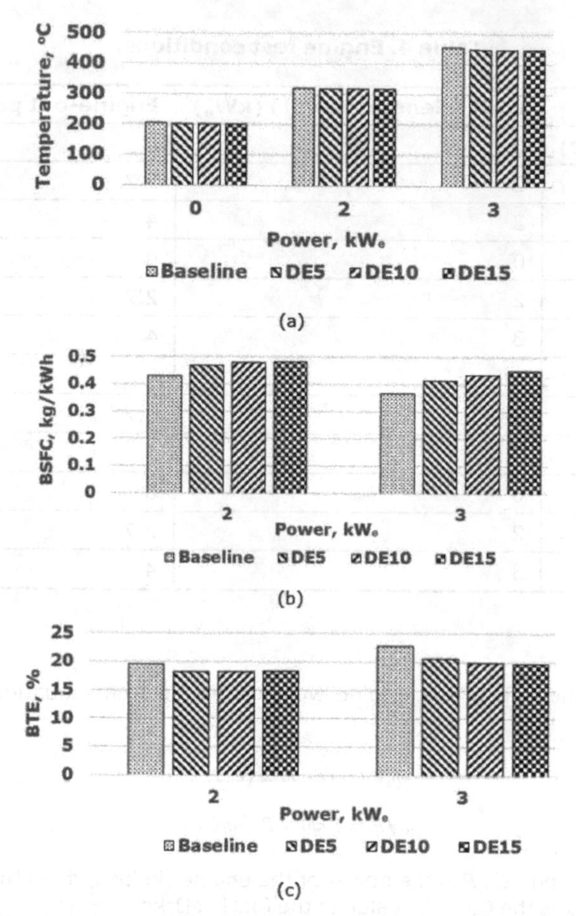


**Figure 1. Effect of the concentration of ethanol on engine performance.**

### 3.1.1 *Flame temperature*

Figure 1 (a) shows that the DE fuel blends led to a slight drop in the temperature of the flame below the baseline. The observed drop in temperature was as a result of the relatively high latent heat of vaporization of ethanol compared to baseline diesel (Table 1). The injected DE fuel blends absorbed more heat from the gases in the combustion chamber to evaporate than the heat absorbed by pure diesel. This led to the observed Low Temperature Combustion (LTC) that occurred when the engine was run on the DE blends. The maximum average drop in the combustion temperature orchestrated by the ethanol-blended fuels was 7 °C at the 3 $kW_e$ load condition.

### 3.1.2 *Brake Specific Fuel Consumption (BSFC)*

The BSFC of the engine increased above the baseline as the concentration of ethanol in the DE blends increased. The Lower Heating Value (LHV) of ethanol is much lower than that of ULSD (Table 1). Consequently, as the percentage of ethanol increased in the DE blends, the DE fuel mass that was injecteded to achieve the same power increased above the baseline (Figure 1 (b)). Rakopoulos et al. (5) also reported that the BSFC increased above the baseline as the concentration of ethanol increased. The observed increase in the BSFC of the engine as the concentration of ethanol increased reflects the relatively low energy density of DE blends compared to pure diesel.

### 3.1.3 *Brake Thermal Efficiency (BTE)*

The BTE of the engine decreased below the baseline as the concentration of ethanol increased in the DE blends. This was due to the relatively low CN and the relatively high latent heat of vaporization of ethanol compared to ULSD. The relatively low CN of ethanol led to longer Ignition Delays (ID) for the DE fuel blends which resulted in relatively short Duration of Combustion (DoC) for the blends compared to diesel baseline. The injection timing of the engine was constant at 13° bTDC, notwithstanding the retarded Start of Combustion (SoC) for the DE blends (due to the increase in the ID caused by the low CN of ethanol). The relatively high latent heat of vaporization of ethanol retarded the evaporation and the SoC of the ethanol in the DE blends. The high heat of evaporation of ethanol decreased the temperature of the flame below the baseline thereby leading to rapid (advanced) quenching of the flame. The combined effect of the relatively low CN and high latent heat of vaporization of ethanol (increase in the ID and the retardation of the evaporation of the injected fuel blend) led to relatively short DoC. Therefore, there was insufficient time to attain a high degree of completeness of combustion when the engine was run on the DE blends. This caused the level of the unburned hydrocarbons to increase as the concentration of ethanol increased in the DE blends (this is further explained in Section 3.2.1.1). As a consequence, the BTE of the engine decreased below the baseline as the concentration of ethanol increased in the DE blends.

## 3.2 Effect of diesel-ethanol (DE) fuel blends on engine-out emissions

The engine-out emissions that were measured included the gaseous emissions and the particulate emissions. Sections 3.2.1 and 3.2.2 present the results for the engine-out gaseous emissions. The results for the particulate emissions are presented in Section 3.2.3 in terms of Particulate Number (PN) and Particulate Matter (PM) distributions.

### 3.2.1 *Regulated gaseous emissions*

The effect of the ethanol-blended fuels on the levels of the regulated emissions (THC, CO, and $NO_x$) are presented in this section. The emission results for the regulated pollutant gases are given in ppm and g/kWh.

#### 3.2.1.1 TOTAL HYDROCARBON (THC) EMISSIONS

Figure 2 (a) and (b) shows that the THC emission levels increased above the baseline at all the tested conditions of power as the concentration of ethanol increased in the DE fuel blends. This was due to two factors. Firstly, the relatively low CN of ethanol which increased ID of the DE blends while the fuel injection timing remained constant at 13° bTDC led to the incomplete combustion of the injected fuel mass. There was insufficient time to achieve a high degree of completeness of combustion of the injected fuel leading to relatively high levels of unburned hydrocarbons when the engine was run on the DE blends. Secondly, the relatively high heat of evaporation of ethanol led to relatively low flame temperatures and the rapid quenching of the flame. Consequently, there was inefficient vaporization of the injected DE fuel masses as well as the formation of local rich zones due to the inefficient mixing of air and fuel. These also led to the incomplete combustion of the injected DE blends and the observed high THC levels above the baseline. However, as the power of the engine increased, the THC levels for all the tested fuels became lower than their corresponding levels at idle. This was because the temperature of the flame increased as the power of the engine increased (Figure 1 (a)).

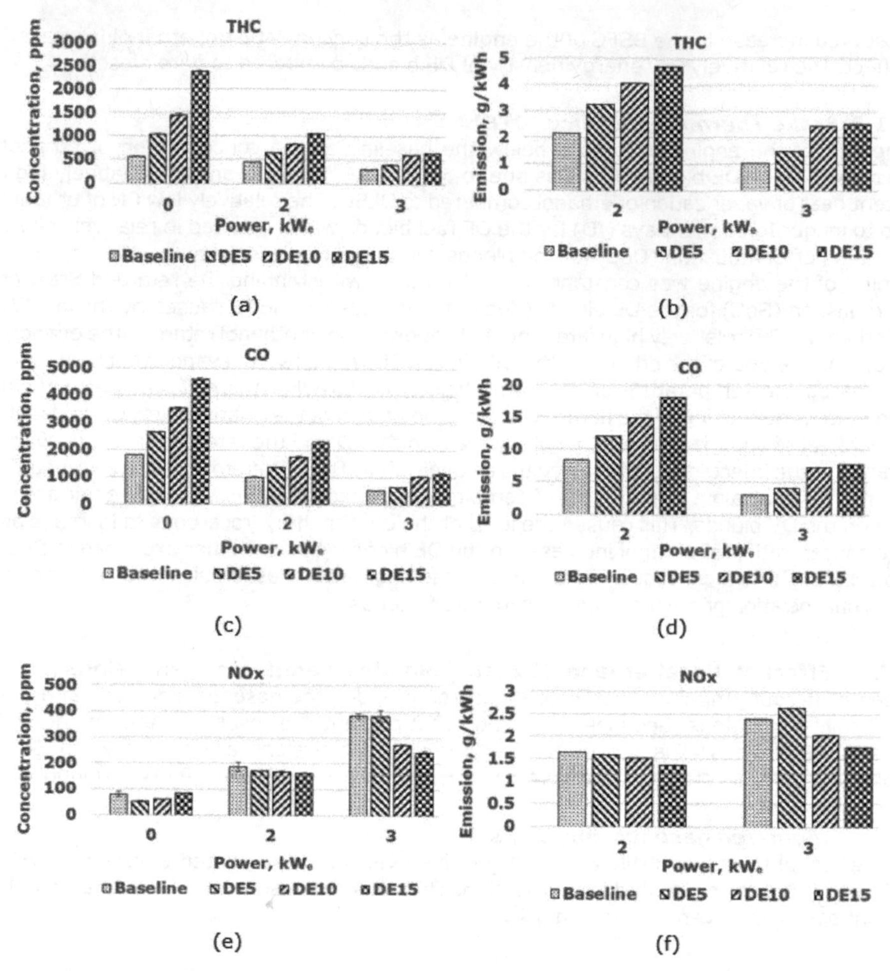


**Figure 2. Effect of the concentration of ethanol on THC, CO, and $NO_x$ emissions.**

The vaporization and combustion of the injected fuel masses as well as the in-cylinder oxidation of the unburned hydrocarbons were enhanced as the temperature of the flame increased. At the idle condition, DE15 increased the THC emissions by a factor of 4 above the baseline while at the the high power condition (3 kW$_e$), the THC emissions were increased by a factor of 2.5 above the baseline. Generally, the engine-out THC emissions for the DE fuel blends at the idle and the 2 kW$_e$ power conditions in the current work were observed to be higher than the value that was reported for diesel engines (600 ppm) in literature (14). This was largely due to the relatively low injection pressure (~20 MPa) of the Gen-set engine (the typical fuel injection pressure in relatively large diesel engines is 160-200 MPa). The low injection pressure of the Gen-set engine led to inefficient atomisation and vaporisation of the injected fuel mass.

#### 3.2.1.2 CARBON MONOXIDE (CO) EMISSIONS

Figure 2 (c) and (d) show that the emission levels for CO increased above the baseline at all the tested conditions of power as the concentration of ethanol increased. The same factors that led to the increase in the THC emissions above the baseline (Section 3.2.1.1) also led to the observed increase in the levels of CO. The same trend was observed for the THC and CO emissions from the engine because both species are products of incomplete combustion. As such, the levels of the engine-out CO also decreased for all the tested fuels as the power of the engine increased. DE15 increased the emission level of CO above the baseline by a factor of 2.5 at the high condition of power (3 $kW_e$). The trend that was observed in the current work for CO contradicts what was reported by Rakopoulos et al. (5).

#### 3.2.1.3 NITROGEN OXIDES ($NO_X$) EMISSIONS

Figure 2 (e) and (f) show that the investigated DE fuel blends decreased the levels of the engine-out $NO_x$ below the baseline at the tested conditions of power. This was due to the relatively low combustion temperature that was achieved when the engine was run on the ethanol-blended fuels. However, as the power of the engine increased from idle to 3 $kW_e$, the emission of $NO_x$ increased for all the tested fuels due to the increase in the temperature of the flame that occurred the high power conditions (the emission of $NO_x$ in diesel engines is enhanced at relatively high flame temperatures). The maximum reduction in $NO_x$ below the baseline was ~37% for DE15 at 3 $kW_e$ Figure 2 (e). 37% reduction in $NO_x$ below the baseline compares well to the 30% reduction that was reported by Salih (4).

### 3.2.2 *Emission levels for specific hydrocarbons and VOCs*

The investigated unregulated engine-out pollutant gases in the current work are ethanol, ethylene, aldehydes, 1,3-butadiene, and benzene. The FTIR analyser that was used in the current work was calibrated to measure fifty (50) species. However, this section presents the emission results for the detectable pollutant gases.

#### 3.2.2.1 ETHANOL

Figure 3 (a) shows that the engine-out emission of ethanol increased above the baseline as the concentration of ethanol increased at each of the tested conditions of power. There was a drastic increase in the level of the emitted ethanol above the baseline at idle for DE10 and DE15. This was due to the relatively low combustion temperature at idle. The observed low combustion temperature at the idle condition neither favoured the vaporization/combustion nor the oxidation of the ethanol fraction of the DE blends. Figure 3 (a) also shows that the levels of ethanol decreased as the power of the engine increased. This was because, at the relatively high flame temperatures that occurred as the load of the engine was increased, the oxidation of ethanol by the oxygen in air and the oxygen in the alcohol was enhanced. Kass et al. (7) reported that the engin-out ethanol was <20 ppm for DE10 and DE15. However, in the current work, the observed engine-out emission levels for DE15 at 2 $kW_e$ and 3 $kW_e$ were 91 ppm and 50 ppm respectively.

#### 3.2.2.2 ETHYLENE

Figure 3 (b) shows that the levels of the engine-out ethylene increased above the baseline as the concentration of ethanol increased in the diesohol fuel blends. However, as the temperature of the flame increased (due to increase in the power of the engine), the levels of ethylene decreased drastically. This was because the drastic increase in the temperature of the flame enhanced the oxidation of the emitted ethylene by the oxygen of air and the oxygen of the -OH group of ethanol.

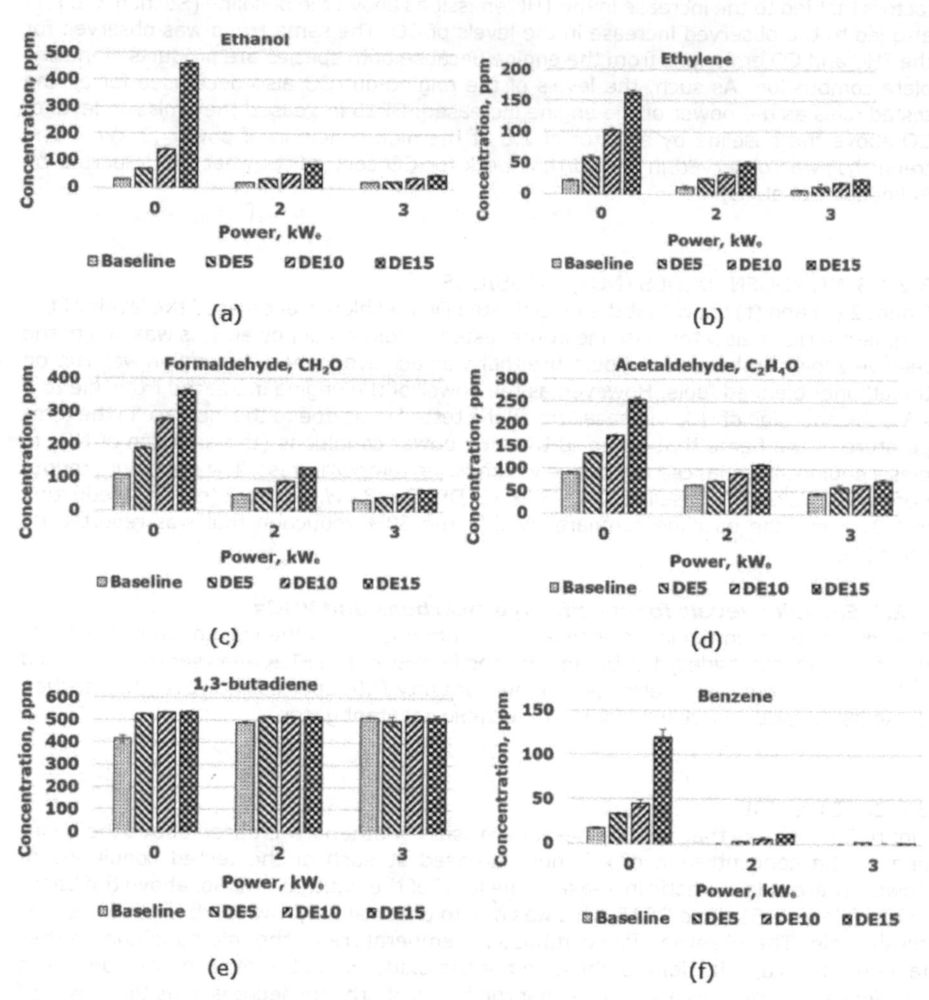


**Figure 3. Effect of the concentration of ethanol on specific engine-out pollutant gases (hydrocarbons and VOCs).**

3.2.2.3 ALDEHYDES

The ethanol-blended fuels led to an increase in the levels of aldehydes emissions ($CH_2O$ and $C_2H_4O$) above the baseline (Figure 3 (c) and (d)). The observed trends for the aldehydes in the current work are in agreement with what was reported by He et al. (8). However, Kass et al. (7) reported that the levels for aldehydes were not significantly affected by DE fuel blends. The emission levels for the aldehydes in the current work were significantly higher at idle for the tested DE blends than the corresponding levels at the higher power conditions. The emission levels of the aldehydes decreased drastically for the DE blends as the power of the engine increased due to the temperature-enhanced oxidation of the aldehydes by the oxygen in the oxygenated fuel blends and the oxygen of air.

3.2.2.4 1,3-BUTADIENE

Figure 3 (e) shows that the emission levels for 1,3-butadiene were quite significant for all the tested fuels. At idle, the DE fuel blends led to ~30% increase in the emission level for 1,3-butadiene above the baseline. As the power of the engine increased, the emission levels for 1,3-butadiene decreased slightly for the DE blends whereas they increased for pure diesel. The effect of the DE fuel blends on the emission levels for 1,3-butadiene was quite insignificant at the high power conditions.

3.2.2.5 BENZENE

The tested DE fuel blends caused a drastic increase in the emission levels for benzene (a PAH) at the idle condition; the increase was by a factor of 6 for DE15 as shown in Figure 3 (f). However, as the power of the engine was increased, the emitted benzene decreased drastically for all the tested fuels. At 3 $kW_e$, the levels for benzene in the engine-out exhaust became negligible. This was because, as the power of the engine increased, the relatively high combustion temperatures that were attained in the combustion chamber enhanced the oxidation of benzene to lighter hydrocarbons.

### 3.2.3 *Particulate emissions*

Figure 4 (a) shows that the peak PN (maximum number concentration of the emitted particles) increased as the concentration of ethanol increased in the DE fuel blends. At the higher concentrations of ethanol (10% and 15%) and at the higher power conditions, the peak PN increased drastically above the baseline. This can be attributed to the increase in the BSFC at the high concentrations of ethanol. The highest increase in the peak PN above the baseline was by a factor of 10 for DE10 at the medium power condition (2 $kW_e$). However, at the high power condition (3 $kW_e$), the peak PN decreased for all the tested fuels compared to the corresponding values at idle and 2 $kW_e$. The observed decrease in the peak PN was due to the enhanced oxidation of soot that occurred at the relatively high combustion temperature that was attained at 3 $kW_e$. At near-stoichiometric conditions, soot oxidation by the -OH group of ethanol is enhanced (14). Leach et al. (15) reported that the PN emissions increased as the load on a highly boosted Gasoline Direct Injection (GDI) engine was increased. The GDI engine was run on EN228 compliant gasoline. In the current work, the PN emissions for the DE blends showed decreased as the load on the engine increased above the medium load (Figure 4 (a)). This confirms the potential of oxygenated fuels/fuel blends such as ethanol-blended fuels to reduce the engine-out PN emissions at relatively high engine loads. The observed decrease in the PN emission levels for the tested DE blends at the 3 $kW_e$ load condition indicated that at loads above the medium load (2 $kW_e$), the temperature of the flame became sufficiently high to enhance the oxidation of soot by the oxygen in the -OH group of ethanol. Therefore, at the 3 $kW_e$ load, the PN emission levels for the DE blends decreased below the corresponding levels at the medlun load (2 $kW_e$). Unlike the case of the DE blends in the current work, the reported PN emissions for the GDI engine increased monotonously (15) because gasoline is not an oxygenated fuel.

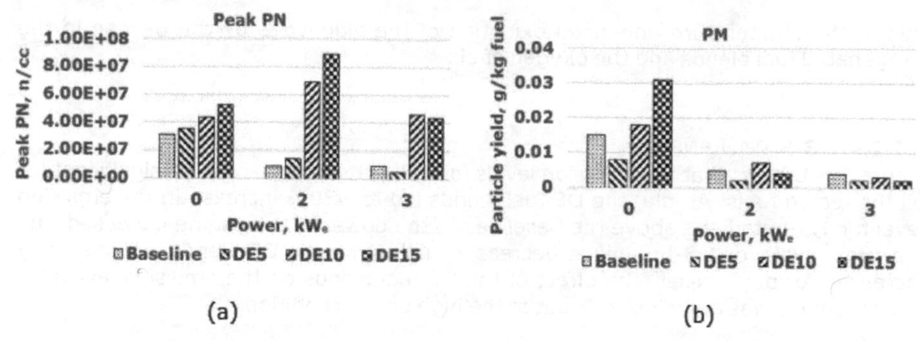


**Figure 4. Effect of the concentration of ethanol on particulate emissions.**

The result for the PM (Figure 4 (b)) shows that at the high power condition (3 $kW_e$), the DE fuel blends led to a decrease in the mass of the emitted particles below the baseline. Generally, the mass of the emitted particles per unit fuel mass decreased drastically as the power of the engine increased. The combustion temperature was highest at 3 $kW_e$. Relatively high combustion temperatures enhanced the oxidation of soot by the oxygen of ethanol. Rakopoulos et al. (5) and Lapuerta et al. (6) also reported that the emission of particulates decreased below the baseline as the load on the engine increased and as the concentration of ethanol increased from 5% to 15%.

The effect of increasing the concentration of ethanol on the diameter of the emitted particles (Dp) at the peak PN is presented in Table 5 for the tested conditions of power. Rows 5 and 6 of columns 3 and 4 in Table 5 show that, at the relatively high concentrations of ethanol (DE10 and DE15) and at the high conditions of power (2 and 3 $kW_e$), the Dp at the peak PN decreased as the concentration of ethanol increased. Table 5 and Figure 4 (a) show that the engine-out PN emissions for DE5 were similar to those for pure diesel. This was due to the relatively low concentration of ethanol in DE5. The disparity in the observed values of the Dp at the maximum PN at 2 $kW_e$ for baseline diesel (65 nm) and DE5 (18 nm) was because the PN distribution for baseline diesel at the 2 $kW_e$ power condition was strongly bimodal, unlike the PN distribution for DE5 (Figure 5). The first PN peak for pure diesel (which was lower than the second) occurred at a Dp of ~24 nm. The observed Dp at the first peaks of the PN distributions for baseline diesel and DE5 occurred in the nanoparticles diameter range (Dp<30 nm). The second PN peaks for pure diesel and DE5 occurred at Dp=65 nm. DE5 had a weak PN peak at Dp=65 nm as shown in Figure 5. Furthermore, Table 5 and Figure 4 (a) show that, at the relatively high concentrations of ethanol (DE10 and DE15) and at the medium and high power conditions, the tested DE fuel blends enhanced the formation of nucleation mode particles (nanoparticles). This indicated that as the load on the engine was increased from idle to 3 $kW_e$, the Dp of the emitted particles at the peak PN decreased. The particulate emissions results in the current work are similar to the results of Lapuerta et al. (6). The authors used DE7.7 in their investigation. Lapuerta et al. (6) reported that at relatively low concentrations of ethanol, the combustion behaviour of DE fuel blends was similar to that of pure diesel. They also reported that DE fuel blends enhanced the formation of nanoparticles. The authors' results for PN and PM are similar to what was observed in the current work.

**Table 5. Effect of the concentration of ethanol on the diameter (Dp) of particles at peak PN.**

| Fuel blend | Dp @ peak PN, nm | | |
|---|---|---|---|
| | Idle | 2 $kW_e$ | 3 $kW_e$ |
| Baseline | 49 | 65 | 56 |
| DE5 | 37 | 18 | 56 |
| DE10 | 65 | 24 | 21 |
| DE15 | 75 | 21 | 18 |

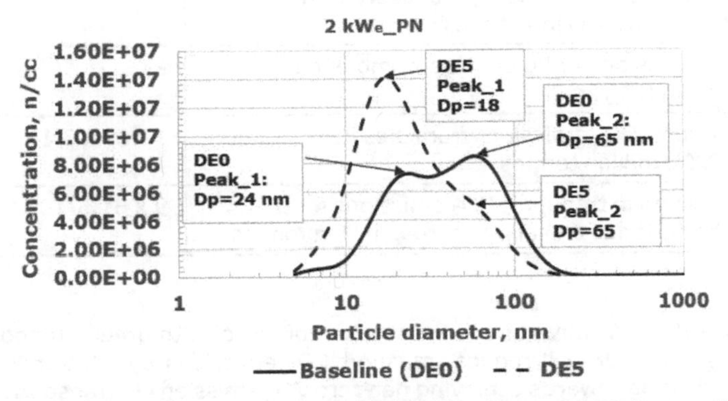


**Figure 5. Bimodal Particle Number (PN) distribution of DE0 (pure diesel) and DE5 at 2 $kW_e$.**

### 3.3 Achievable reduction in transport $CO_2$ emissions by the use diesel-ethanol fuel blend

The current work confirms that diesel-ethanol fuel blends containing up to 15% of (anhydrous) green ethanol can be utilised in diesel engines. 15% by volume of zero-carbon ethanol in diesel (DE15) is equivalent to an ethanol substitution of diesel of ~9% by energy. This implies that, if 15% by volume of diesel is replaced substituted with green ethanol in diesel engines in the UK, the transport $CO_2$ emissions will reduce by 9%. Table 6 summarises the analysis that was carried out to determine the equivalent potential reduction in transport $CO_2$ emissions in million tonnes per year (based on the consumption of diesel for transport in the UK in 2020). The efficiency of the combustion of the ethanol fraction of the DE blend was assumed equal to that of ULSD.

**Table 6. Possible $CO_2$ savings from the substitution of diesel with 15% by volume of green ethanol.**

| S/n | Item | Calculation | Value |
|---|---|---|---|
| 1 | Energy density of diesel, MJ/litre | - | 36 |
| 2 | Energy density of ethanol, MJ/litre | - | 21.4 |
| 3 | Green ethanol substitution of diesel by volume, % | - | 15 |
| 4 | Green ethanol substitution of diesel by energy, % | (15x21.4)/36 | ~9 |
| 5 | Consumption of diesel by transport in the UK (2020) (16), million tonnes | - | 19.69 |
| 5 | $CO_2$ emission per kg of diesel combusted (17), kg $CO_2$/kg diesel | - | 3.1 |
| 6 | Transport $CO_2$ emissions from diesel (2020), million tonnes | 19.69 x 3.1 | 61.05 |
| 7 | Reduction in transport $CO_2$ emissions for 9% substitution of diesel by energy, million tonnes | 9 x 61.5/100 | ~5.5 |

Table 6 shows that 15% by volume substitution of diesel with green ethanol in transport vehicles in the UK will reduce transport $CO_2$ emissions by ~5.5 million tonnes thereby contributing towards achieving net-zero $CO_2$ emissions in transport.

## 4 CONCLUSION

The effect of ethanol-blended fuels on the performance and the emissions of a 5.7 kW engine output diesel engine was investigated in the current work. A FTIR analyser was utilised to speciate the engine-out exhaust gases such that emission results for regulated pollutant gases and unregulated pollutant gases were reported. The particle size distributions of the engine-out particulate emissions were measured in the current work by the DMS500. The current work showed that increasing the concentration of ethanol in the DE blends caused the BSFC of the engine to increase above baseline diesel while the BTE decreased below diesel baseline. The emission levels for THC and CO increased by a factor of 2.5 above the baseline at the high load condition (3 $kW_e$) for DE15 (the highest concentration of ethanol in the current work). The maximum reduction in the engine-out $NO_x$ that was achieved in this work was 37% at an ethanol concentration of 15% and and an engine load of 3 $kW_e$ (the highest load). The results of the work also showed that the engine-out emission levels for ethanol, ethylene, aldehydes, and benzene increased above the baseline as the concentration of ethanol increased in the DE fuel blends. However, increasing the power of the engine caused the emission levels of the pollutant gases to reduce drastically. The emission levels for 1,3-butadiene were not significantly affected by the ethanol-blended fuels at the medium and high power conditions. This indicated that at the relatively high loads, the tested DE blends had insignificant effect on the emission levels for 1,3-butadiene. It was also shown in the current work that the tested DE fuel blends reduced the engine-out Particulate Matter (PM) emissions below the baseline at the high conditions of power. However, ethanol-blended fuels enhanced the production of nanoparticles. At each of the tested loads, the peak Particulate Number (PN) concentrations of the emitted particles were observed to increase above the baseline as the concentration of ethanol increased in the DE blends. The highest increase in the peak PN above the baseline was by a factor of 10 for DE10 at the medium power condition (2 $kW_e$). The particulate emission results for DE5 were similar to those for pure diesel. The current work also showed that using 15% by volume of ethanol in diesel will reduce transport $CO_2$ emissions in the UK by 9% which is equivalent to $CO_2$ savings of ~5.5 million tonnes.

**ACKNOWLEDGEMENT**

The current work was funded by the Petroleum Technology Development Fund (PTDF), Nigeria and supported by the National Agency for Science and Engineering Infrastructure (NASENI), Nigeria.

**REFERENCES**

[1] Likos, B., Callahan, T. J. & Moses, C. A. (1982) Performance and Emissions of Ethanol and Ethanol-Diesel Blends in Direct-Injected and Pre-Chamber Diesel Engines. SAE technical paper 821039. doi:10.4271/821039
[2] Yu, S. & Zheng, M. (2016) Ethanol–diesel premixed charge compression ignition to achieve clean combustion under high loads. *Proceedings of the Institution of Mechanical Engineers, Part D: Journal of Automobile Engineering*. 230, 527–541. doi:10.1177/0954407015589870
[3] Divekar, P., Han, X., Tan, Q., Asad, U., Yanai, T., Chen, X., Tjong, J. & Zheng, M. (2017) Mode Switching to Improve Low Load Efficiency of an Ethanol-Diesel Dual-Fuel Engine. SAE technical paper 2017-01-0771. doi:10.4271/2017-01-0771
[4] Salih, F. M. (1990) *Automotive fuel economy measures and fuel usage in Sudan*. Ph.D Thesis, University of Leeds
[5] Rakopoulos, D. C., Rakopoulos, C. D., Kakaras, E. C. & Giakoumis, E. G. (2008) Effects of ethanol–diesel fuel blends on the performance and exhaust emissions of heavy duty DI diesel engine. *Energy Conversion and Management*. 49, 3155–3162. doi:10.1016/j.enconman.2008.05.023
[6] Lapuerta, M., Armas, O. & García-Contreras, R. (2009) Effect of Ethanol on Blending Stability and Diesel Engine Emissions. *Energy & Fuels*. 23, 4343–4354. doi:10.1021/ef900448m
[7] Kass, M. D., Thomas, J. F., Storey, J. M., Domingo, N., Wade, J. & Kenreck, G. (2001) Emissions From a 5.9 Liter Diesel Engine Fueled With Ethanol Diesel Blends. SAE technical paper 2001-01-2018. doi:10.4271/2001-01-2018
[8] He, B.-Q., Shuai, S.-J., Wang, J.-X. & He, H. (2003) The effect of ethanol blended diesel fuels on emissions from a diesel engine. *Atmospheric Environment*. 37, 4965–4971. doi:10.1016/j.atmosenv.2003.08.029
[9] Li, D., Zhen, H., Xingcai, L., Wu-Gao, Z. & Jian-Guang, Y. (2004) Physico-chemical properties of ethanol-diesel blend fuel and its effect on performance and emissions of diesel engines. *Renewable Energy*. 30, 967–976. doi:10.1016/j.renene.2004.07.010
[10] Hansen, N., Miller, J. A., Kasper, T., Kohse-Höinghaus, K., Westmoreland, P. R., Wang, J. & Cool, T. A. (2009) Benzene formation in premixed fuel-rich 1,3-butadiene flames. *Proceedings of the Combustion Institute*. 32, 623–630. doi:10.1016/j.proci.2008.06.050
[11] AVL (2015) *GCA Gas exchange and combustion analysis*, Germany, AVL.

[12] Rakopoulos, D. C., Rakopoulos, C. D., Papagiannakis, R. G. & Kyritsis, D. C. (2011) Combustion heat release analysis of ethanol or n-butanol diesel fuel blends in heavy-duty DI diesel engine. *Fuel*, 90, 1855–1867. doi:10.1016/j.fuel.2010.12.003
[13] Cambustion (2011) *DMS500 user manual*. Cambridge, Cambustion Ltd. 45
[14] Heywood, J. B. (1988) *Internal combustion engine fundamentals*, New York, McGraw-Hill, Inc. 567
[15] Leach, F., Stone, R., Richardson, D., Lewis, A., Akehurst, S., Turner, J., Remmert, S., Campbell, S. & Cracknell, R. F. (2018) Particulate emissions from a highly boosted gasoline direct injection engine. *International Journal of Engine Research*, 19, 347–359. doi:10.1177/1468087417710583
[16] DUKES (2021) Digest of UK energy statistics [Online]. [Accessed 6 August 2021]. Available from: https://www.gov.uk/government/statistics/
[17] FR (2021) Carbon emissions of different fuels [Online]. [Accessed 7 August 2021]. Available from: https://www.forestresearch.gov.uk/

*International Conference on Powertrain Systems for Net-Zero Transport*
*Institution of Mechanical Engineers, ISBN 978-1-032-11281-7*

# Magma xEV and sustainable liquid fuels – steps towards net-zero propulsion

**R.J. Osborne, A. Saroop, J. Stokes, L. Valenta, R. Penning**
Ricardo UK

**D. Richardson, A. Gimmini, S.M. Sapsford**
Coryton Advanced Fuels

## ABSTRACT

The electrification of powertrains provides a critical opportunity to change the way that engines are designed and developed, allowing their efficiency to be increased and their cost reduced. Alongside this, sustainable liquid fuels offer the potential for significant greenhouse gas reductions, while using current fuelling infrastructure and addressing the existing vehicle fleet.

This paper draws on ongoing Ricardo projects in the field of dedicated hybrid engines (DHEs). The Magma xEV combustion concept employs very high compression ratio, long stroke architecture, and advanced ignition and knock mitigation technologies, for DHEs requiring the highest efficiency. In the latest research project, a pre-chamber combustion system (with both active and passive operation) has been applied to the Magma xEV engine, to enable the highest levels of charge dilution and further increase brake thermal efficiency. The combustion concept has been developed using virtual product development approaches and validated with a single-cylinder research engine.

A sustainable biomass-to-liquid (BtL) gasoline fuels has also been tested with this engine, and the impact of these fuels on the combustion process has been reported.

## 1 INTRODUCTION

The automotive industry is experiencing a period of unprecedented change. The primary challenge is the urgent need to defossilise transport to limit global temperature rises to the 1.5 °C set out in the Paris Agreement. At the same time as meeting the climate challenge, the industry must also continue to reduce air-quality emissions, provide a larger range of vehicle architectures, and bring vehicles to the market more quickly.

Some policymakers and commentators have concluded that there is only one technical solution for road transport, the battery electric vehicle (BEV). This has resulted in proposals for the prohibition of the sale of internal combustion engine (ICE) vehicles in some countries and cities. This approach does not consider the life-cycle greenhouse gas emissions of different vehicles, and critically also fails to address the 1.4 bn passenger vehicles already on the road worldwide.

However, the electrification of vehicles also presents a major opportunity for the ICE. Conventional engines operate over a large range of engine speed and load, and must meet the contrasting requirements for rated power and idle operation. A dedicated hybrid engine (DHE), one which is designed to run only in a specific hybrid electric vehicle (HEV) architecture, can operate over a much narrower operating range. For series-hybrid architectures, the engine can run at only two operating points, as shown in Figure 1. This allows design parameters to be selected which favour these operating conditions.

DOI: 10.1201/9781003219217-8

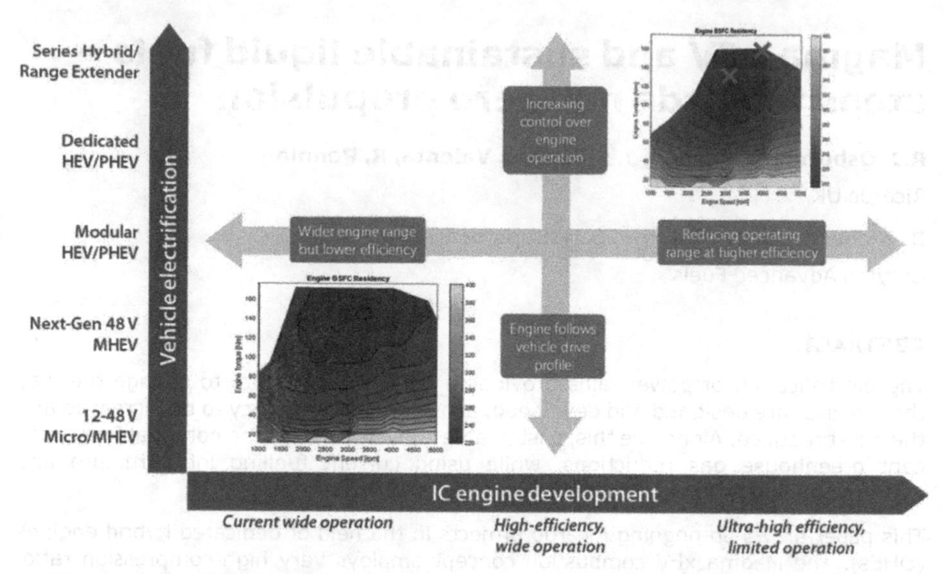


**Figure 1. Electrification of vehicles as an opportunity for the IC engine.**

It is generally considered that Sir Harry Ricardo developed the first pre-chamber engines in work conducted up to 1922 [1]. Since then many pre-chamber designs have been proposed, and these ignition systems have been widely employed in large marine and stationary power-generation gas engines. More recently a new generation of pre-chamber gasoline engines have been developed for passenger-car applications, generally focussed on extending dilution limits to improve efficiency [2-6]. The engines have employed passive, active (fuel-fed) and scavenged or air-assisted pre-chambers. Passive pre-chambers have subsequently been employed in Formula 1 engines [7], and first passive pre-chamber ignition system has been launched in a road car (this engine has a second spark plug in the main chamber) [8].

A wide range of sustainable fuels for IC engines are under development. Considering sustainable liquid fuels, the two key groups are biofuels, produced from biomass via a variety of production routes, and electrofuels (or e-fuels) which are synthesised from $CO_2$ and hydrogen. It is likely that many or all of these different sustainable fuels will be needed in different markets and applications to meet the climate challenge. In this study a sustainable biomass-to-liquid fuel produced from forestry and agricultural wastes has been tested.

## 2 ENGINE CONCEPT

### 2.1 Ricardo DHE

Ricardo has developed a dedicated hybrid engine (DHE) concept for application to full HEVs, plug-in hybrid electric vehicles (PHEVs), series HEVs and range-extended electric vehicles (REEVs). The concept is focussed on C-segment vehicles and is based around a three-cylinder 1.5 litre direct-injection gasoline engine.

The key elements of the Ricardo DHE are shown in Figure 2. A lightweight cylinder block structure is employed with 'clamshell' engine covers providing the basis for encapsulation for NVH and efficiency benefits. A three-cylinder engine with 1.5 litre displacement would normally require a primary balancer shaft in order to achieve acceptable NVH characteristics. In a DHE application it is not necessary to run at problematic low engine speeds and so the balancer shaft can be deleted. A front-end accessory drive (FEAD) is also not needed, and the valvetrain and boosting system have been simplified.

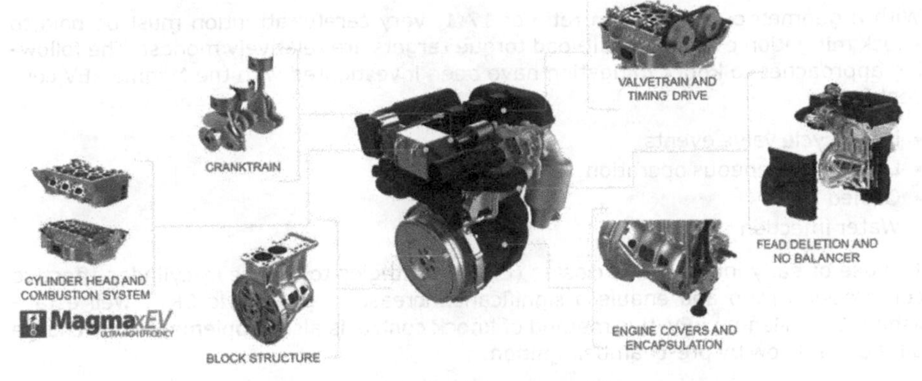


**Figure 2. Elements of Ricardo DHE concept.**

A range of DHE specifications have been developed to address the requirements of different hybrid-electric vehicle architectures. For plug-in vehicles with large battery capacity the engine thermal efficiency is not the most important feature, and the engine specification is targeted at lowest cost. In some series-hybrid applications by contrast the engine BTE is critical, and this is the purpose of the Magma xEV combustion system.

## 2.2 Magma xEV combustion system

The effect of geometric compression ratio on part-load fuel consumption is well known. Increasing compression ratio improves ideal cycle efficiency but increases the surface area to volume (S/V) ratio and, therefore, in-cylinder heat losses [9]. Combining compression ratio and bore-to-stroke (B/S) ratio effects gives the characteristic shown in Figure 3. For a square engine (B/S = 1) there is little benefit in raising compression ratio above 13 - 14:1, as increased S/V ratio leads to higher heat losses. As B/S ratio reduces, further increases in compression ratio continue to provide benefit. Based on this analysis, the bore-to-stroke ratio for Magma xEV was set at 0.70, and the target geometric compression ratio at 17:1.

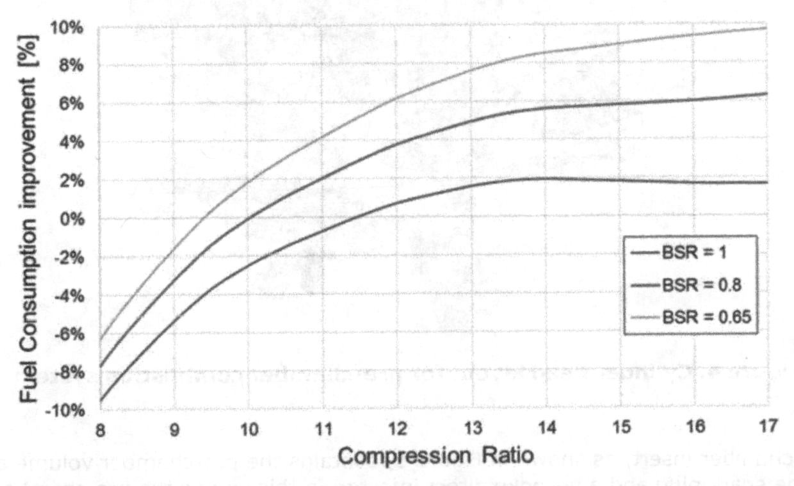


**Figure 3. Influence of compression ratio and bore-to-stroke ratio (BSR) on part-load BSFC.**

With a geometric compression ratio of 17:1, very careful attention must be paid to knock mitigation even if the full-load torque targets are relatively modest. The following approaches to knock mitigation have been investigated with the Magma xEV concept [10]:

- Miller cycle valve events
- Lean homogeneous operation
- Cooled EGR
- Water injection

The use of early intake-valve closing (EIVC) strategies to reduce in-cylinder effective compression ratio and enable a significant increase in geometric CR is well established. With Magma xEV, this method of knock control is also supplemented by charge dilution and now by pre-chamber ignition.

### 2.3 Pre-chamber combustion system

The objective of the present study was to develop a pre-chamber combustion system capable of both active (fuel-fed) and passive operation applied in the Magma xEV architecture. The overall layout of the combustion system and cylinder head is shown in Figure 4.

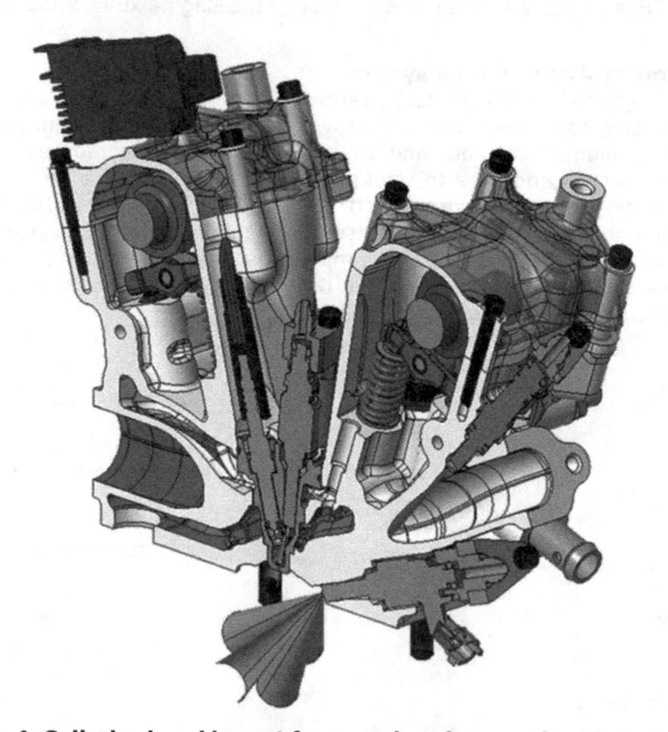

**Figure 4. Cylinder head layout for pre-chamber combustion system.**

A pre-chamber insert, as shown in Figure 5, contains the pre-chamber volume along with the spark plug and a bespoke direct injector. In this engine the pre-chamber has a volume of just above 4% of the overall clearance volume.

**Figure 5. Pre-chamber insert.**

In pre-chamber combustion systems turbulence in the main chamber is dominated by the jets that emerge from the pre-chamber nozzle. As a result, the high-tumble intake ports developed for earlier Magma xEV combustion systems ($R_t$ of 2.1) were replaced with more conventional intake ports with tumble ratio of 1.0. The combustion chamber is shown in Figure 6, where the pre-chamber nozzle can be observed along with the absence of shrouding around the intake valves (a feature of the higher tumble combustion systems).

**Figure 6. Magma xEV combustion chamber with pre-chamber nozzle.**

## 3 MODEL-BASED DEVELOPMENT OF THE COMBUSTION SYSTEM

A programme of integrated model-based development (IMBD) was undertaken to develop the pre-chamber combustion system. 3-D computational fluid-dynamics (CFD) using Ricardo VECTIS was used to develop the following aspects of the combustion system design:

- Pre-chamber volume
- Pre-chamber nozzle geometry

- Pre-chamber injector position and spray pattern
- Start of injection timing for the pre-chamber injector
- Main chamber injector spray pattern

A large number of pre-chamber nozzle designs were initially simulated, to select suitable options for full-cycle simulation. The pre-chamber shape and volume were then developed along with the injection timing approach for active pre-chamber operation.

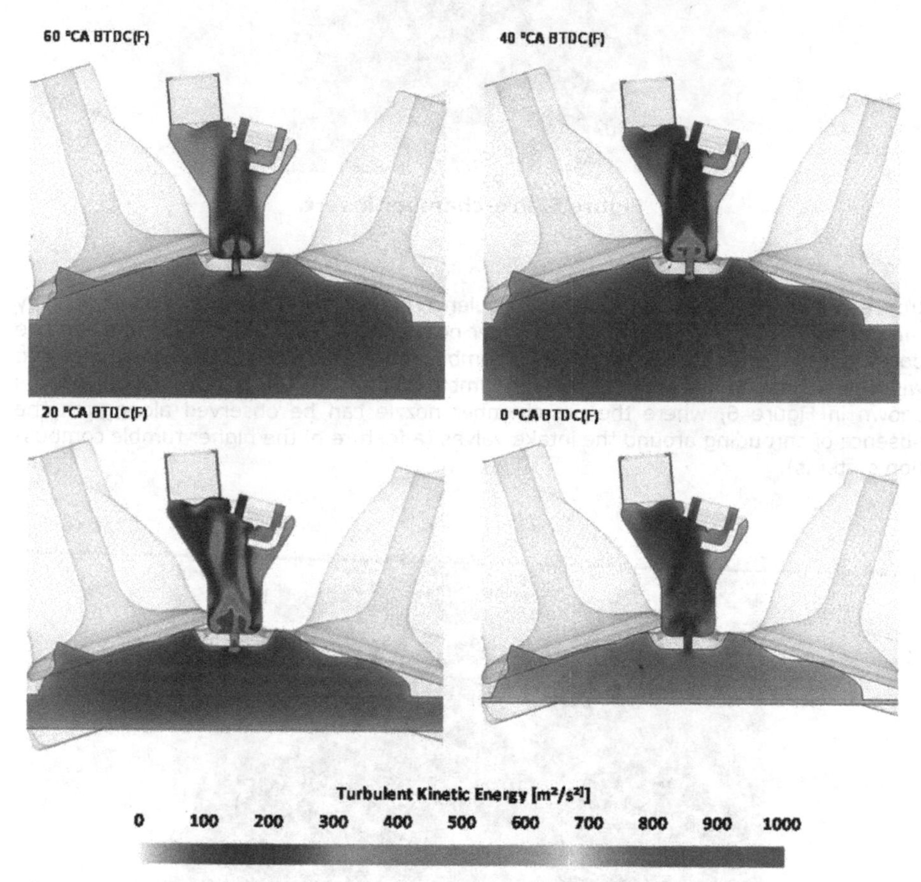


**Figure 7. Turbulent kinetic energy (TKE) levels over the 60 °CA period prior to TDC for passive pre-chamber operation.**

It is well known that scavenging of pre-chambers produces high levels of turbulence, but what is important are the levels of turbulence at ignition timing. Turbulent kinetic energy (TKE) values for the final pre-chamber design are shown in Figure 7, for an operating condition of 3000 rev/min, 11 bar NIMEP. The TKE levels observed in the pre-chamber are high compared with a conventional combustion system at an equivalent operating condition.

The impact of pre-chamber injection timing for active pre-chamber operation was also analysed. This is illustrated in Figure 8 which compares the equivalence ratio distribution (probability density function) for two different start of injection timings, with the objective of stoichiometric mixture in the pre-chamber. With an injection timing of 320 °CA BTDC(F), pre-chamber injection starts at the same time as the main chamber injection. The relatively narrow equivalence ratio distribution indicates good mixing in the pre-chamber, but the overall mixture is lean due to fuel flowing out of the pre-chamber. Delaying the start of injection to 120 °CA BTDC(F) prevents this fuel flow producing better control over pre-chamber lambda, which is distributed around stoichiometry. The later injection does reduce the time available for mixing, which is illustrated by a broader distribution of equivalence ratio.

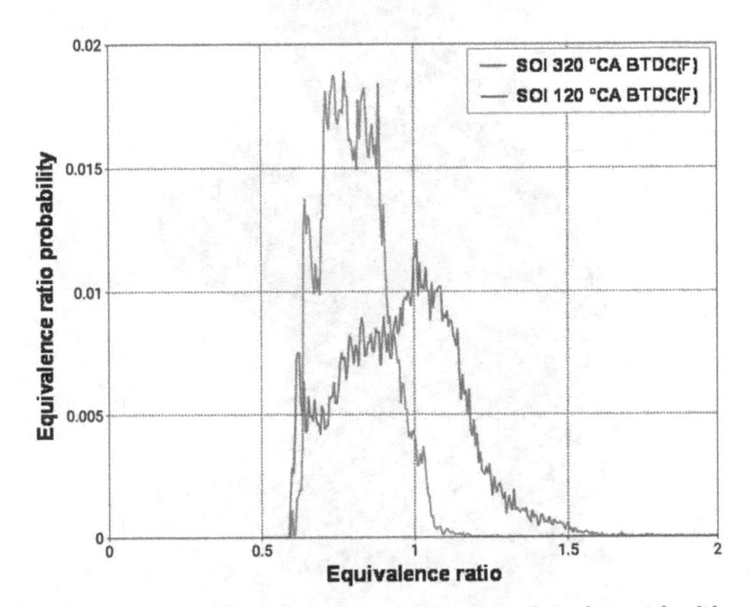


**Figure 8. Equivalence ratio distribution in the pre-chamber at ignition timing (11 °CA BTDC(F)) for two different start of injection timings.**

## 4 ENGINE TESTING

### 4.1 Single-cylinder engine specification

An experimental programme was conducted using a Ricardo Hydra single-cylinder research engine with the specification shown in Table 1. The cylinder head, piston crown and cam profiles were developed specifically for the Magma xEV engine.

**Table 1. Key parameters of the test engine.**

| Displacement volume/no. cylinders | 0.5 litres/1 cylinder |
|---|---|
| Bore | 76.6 mm |
| Stroke | 109.1 mm |
| Number of valves | 4 |
| Compression ratio | 17:1 |
| Maximum fuel pressure | 200 bar |

The engine, as shown in Figure 9, was installed in a research facility with an external high-pressure air supply and the necessary measurement systems, sensors, and actuators.

**Figure 9. Magma xEV single-cylinder research engine.**

A schematic of the engine test facility is shown in Figure 10. The pre-chamber was fitted with a pressure transducer, and the main chamber with two pressure transducers. An exhaust orifice was fitted to simulate the damping characteristics of a turbine and waste-gate.

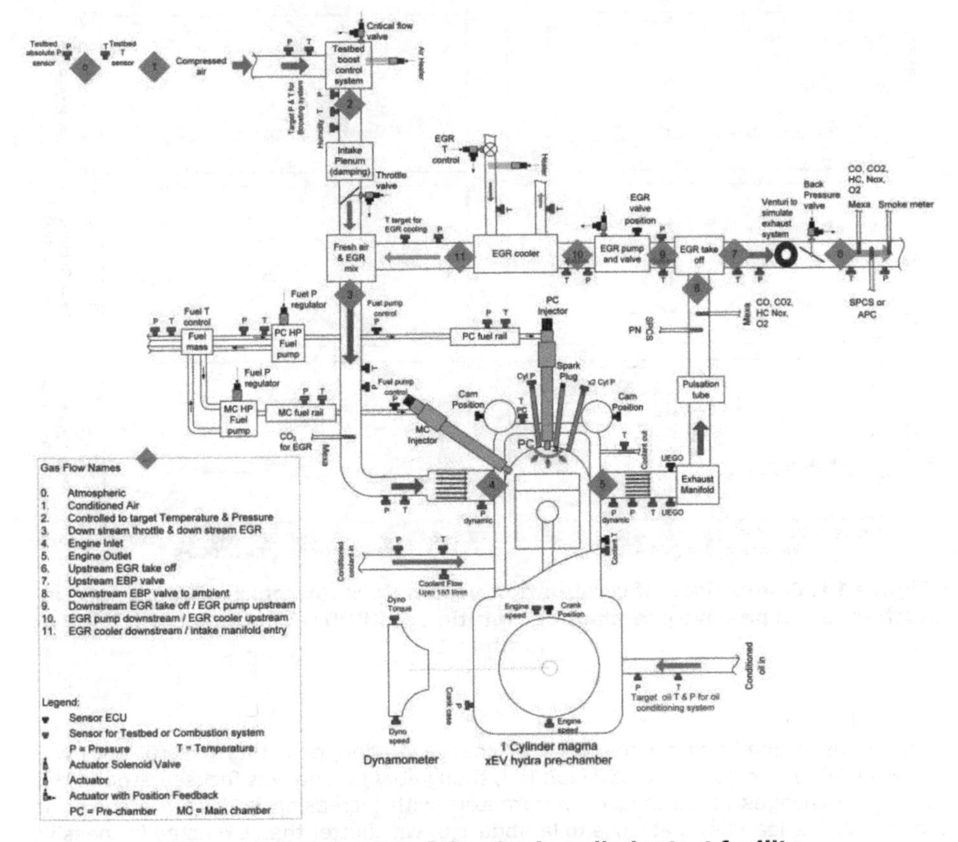


**Figure 10. Schematic of the single-cylinder test facility.**

## 4.2 Passive pre-chamber results

Pre-chamber ignition has a striking influence on combustion parameters, as shown in Figure 11 At an operating condition of 2000 rev/min and 5 bar net IMEP, stoichiometric combustion with conventional spark ignition, corona-discharge ignition [10] and passive pre-chamber ignition are compared for a range of start of injection (SOI) timing.

At this low-load operating point, the ignition delay angle (ignition – 10% mass fraction burned) is quite long for the conventional ignition system, at more than 30 °CA. For the two distributed ignition processes (corona discharge and passive PC) there is a dramatic reduction in ignition delay angle to around 12 °CA. Turning to burn angle (10 – 90% MFB), again the conventional ignition system has the slowest combustion at around 20 °CA. Corona-discharge ignition produces slightly shorter burn angles, but for the passive pre-chamber operation these are reduced by around half, to 10 °CA. The overall very rapid heat release contributes to increased thermal efficiency and dilution tolerance and increases peak cylinder pressure and rate of pressure rise.

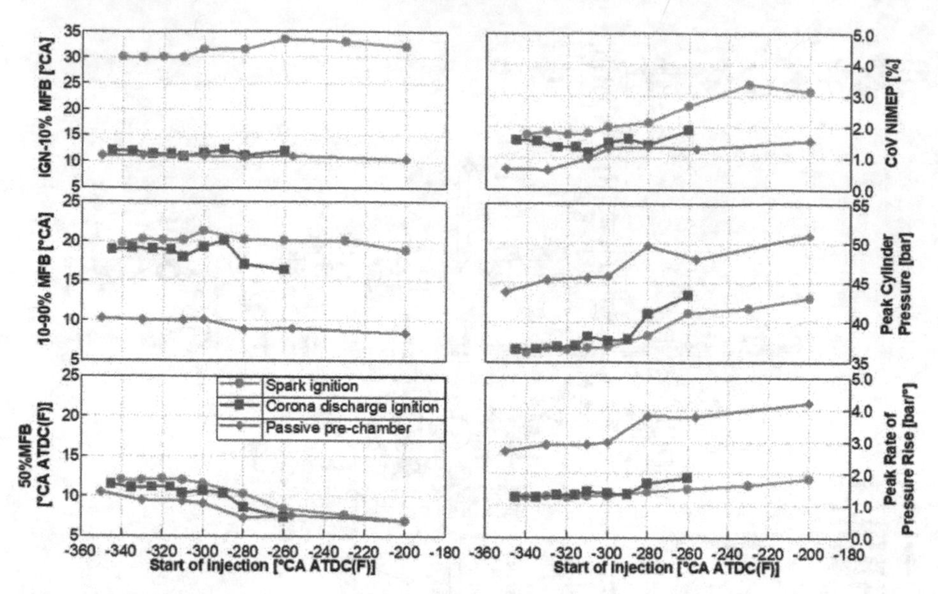


**Figure 11. Comparison of combustion parameters for spark ignition, corona discharge and passive pre-chamber ignition at 2000 rev/min, 5 bar NIMEP at lambda 1.**

In previous engine testing with corona-discharge ignition, peak thermal efficiency was found at 3000 rpm 11 bar NIMEP load [9], thus initial testing was focused around this area. The changes in combustion parameters with increasing lambda are shown in Figure 12. The lean limit, at close to lambda 1.8, was better than expected for passive pre-chamber operation, and combustion remains rapid even at the lean limit with 10-90% MFB of about 15 °CA.

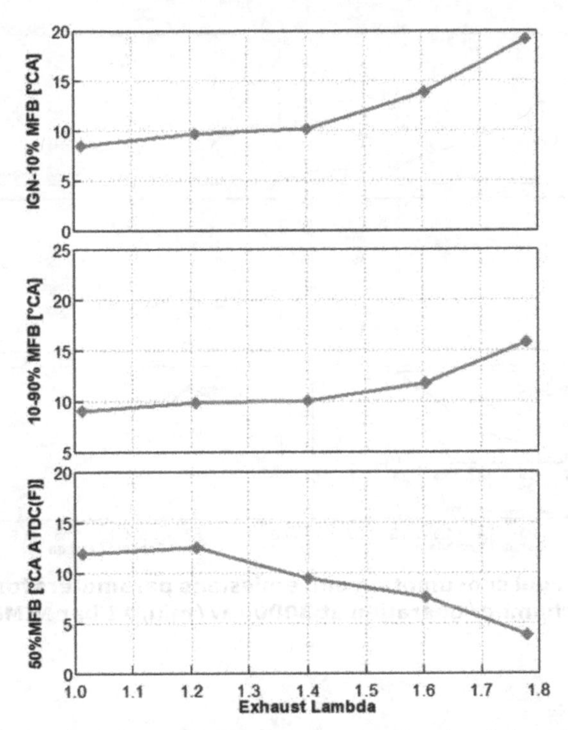


**Figure 12. Combustion parameters for varying lambda with passive pre-chamber operation at 3000 rev/min, 11 bar NIMEP.**

Lean operation results in improved fuel consumption with acceptable combustion stability and with reduced emissions including PN, Figure 13. All specific fuel consumption values have been corrected to a fuel lower calorific value of 42.5 MJ/kg throughout the paper. The engine out PN (> 23 nm) observed even at lambda 1 is in the range of a typical automotive gasoline engine application. The volume fraction of unburned hydrocarbons (HC) increases by more than half between stoichiometric mixture and lambda 1.8 but still remains in the range of a typical gasoline engine with stochiometric operation.

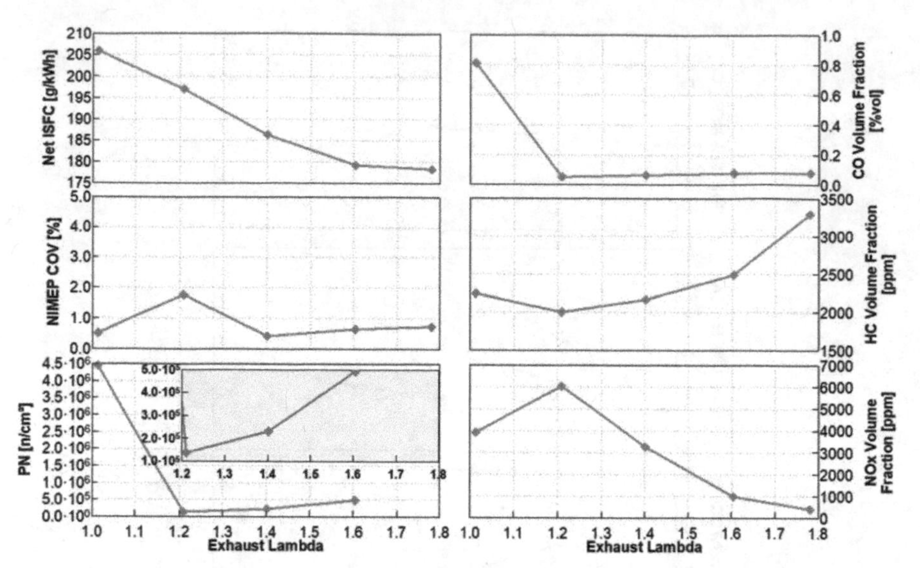


**Figure 13. Fuel consumption and emissions parameters for passive pre-chamber operation at 3000 rev/min, 11 bar NIMEP.**

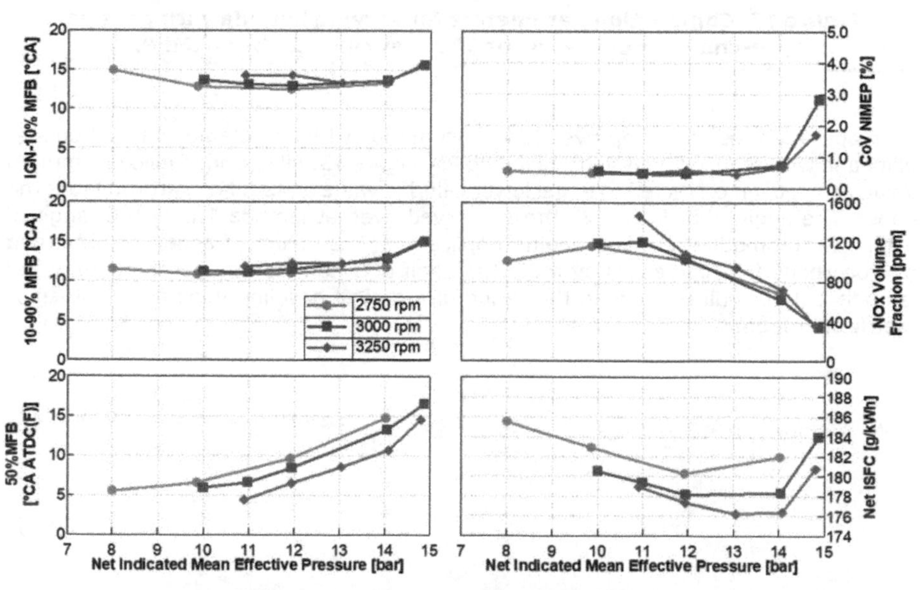


**Figure 14. Load range tests at 2750, 3000 and 3250 rev/min at constant exhaust lambda of 1.6 in the region of peak thermal efficiency with passive pre-chamber combustion.**

Load, lambda, and ignition timing swings were undertaken to identify the peak efficiency condition for the pre-chamber engine. The results from load range tests at three engine speeds for lambda 1.60 are shown in Figure 14. Ignition delay angle (ignition – 10% MFB) and burn angle (10 – 90% MFB) are similar with different speeds and loads. Position of 50% MFB is retarded, NOx and fuel consumption are reduced with increasing load. At highest loads of approximately 15 bar NIMEP the engine reaches a stability limit and an increase in fuel consumption is observed. An engine speed of 3250 rev/min speed and load of 13.5 bar NIMEP was identified as the best efficiency condition. Peak efficiency has moved to higher load with pre-chamber operation due to the positive impact of higher burn-rate on combustion knock.

Combustion behaviour across a range of speed and load was characterised with both stoichiometric and lean passive pre-chamber combustion. Maps of net ISFC comprised of more than 50 test-points each are shown in Figure 15, for stoichiometric operation, and Figure 16 for lean operation. Combustion stability was observed to be good for low speed and low load operation. Minimum NISFC was below 180 g/kWh for lean operation and a wide operating area (which is also a key operating area for dedicated-hybrid engines) is below 185 g/kWh.

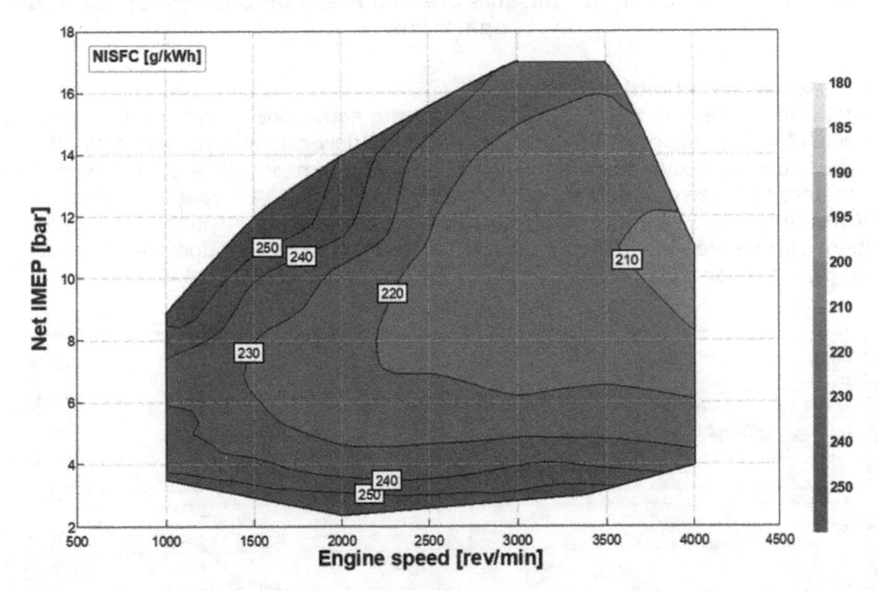


**Figure 15. Net ISFC map for passive pre-chamber combustion at exhaust lambda 1.**

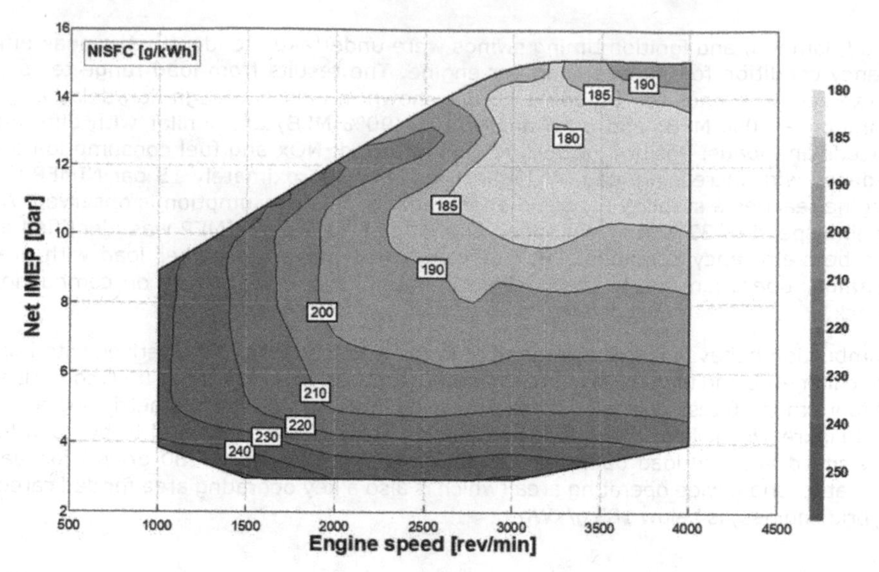


**Figure 16. Net ISFC map for passive pre-chamber combustion for lean-burn operation.**

## 4.3 Active pre-chamber results

A programme of testing was then undertaken with active pre-chamber operation, with injection of fuel in the pre-chamber via a second low flow-rate injector with separate fuel pressure control. Active pre-chamber ignition has a similar influence on combustion parameters to passive operation, as shown in Figure 17. At an operating condition of 3000 rev/min and 11 bar net IMEP, combustion characteristics with corona-discharge ignition, passive pre-chamber ignition and active pre-chamber ignition are compared for varying exhaust lambda.

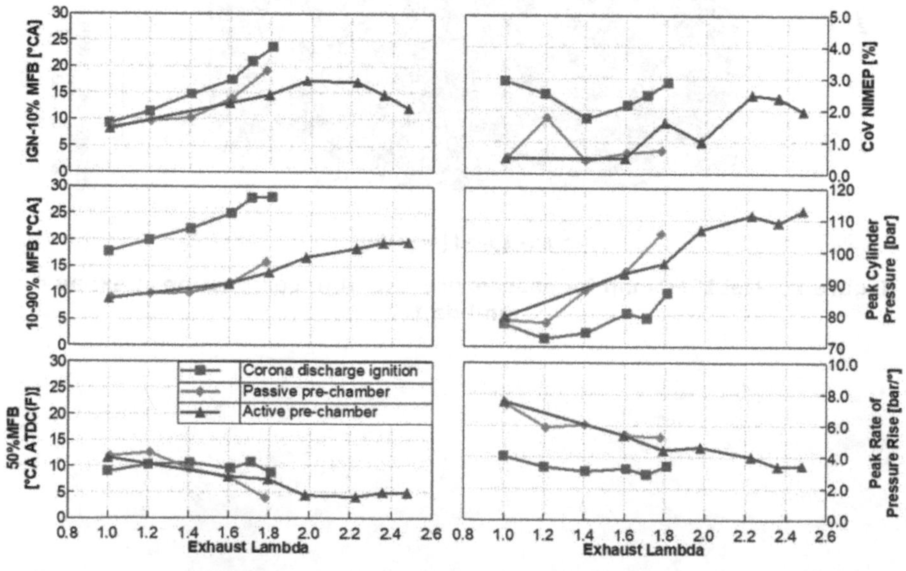


**Figure 17. Comparison of combustion parameters for corona discharge, passive and active pre-chamber ignition at 3000 rev/min, 11 bar NIMEP.**

Ignition delay angle is shorter for both passive and active pre-chamber operation compared with corona-discharge ignition. However, for burn angle (10 – 90% MFB) there is a much larger reduction: for both pre-chamber operating modes, burn angle is reduced by more than 10 °CA when compared to corona-discharge ignition.

This very rapid heat release presents a challenge for rate of pressure rise. The peak rate of pressure rise at exhaust lambda 1 (without any EGR) is 8 bar/°CA with pre-chamber ignition which is double that for corona-discharge. For lean conditions, peak cylinder pressure is increased by more than 10 bar with pre-chamber ignition and reaches 115 bar at lambda 2.5.

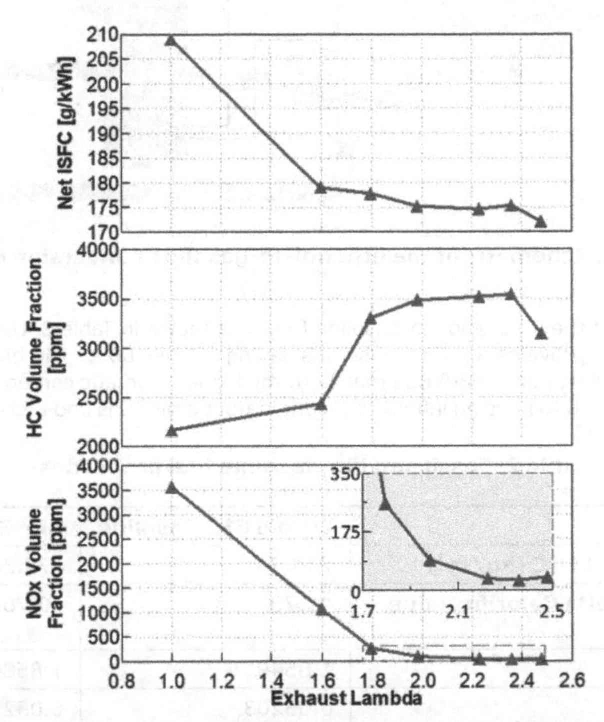


**Figure 18. Fuel consumption and emissions parameters for active pre-chamber operation at 3000 rev/min, 11 bar NIMEP.**

Net ISFC, HC and NOx are observed to flatten out beyond lambda 2.0, and NOx reduces below 50 ppm at higher lambda as shown in Figure 18.

## 4.4 Sustainable fuel testing

A series of tests was also completed with a sustainable biomass-to-liquid fuel developed by Coryton. The sustainable gasoline used in this study was produced in two main stages. The first stage is to generate a second generation advanced bio-ethanol from waste biomass; in this case agricultural waste such as straw. This lignocellulosic biomass is firstly pre-treated to improve the accessibility of enzymes. After pre-treatment, the biomass undergoes enzymatic hydrolysis for conversion into sugars which are then subsequently fermented to ethanol by the use of various microorganisms. This bio-ethanol is then dehydrated into ethylene and 'grown' into longer chain hydrocarbons in the presence of a zeolite catalyst at 300–400 °C (Figure 19).

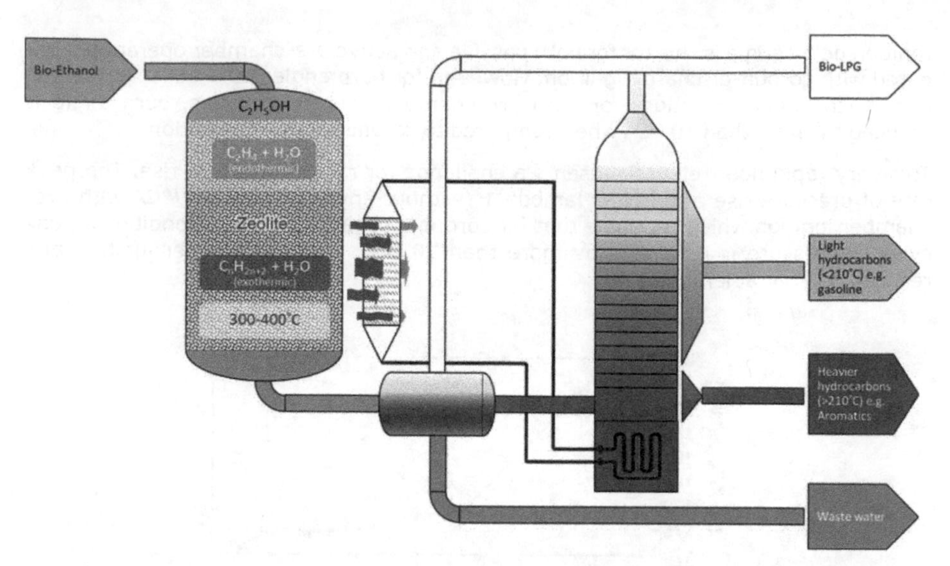


**Figure 19. Schematic of the ethanol-to-gasoline conversion process.**

The properties of the fossil and bio-gasoline fuels are shown in Table 2. Using REDII definitions, this fuel provides a greenhouse gas saving of over 80%. The bio-gasoline has a higher final boiling point (FBP) due mainly to the higher aromatic content, and this was subsequently found to have an influence on engine-out emissions and fuel consumption.

**Table 2. Fossil and bio-gasoline fuel properties.**

| **Parameter** | **Euro 6 E10 Gasoline** | **Bio-Gasoline** |
|---|---|---|
| **Density @ 15 °C [kg/m$^3$]** | 744.70 | 775.30 |
| **Lower (Nett) Calorific Value [MJ/kg]** | 41.70 | 40.76 |
| **H:C Ratio** | 1.8549 | 1.8584 |
| **O:C Ratio** | 0.03203 | 0.03278 |
| **Analysed RON** | 97.0 | 96.1 |
| **E150 [%]** | 90.1 | 68.1 |
| **Final boiling point (FBP) [°C]** | 181.4 | 207.4 |
| **Olefins [%v/v]** | 7.8 | 1.3 |
| **Aromatics [%v/v]** | 26.3 | 38.8 |
| **GHG savings (RED II) [%]** | 0 | 80 |

Passive pre-chamber ignition results for Euro 6 and bio-gasoline fuel are compared at 3250 rpm and 13.5 bar NIMEP load (which was found to be the peak efficiency point), Figure 20. Bio-gasoline has a lower RON which impacts the position of 50% MFB and hence also burn duration (10-90% MFB). It is likely that the increased HC observed with bio-gasoline is due to higher final boiling point (FBP). The increase of approximately 5 g/kWh in net ISFC with bio-gasoline results from both lower RON and increased HC. All specific fuel consumption values have been corrected to a fuel lower calorific value of 42.5 MJ/kg.

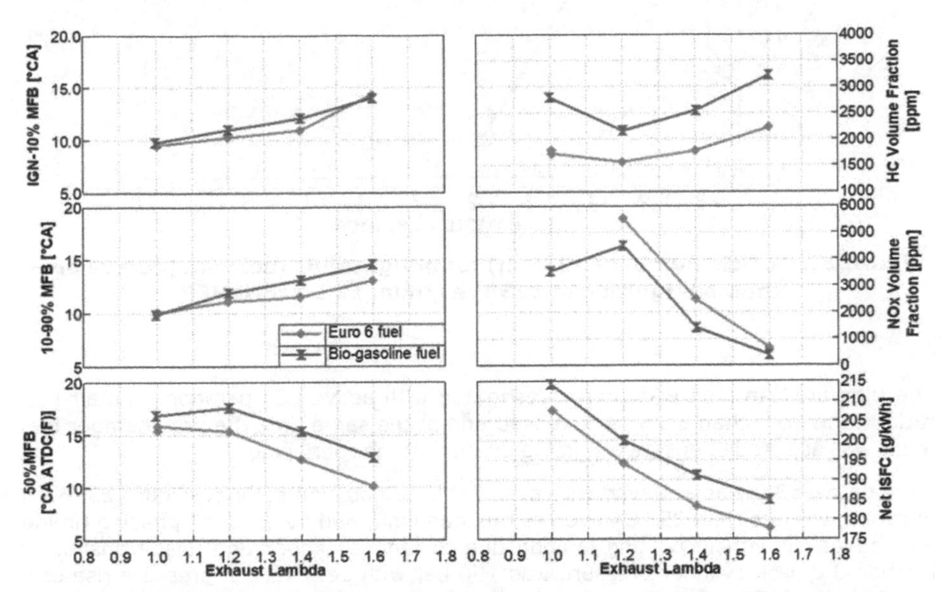


**Figure 20. Comparison of combustion parameters, emissions and net ISFC for fossil and bio-gasoline fuels with passive pre-chamber ignition at 3250 rev/min, 13.5 bar NIMEP.**

On comparison of PN data (>23 nm) for both the fuels it was observed that there is an increase in PN with bio-gasoline, but it remains in the typical range for 200 bar fuel pressure, Figure 21.

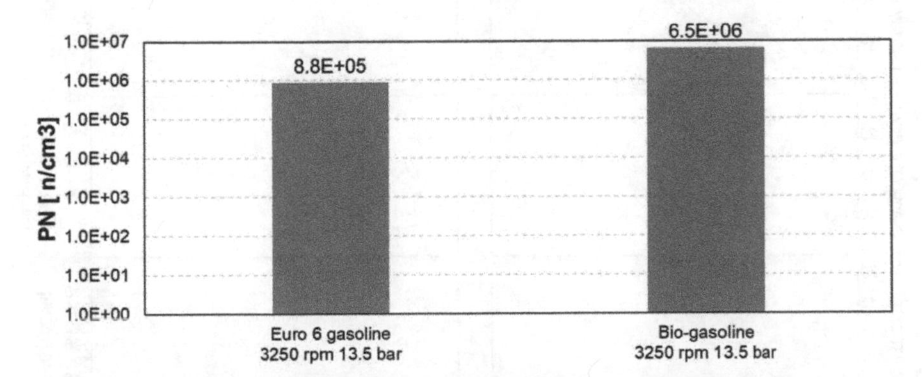


**Figure 21. Particle number (>23 nm) for Euro 6 gasoline and bio-gasoline fuel with passive pre-chamber ignition at lambda 1.**

PN data for a full lambda swing with passive pre-chamber ignition at 3250 rpm and 13.5 bar NIMEP shows that the results are in the typical range for a gasoline engine and reduce with increased lambda, as shown in Figure 22.

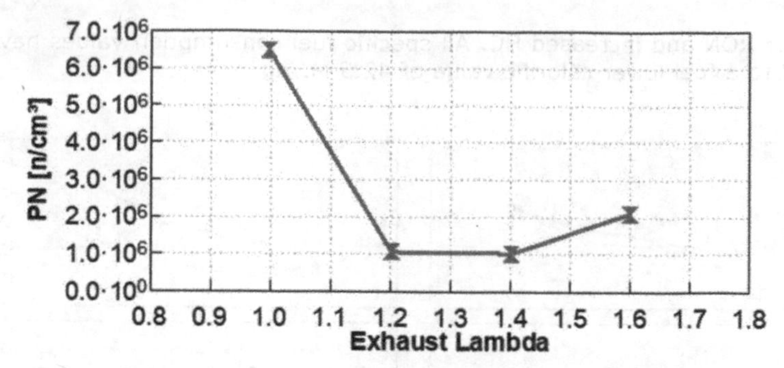


**Figure 22. Particle number (>23 nm) for bio-gasoline fuel with passive pre-chamber ignition at 3250 rev/min, 13.5 bar NIMEP.**

The ultimate lean limit was then investigated with active PC operation. The air-fuel ratio in the main chamber was increased and at the same time the pre-chamber fuel mass was adjusted to ensure a suitable mixture at the spark plug.

Lambda over 3.0 was achieved with acceptable combustion stability, Figure 23. Burn duration was less than 25 °CA even at the lean limit, and combustion phasing timing was advanced with increasing lambda due to reduced knock and heat transfer. At lambda 3.0 peak cylinder pressure was 150 bar with peak rate of pressure rise of 5 bar/°CA. Net ISFC shows improvement up to lambda 2.6 and flattens out beyond this point. This behaviour is slightly different to the Euro 6 fossil fuel where net ISFC had flattened out at a lambda level of 2.0.

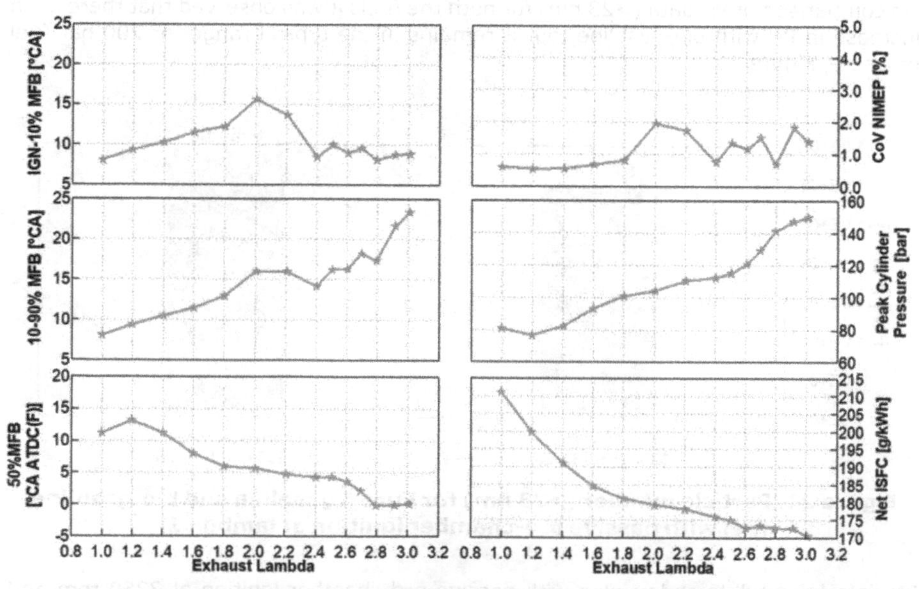


**Figure 23. Combustion parameters and fuel consumption for active pre-chamber operation with bio-gasoline fuel at 3000 rev/min, 11 bar NIMEP.**

Volume fraction and net indicated specific emissions presented in Figure 24 show that CO, HC, and NOx are reduced with lean operation. CO is minimum at lambda 1.2 (where NOx is highest) and starts to increase at lambda beyond 1.2. This rise in CO is due to reduced in-cylinder temperature leading to partial burn. HC volume fraction is observed to increase up to lambda 2.0, due to increased cylinder pressure (causing increased charge in the crevices), increased quenching distance and reduced combustion efficiency. Beyond lambda 2.0, the drop in HC is due to further dilution, but net ISHC is observed to flatten out. NOx has flattened out beyond lambda 2.0.

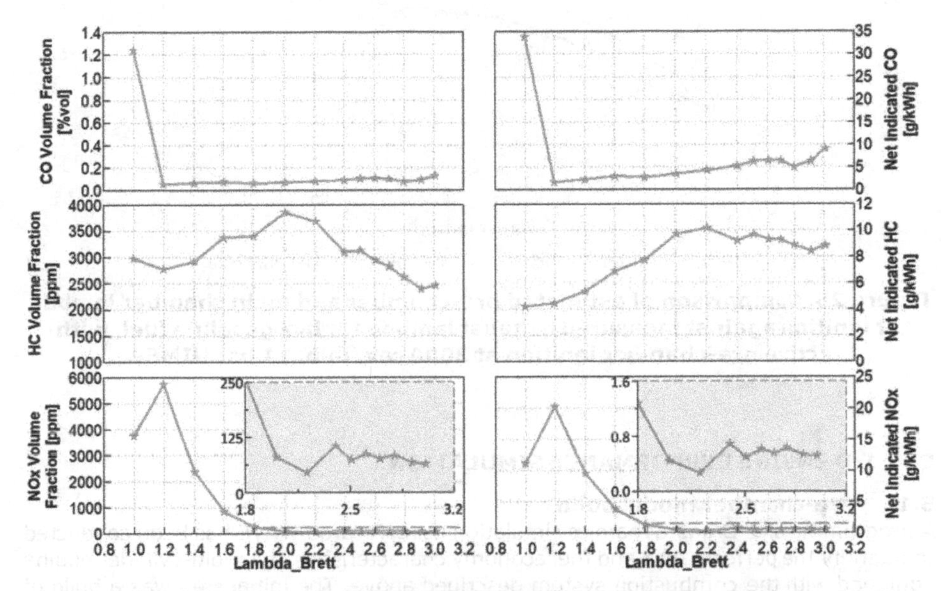


**Figure 24. Volumetric and specific emissions for bio-gasoline fuel, with active pre-chamber ignition at 3000 rev/min, 11 bar NIMEP.**

The estimated pre-chamber and main chamber lambda at spark timing is compared to exhaust lambda in Figure 25. It is observed that as exhaust lambda is increased the main chamber lambda also increases but the pre-chamber lambda had to be made rich to initiate combustion in the pre-chamber particularly beyond lambda 2.0. The drop in ignition delay (ignition – 10% MFB) beyond lambda 2.0 and flattening of burn duration (10-90% MFB) between lambda 2.0 to 2.6 observed in Figure 23 can be linked to the pre-chamber lambda in Figure 25.

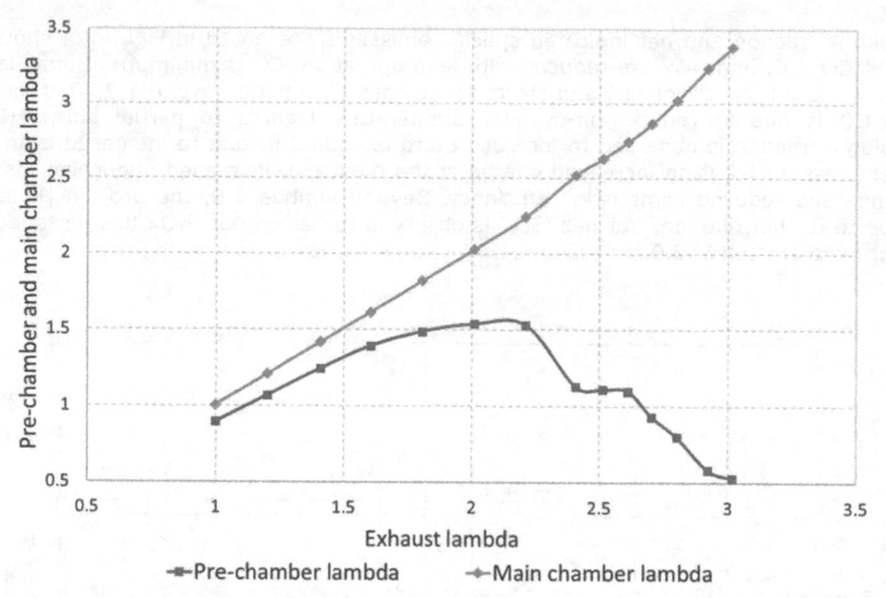


**Figure 25. Comparison of estimated pre-chamber and main chamber lambda at ignition against measured exhaust lambda for bio-gasoline fuel, with active pre-chamber ignition at 3000 rev/min, 11 bar NIMEP.**

## 5 1-D ENGINE PERFORMANCE SIMULATION

### 5.1 Pre-chamber model build

A programme of 1-D gas-dynamics simulation using Ricardo WAVE has been conducted to quantify the performance and fuel economy characteristics of a multi-cylinder engine equipped with the combustion system described above. The initial step was a build of a 1-D model with the pre-chamber.

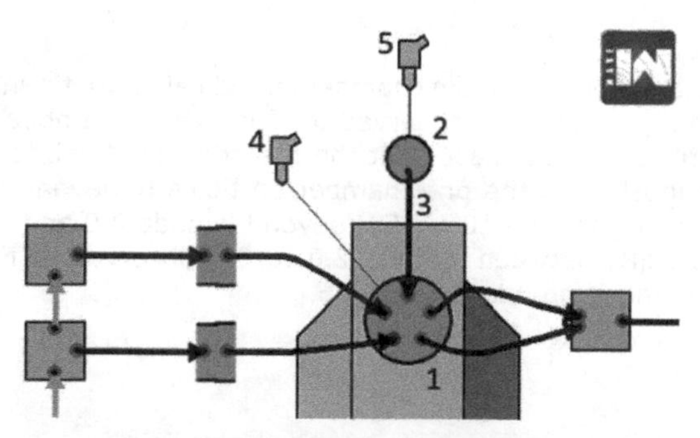


**Figure 26. Layout of the Magma xEV pre-chamber combustion system in Ricardo WAVE.**

Compared to the traditional engine models, the pre-chamber 1-D models contains along with the main cylinder (1) an additional cylinder representing the pre-chamber (2), as shown in Figure 26. These are connected by a single duct (3) representing the jets of the pre-chamber. The model is equipped with two injectors (4, 5) which allows passive as well as active pre-chamber modes to be simulated.

The pre-chamber geometry was defined to represent the same volume and surface area as does the pre-chamber in reality. To reach convergence it was necessary to reduce the simulation timestep. This in turn led to significantly longer computational times. To compensate for that, seven jets as used in the engine were replaced by one jet in the 1-D model. A study using a steady-state flow condition comparing seven jets with single jet to define required friction and heat transfer settings of the single jet was conducted to ensure that this step does not affect the model behaviour.

## 5.2 Pre-chamber model validation

The validation of the model was carried out using test data obtained from single cylinder engine. The validation was undertaken at three key-points:

- Passive pre-chamber combustion with exhaust lambda 1.6
- Active pre-chamber combustion with exhaust lambda 2.0
- Active pre-chamber combustion with exhaust lambda 2.2

In addition to the standard validation process used for modelling engines with a traditional combustion system, the pre-chamber combustion and gas exchange events must also be correlated. Jet discharge coefficients were adjusted along with pre-chamber heat release and combustion efficiency to match test data in-cylinder pressure traces, as shown in Figure 27.

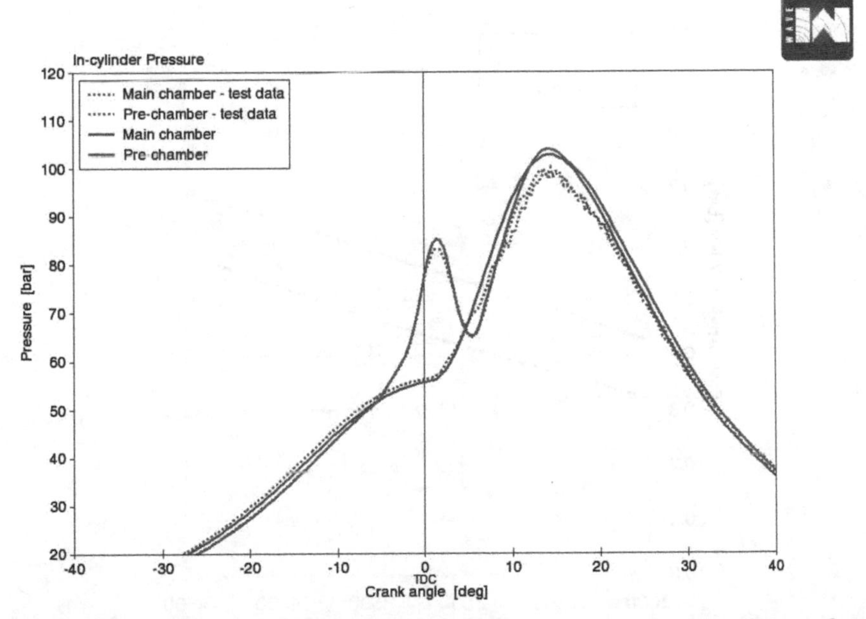


**Figure 27. In-cylinder pressure match at passive pre-chamber key-point.**

The model validation was an enabler to characterize the combustion system in terms of the trends of combustion duration, combustion efficiency, ignition delay and knock limitation depending on in-cylinder lambda. The validation therefore provided a solid base for the following multi-cylinder simulations.

### 5.3 Multi-cylinder engine simulation

A three-cylinder 1.5 litre Magma xEV dedicated hybrid engine (DHE) equipped with the pre-chamber combustion system was then simulated. A friction model for the engine was developed using the Ricardo Friction Analysis Tool (FAST). The FAST methodology uses a semi-empirical equation approach to build a friction model for engine architectures in a time efficient way. FAST builds on the methods of Sandoval, Heywood et al. [11, 12], but with the following enhancements:

- Correlation to measured data from modern engines
- Ability to calculate friction under fired conditions as well as motored
- Ability to deal with systems not otherwise considered (e.g. balancer shafts, vacuum pump)
- Physics-based treatment of piston and rings (enables user to examine effects of ring tension, connecting rod length etc.)

Since the first Magma xEV engine was designed, further friction reduction measures have been undertaken. The final firing friction mean effective pressure (FMEP) curve for the Magma xEV multi-cylinder pre-chamber engine is shown in Figure 28, for both full-load and no-load conditions. This characteristic embodies the low-friction approach that was taken in designing key engine features such as bearing sizes, the low rated engine speed, and the absence of a primary balance shaft. However, it also reflects the friction penalty that arises from the long-stroke architecture.

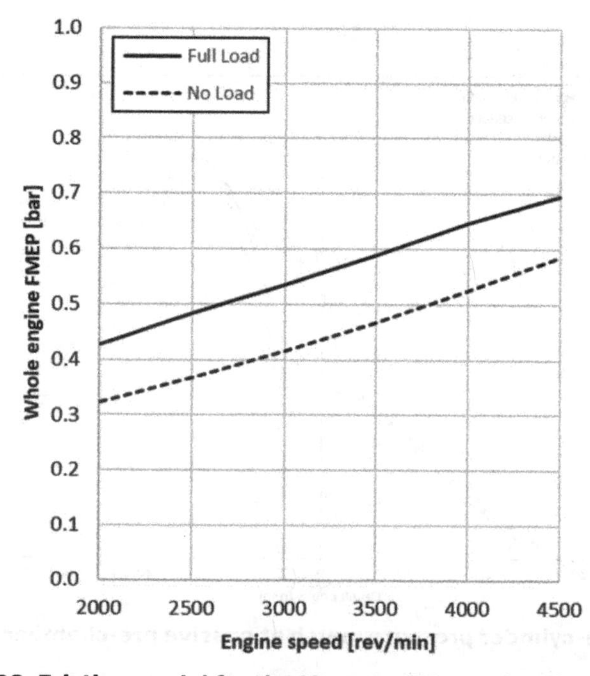


**Figure 28. Friction model for the Magma xEV pre-chamber engine.**

The multi-cylinder model was equipped with single stage turbocharger with waste-gate and a water charge air cooler (WCAC) and further optimised. The optimisation of load and lambda led to the solutions summarized in Table 3. The maximum BTE was found with main chamber lambda at SOC of 1.62, BMEP of 14.1 bar and engine speed of 3250 rev/min. The model was also run at stoichiometric conditions (without EGR) to enumerate the benefit of lean combustion. For the investigated key-point BTE is increased by 3.5 percentage points in lean-burn mode compared with stoichiometric operation.

**Table 3. Multi-cylinder engine simulation results.**

| Parameter | Lean combustion with passive pre-chamber | Stoichiometric combustion with passive pre-chamber |
|---|---|---|
| **Engine speed [rev/min]** | 3250 | 3250 |
| **BMEP [bar]** | 14.1 | 14.1 |
| **Main chamber lambda at SOC [-]** | 1.62 | 1.00 |
| **BTE [%]** | 45.4 | 41.9 |

## 6 CONCLUSIONS

- The Magma xEV engine concept has been further enhanced through the development of a pre-chamber combustion system
- The pre-chamber combustion system has been simulated and tested in both active and passive modes, with the focus on lean-burn combustion
- Compared with both conventional spark ignition and corona-discharge ignition,-chamber ignition increases burn-rates, which is beneficial for both combustion knock and dilution tolerance
- In passive mode a lean limit of 1.8 was observed, and with active pre-chamber fuelling lambda values beyond 3.0 are possible
- Under lean conditions large reductions in specific fuel consumption and NOx emissions are demonstrated
- The engine has also been tested with a sustainable biomass-to-liquid fuel with similar combustion characteristics and lean limit
- Although unburned hydrocarbons and fuel consumption were higher with the bio-gasoline fuel, this is outweighed by the well-to-wheels $CO_2$ benefit of this fuel
- A multi-cylinder engine equipped with this combustion system would achieve a peak brake thermal efficiency (BTE) of 45.4% with passive pre-chamber operation

## ACKNOWLEDGMENTS

The authors would like to thank the directors of Ricardo plc and Coryton for permission to publish this paper. We would also like to acknowledge the significant contributions to the work made by Jon Davis, Phil Carden, Simon Vella, Chris Critchley and Adam Finnerty.

**DEFINITIONS/ABBREVIATIONS**

| | |
|---|---|
| BDC | Bottom dead centre (piston) |
| BEV | Battery electric vehicle |
| BMEP | Brake mean effective pressure |
| BSFC | Brake specific fuel consumption |
| BTDC(F) | Before top dead centre (firing) |
| BTE | Brake thermal efficiency |
| CA | Crank angle |
| DHE | Dedicated hybrid engine |
| EGR | Exhaust gas recirculation |
| EIVC | Early inlet valve closing |
| FBP | Final boiling point |
| FMEP | Friction mean effective pressure |
| GIMEP | Gross indicated mean effective pressure |
| HC | Hydrocarbon |
| HEV | Hybrid electric vehicle |
| ICE | Internal combustion engine |
| IMEP | Indicated mean effective pressure |
| ISFC | Indicated specific fuel consumption |
| LIVC | Late inlet valve closing |
| MFB | Mass fraction burned |
| NIMEP | Net indicated mean effective pressure |
| NISFC | Net indicated specific fuel consumption |
| PHEV | Plug-in hybrid electric vehicle |
| PMEP | Pumping mean effective pressure |
| $R_t$ | Tumble ratio |
| REEV | Range-extended electric vehicle |
| SOC | Start of combustion |
| TDC | Top dead centre (piston) |
| TKE | Turbulent kinetic energy |
| WCAC | Water charge air cooler |

## REFERENCES


[1] Ricardo, H. R.: "Recent Research Work on the Internal-Combustion Engine", SAE Annual Meeting, 10-13 January 1922.

[2] Bunce, M. and Blaxill, H., "Sub-200 g/kWh BSFC on a Light Duty Gasoline Engine," SAE Technical Paper 2016-01-0709, 2016, doi:10.4271/2016-01-0709.

[3] Cooper, A., Harrington, A., Bassett, M., Reader, S. et al., "Application of the Passive MAHLE Jet Ignition System and Synergies with Miller Cycle and Exhaust Gas Recirculation," SAE Technical Paper 2020-01-0283, 2020, doi:10.4271/2020-01-02839.

[4] Sens, M., Binder, E., Reinicke, P.-B., Rieß, M., Stappenbeck, T., and Woebke, M., "Pre-Chamber Ignition and Promising Complementary Technologies," in 27th Aachen Colloquium Automobile and Engine Technology, 2018.

[5] Kobayashi, H., Komura, K., Nakanishi, K., Ohta, A., Narumi, H., Takegata, N., "Technology for Enhancing Thermal Efficiency of Gasoline Engine by Pre-Chamber Jet Combustion". Honda R&D Technical Review, Vol. 30, No. 2, October 2018.

[6] Serrano, D., Zaccardi, J.-M., Müller, C., Libert, C. et al., "Ultra-Lean Pre-Chamber Gasoline Engine for Future Hybrid Powertrains," SAE Int. J. Advances & Curr. Prac. in Mobility 2(2):607–622, 2020, doi:10.4271/2019-24-0104.

[7] Campbell-Brennan, J. "Feel The Burn", Racecar Engineering, June 2020.

[8] "How Maserati's New V6 is More Potent Yet Also More Efficient", Autocar, 29 Jul 2020.

[9] Pendlebury, K., Stokes, J., Dalby, J. and Osborne, R. J.: "The Gasoline Engine at 2020", 23rd Aachen Colloquium, October 2014.

[10] Sellers, R., Osborne, R., Cai, W., and Wang, Y., "Designing and Testing the Next Generation of High-Efficiency Gasoline Engine Achieving 45% Brake Thermal Efficiency," presented at in the 28th Aachen Colloquium Automobile and Engine Technology, 2019.

[11] Sandoval, D. and Heywood, J.,"An Improved Friction Model for Spark-Ignition Engines," SAE Technical Paper 2003-01-0725, 2003, doi: 10.4271/2003-01-0725.

[12] Patton, K., Nitschke, R., and Heywood, J., "Development and Evaluation of a Friction Model for Spark-Ignition Engines," SAE Technical Paper 890836, 1989, doi: 10.4271/890836.

*International Conference on Powertrain Systems for Net-Zero Transport*
*Institution of Mechanical Engineers, ISBN 978-1-032-11281-7*

# Thermal swing coatings for future sustainable heavy-duty IC engines

**A. Hegab[1], K. Dahuwa[1], A. Cairns[1], A. Khurana[2], R. Francis[2]**

[1]Powertrain Research Centre, University of Nottingham, UK
[2]Keronite International Ltd, UK

## ABSTRACT

The experimental work examined the use of a novel plasma electrolytic oxidation (PEO) thermal barrier coating (TBC) for reduced wall heat transfer and increased thermal efficiency in future diesel engines. The aim was to insulate the heat flow from the working gas in the combustion chamber to the cylinder wall by reducing local thermal conductivity of the material during the combustion process, hence reducing the heat loss. The work involved the development of the novel PEO piston coating, and detailed a comparison of engine performance parameters and exhaust emissions characteristics with PEO-coated piston, to those obtained with standard uncoated piston.

## 1 INTRODUCTION

Transport is the third largest greenhouse gas (GHG) emitting sector globally, and accounts for 27% of total GHG emissions in the UK, where 91% of these come from road transport (1). Global concern regarding vehicle emissions has notably escalated in recent years, and major cities around the world, including Paris, Madrid, Mexico City, and Athens have declared their intentions to ban diesel vehicles from city centres by 2025 (2). In early 2020, the UK government announced an aggressive stance and stated that the sales of new petrol, diesel and (potentially) hybrid cars will be banned from 2035, five years earlier than initially planned, in an attempt to reduce air pollution and attain net zero-emissions by 2050 (3). While the global electric car fleet has almost doubled in number over the past two years with more significant growth to come (4), electric vehicles still only represent a small percentage of vehicles sales, and issues over charging infrastructure, customer acceptance and battery supply may mean that complete conversion takes decades (4). Furthermore, the electrification of transport currently primarily targets passenger cars, and internal combustion (IC) engines are projected to remain in wide use beyond 2040 in heavy duty applications (road, rail, marine, power generation and hybrids) (5). As the transition to the use of low-carbon fuels and hybridization in heavy duty transport is regarded as a medium-term strategy for a more sustainable transport, efforts to develop more energy-efficient and eco-friendly IC engines must continue (6).

One fundamental way to develop high efficiently IC engines is to reduce the losses associated with the process of converting the fuel chemical energy into useful work. Typical modern automotive diesel engines reject about 60% of the fuel energy as a waste heat, mainly to the coolant and to the exhaust gas, in roughly equal shares (7). In an attempt to mitigate this, the concept of adopting thermal barrier coatings (TBC) has been employed in IC engines since the 1980s (8), in which typically a ceramic-based material coating of low thermal conductivity is applied to the whole combustion chamber surface, or to certain parts such as the piston crown, cylinder head, liner or valves (9). The purpose has been to insulate the heat flow from the working gas to the combustion cylinder wall by keeping the wall surface at high

DOI: 10.1201/9781003219217-9

temperature or reducing local material thermal conductivity during the combustion process, hence reducing the heat loss to the coolant (10). This would mean that more fuel energy could be converted to useful work output; improving the engine thermal efficiency (11). Reducing heat transfer also increases the exhaust gas temperatures, which provides greater potential for energy recovery through the use of, for example, electric turbo-compounding or thermoelectric generators (12). Other benefits include protecting the combustion chamber components from thermal stresses, less cooling requirements and faster catalytic light-off to reduce harmful emissions following cold start (13), which may become a key priority in future hybrid trucks.

Several ceramic materials have been used for TBC applications over the years, with different techniques for coating deposition investigated (14). Yet, traditional ceramic coating keeps the cylinder wall temperature at high levels even during the intake and compression strokes, due to the large heat capacity of the coating material (15). This would decrease the volumetric efficiency and increase the working gas temperature; reducing the work output and worsening the exhaust emissions (16). A breakthrough in TBC technologies was attained when Toyota developed the Silica Reinforced Porous Anodised Aluminum (SiRPA) coating (17). In this technology, also known as Thermo-Swing Wall Insulation Technology (TSWIN), the coating material has the capacity to rapidly change the wall temperature in a more dynamic way to follow the transient gas temperature (18). This was attributed to the low-thermal-conductivity and low-heat-capacity of the TSWIN coating, which causes its surface temperature to change greatly even during an extremely short cycle time (19). A comparison of conventional aluminum piston temperature, traditional TBCs and Thermo-Swing coating through a cycle relative to gas temperature is shown in Figure 1 (15). It can be seen from the figure that the surface temperature with Thermo-Swing coating increases during the combustion event, and decreases during the exhaust and intake strokes; dynamically with the gas temperature. This decreases the heat loss (compared to an uncoated piston) and prevents the intake air heating encountered with traditional TBC coatings.

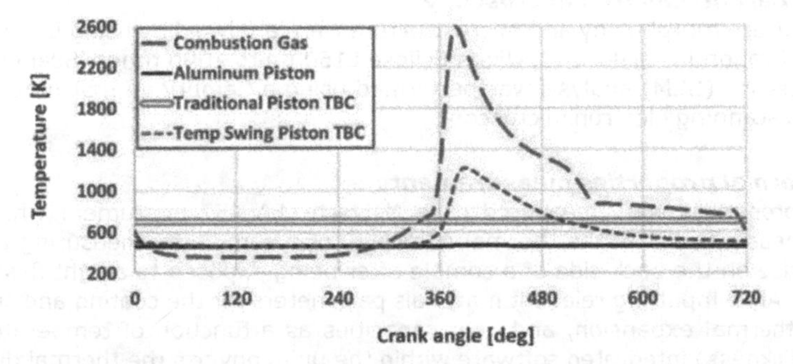


**Figure 1. Comparison of conventional aluminium piston, traditional TBC and thermos-swing coating temperature relative to gas temperature. Adapted from Ref. (15).**

Yet, the high surface roughness that is typical for anodised TBCs, including TSWIN, was found to increase heat transfer, slow down combustion and increase THC emissions (20). In addition to the high surface roughness, porosities and imperfection of the coating process limited the benefits of the TSWIN coating to some extent. One innovative surface coating technique that overcomes many of these problems is the Plasma Electrolytic Oxidation (PEO) (21). PEO creates a thin coating layer with ultra-low thermal conductivity and heat capacity, with an improved homogeneity in coating porosity and high surface finish (22). In this process, the coating grows both inward to the alloy substrate and outward to the coating surface simultaneously which results in excellent adhesion to the substrate metal (23). The microstructural characteristics of the PEO coatings depend on the operational conditions (e.g. voltage level, process duration and electrolyte composition) that can be tailored to accomplish the desired thermal properties, porosity and thickness (24). These merits give PEO advantages over more conventional coating techniques, and hence PEO has penetrated specialist markets in surface engineering, including automotive engines as well as many other industrial applications.

The aim of the present work has been to investigate the potential benefits of applying a novel PEO coating to a diesel engine piston. The work involved the development of the piston coating, where the main objectives were to produce a thin layer coating of low specific heat capacity and low thermal conductivity, followed by the on-engine testing. Detailed comparative studies between standard uncoated piston, PEO coated pistons and traditionally anodised (ANO) piston were steered for engine performance, fuel economy and exhaust gas emissions.

## 2 EXPERIMENTAL APPROACH

### 2.1 Piston coating development

#### 2.1.1 *Plasma electrolytic oxidation coating*

Coatings were prepared using a 25 kW Keronite processing rig and an electrolyte consisting of a dilute alkaline electrolyte containing sodium silicate. The electrolyte was maintained at a temperature of approximately 15°C by re-circulation through a heat exchanger and a bipolar waveform was applied. The approximate current density was 10 $A/dm^2$. Following processing the parts were thoroughly rinsed with de-ionised water. In some cases, a high silicate sealer was applied to the top surface post coating.

#### 2.1.2 *Visual and electron microscopy*

Coating surface morphology and microstructure were analysed by using Leica MZ6 for low magnification up to 4 times, Nikon Eclipse L150 up to 1000 times. Scanning Electron Microscopy (SEM) analysis was performed using a Cambridge Instruments SteroScan 240 scanning electron microscope.

#### 2.1.3 *Thermal properties measurements*

Thermal properties were measured on a Netzsch LFA 457 instrument which gives a direct measurement of the thermal diffusivity of a sample by measuring the temperature rise on the back side of a sample after being exposed to a light flash on the front side. After inputting relevant materials parameters for the coating and substrate (density, thermal expansion, and heat capacities as a function of temperature plus coating thickness) integrated software within the unit converts the thermal diffusivity value to a thermal conductivity of the coating.

### 2.2 Experimental setup

#### 2.2.1 *Engine test facility*

The experiments were carried out on a Ricardo Hydra, single cylinder Direct Injection (DI) diesel engine, with the cylinder head, piston assembly and controls adapted from a Ford Puma production engine. The engine was connected to a swinging frame DC universal dynamometer. The main specifications of the engine are listed in Table 1. A schematic diagram of the experimental setup is shown in Figure 2.

**Table 1. Main specifications of the test engine.**

| Model | Ricardo Hydra |
|---|---|
| No. of cylinders | 1 |
| Capacity | 0.55 litre |
| Cooling system | Water cooled |
| Induction system | Naturally aspirated |
| EGR system | Electronically controlled External loop |
| Bore | 86.0 mm |
| Stroke | 94.6 mm |
| Compression ratio | 15.5:1 |
| Fuel injection system | High-pressure common-rail (HPCR) |

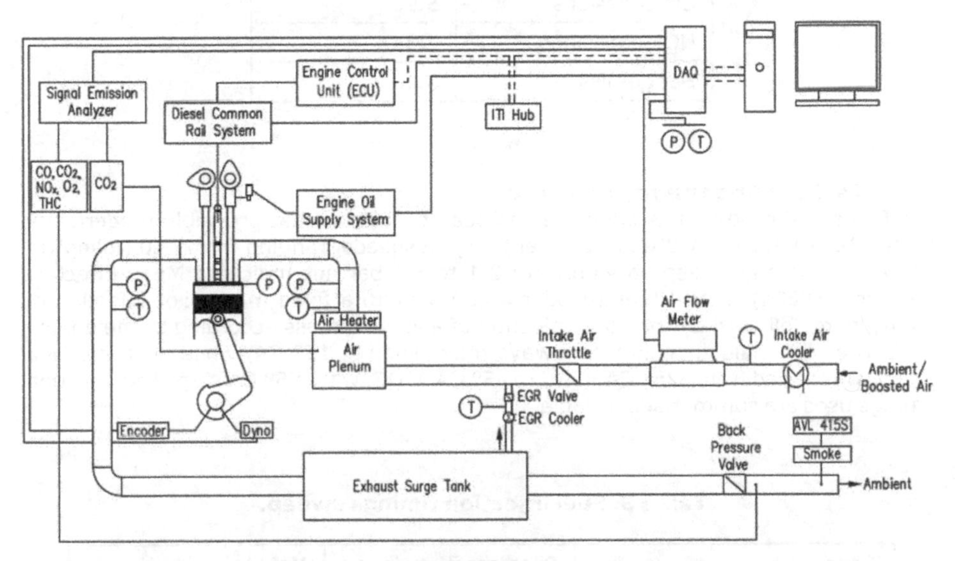


**Figure 2. A Schematic diagram of the experimental setup.**

### 2.2.2 *Data acquisition and instrumentation*

All experimental data was captured in a desktop PC via an integral data acquisition system. Engine operating parameters were monitored and controlled through the Engine Control Unit (ECU). This includes fuel injection parameters such as injection quantity, injection timing (and therefore separation), number of injections and rail pressure. Engine torque was measured using a load cell, while its speed and crank position were monitored using an incremental optical shaft encoder, with an accuracy of 0.5 degree. The in-cylinder pressure was measured using piezoelectric combustion pressure sensor. In-cylinder pressure and crank position signals were fed into high-speed National Instruments (NI) CompactDAQ system, to obtain the crank angle-synchronized in-cylinder pressure data in LabVIEW environment. Temperatures were measured throughout using type-k thermocouples, while oil pressure, intake Manifold

Absolute Pressure (MAP), and Exhaust Manifold Pressure (EXP) were measured using pressure traducers. Engine exhaust gas emissions (NOx, THC, CO, and CO2), and O2 concentration were measured using a Signal gas analyser kit, while the Filter Smoke Number (FSN) was measured using an AVL filter-type smoke meter. The uncertainties of the main measured parameters are summarised in Table 2.

**Table 2. Uncertainties of the main measured parameters.**

| Measured parameter | Uncertainty (%) |
|---|---|
| Diesel fuel flow rate | 0.86 |
| Torque | 1.05 |
| In-cylinder pressure | 0.62 |
| Manifold pressure | 0.68 |
| THC emissions | 1.57 |
| CO emissions | 5.92 |
| NOx emissions | 2.34 |
| Soot emissions | 3.10 |

### 2.2.3 *Test conditions and procedure*

The focus of this study has been the part load conditions that prevail in modern city traffic. Tests were conducted at different engine speeds, ranging from 750 (idling) to 2000 rpm, at load sweep varying from 2.1 to 5.1 bar net Indicated Mean Effective Pressure (IMEPn). Diesel fuel injection was split into a fixed mass pilot injection of 1mg/stroke, followed by the main injection of a varying mass according to the engine load. The pilot-main spacing was always maintained at 15° CA, while pilot injection timing was varied from -25° CA ATDC to -5° CA ATDC, with a 5° CA step. The injection timings used are summarised in Table 3.

**Table 3. Fuel injection timings sweep.**

| Case | Pilot injection timing (° CA ATDC) | Main injection timing (° CA ATDC) |
|---|---|---|
| Timing A | -25 | -10 |
| Timing B | -20 | -5 |
| Timing C | -15 | 0 |
| Timing D | -10 | 5 |
| Timing E | -5 | 10 |

The external EGR loop was deactivated so that the effects of piston coating were not masked by the variation of EGR rates. Nevertheless, the intake air temperature was maintained at 40° ± 1°C to emulate the temperature effect of the EGR that would be encountered in real driving conditions (25).

The net heat release rate (HRR) was calculated using the first law equation (7):

$$\frac{dQ_{net}}{d\theta} = \frac{\gamma}{\gamma - 1} p \frac{dV}{d\theta} + \frac{1}{\gamma - 1} V \frac{dp}{d\theta} \tag{1}$$

where ($\theta$) is the crank angle, ($p$) is the in-cylinder pressure at a given crank angle, ($V$) is the cylinder volume at that point, and ($\gamma$) is the specific heat ratio calculated from a polynomial function of bulk gas temperature at the corresponding crank angle (26). Integrating the HRR as a function of crank angle provides a representation of the total energy released up to a specified angle (aka cumulative heat release). The crank angle at which 50% of heat release occurs (CA50) is used to present the combustion phasing (27); characteristically changing with the injection timing.

Engine energy balance is calculated by applying the first law of thermodynamics to the control volume surrounding engine (10). In the conditions where no EGR is used, a steady-flow energy conversion equation may be expressed as (28):

$$P_b = \dot{m}_f\, LHV_f - P_f - P_p - \dot{Q}_{HT} - \dot{Q}_{Exh} - \dot{Q}_{l,comb} \tag{2}$$

Where ($P_b$) is the brake power, (Pf) is the mechanical friction loss, ($P_p$) is the pumping power loss, ($Q_{HT}$) is the heat transfer loss, ($Q_{Exh}$) is the heat loss in the exhaust and ($Q_{comb}$) is the combustion inefficiency loss.

Engine tests were carried out using four different types of pistons; standard uncoated piston that has been used as a baseline for comparison, two PEO coated pistons with different coating forms and one ANO coated piston. The four pistons used in the tests are identified in Table 4. Set out in Figure 3 are photographs of the pistons used.

**Table 4. Summary of different pistons used.**

| Piston designation | Description |
|---|---|
| BSLN | Baseline, standard uncoated piston as supplied by the manufacturer |
| PEO1 | PEO crown-and-bowl coated piston, extra-smooth bowl, coating thickness 70 μm |
| PEO4 | PEO crown-and-bowl coated piston, coating thickness 70 μm |
| ANO1 | ANO coated piston, coating thickness 50 μm, no sealant |

**Figure 3. Photographs of the typical pistons used.**

The baseline testing was repeated after each engine rebuild to ensure there was no drift. Cylinder head height relative to the piston was adjusted after every rebuild using a series of shims, in order to maintain the geometric compression ratio (CR) within ±0.01 with different piston coating thicknesses.

## 3 RESULTS AND DISCUSSION

### 3.1 Microstructural and thermal property investigation

#### 3.1.1 *Coupon preparation and microstructure*

A thermal swing coating should have as low a thermal conductivity and volumetric heat capacity as possible, ideally < 1 W/mK and <1,500 kJ/K/m3 respectively. However, for practical application further properties are needed, namely excellent adhesion to the substrate, durability under conditions found in-cylinder, and low surface roughness such as not to interfere with combustion.

Keronite's PEO coating process converts the surface of aluminum parts to a ceramic layer, principally comprising of a phase of alumina. However, other species such as silicate can be incorporated from the electrolyte. Since the PEO layer is formed at least partially via conversion of the substrate, adhesion is extremely good (22) and the tolerance of PEO layers to thermal cycling is deemed excellent (29). PEO layers are inherently porous due to the nature of the process (21).

While the thermal conductivity of alumina is significantly lower than aluminum at ~30 W/mK, this is still considerably above the target of 1 W/mK. Incorporation of other very low thermal conductivity phases such as mullite, an aluminosilicate with a thermal conductivity of ~2-4 W/mK, is therefore advantageous. Furthermore, since air has both a low thermal conductivity and heat capacity, maximizing porosity is a critical element of developing thermal swing coating. Yet, highly porous coatings tend to be rougher which may have negative impacts on combustion (30). For that reason, it is important to maximize porosity while minimizing the effect on surface roughness. Keronite has specifically engineered its thermal swing coatings to contain high quantities of low thermal conductivity phases such as mullite, and to be highly porous, yet smooth (Ra<3 μm as coated, <1 μm for sealed samples).

Initial development and characterization of the coatings was performed on coupons of 5 cm diameter, made from an alloy chosen to be similar to that of the pistons. The alloy has a nominal composition of 13.5Si, 4Cu, 2.8-3.0 Ni Bal Al Zr. A number of variants of electrolyte and electrical regime were investigated to optimize the coatings. The selected optimal coating had a nominal thickness of 60-65 μm and surface roughness (Ra) of ca. 2.7-3.0 μm. The free surface and cross sections of coated samples were examined by visual microscopy and SEM. Figure 4(a) is a SEM micrograph of the free surface, while Figure 4(b) is SEM micrograph of a cross section. Significant porosity is evident, including down to the substrate interface. The coating and pore structure were highly uniform across the samples. Some samples were additionally sealed with a silicon rich sealer. A free surface micrograph of the surface post sealing is shown in Figure 4(c). Some cracking of the surface is evident, but the sealer layer was well adhered with the PEO layer.

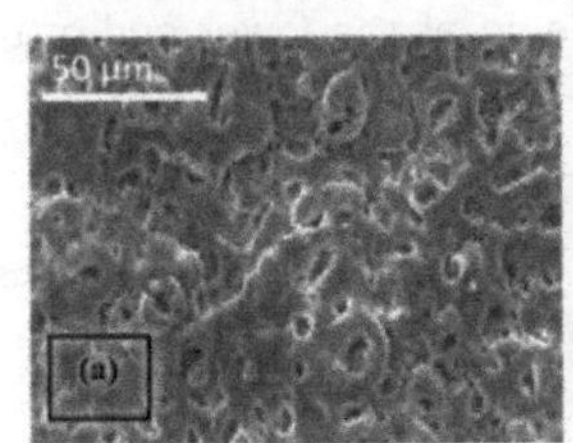

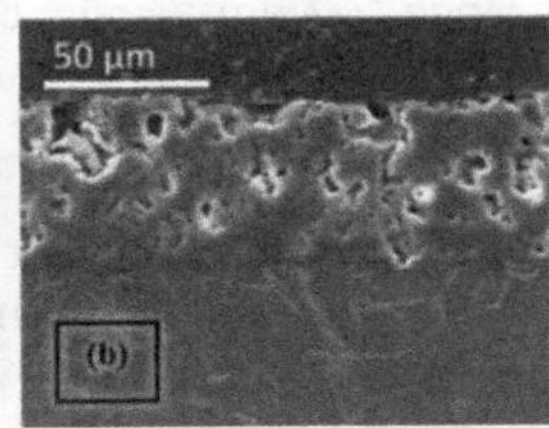

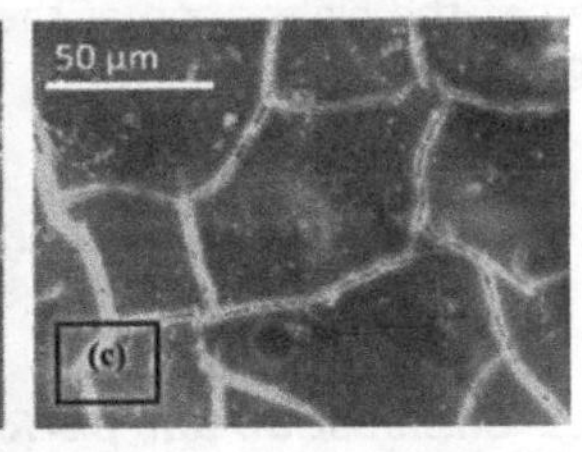


**Figure 4. (a) Free surface SEM of the PEO surface, (b) cross-sectional SEM of the PEO coating, and (c) free surface SEM of the PEO coating after application of the sealer.**

A comparator hard anodised coating was also examined by visual microscopy and SEM. Free surface and cross-sectional SEMs are shown in Figures 5(a) and 5(b), respectively. Significant porosity is again evident and extends down to the surface in some places. However, the overall level of porosity somewhat less than in the PEO coating. The porosity and microstructure are also less uniform than for the PEO sample with noticeable cracks of connected porosity. A number of brighter contrast (Si rich) particles were visible within the coating.

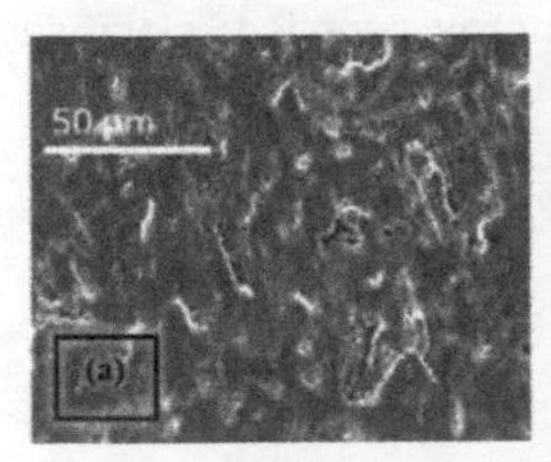

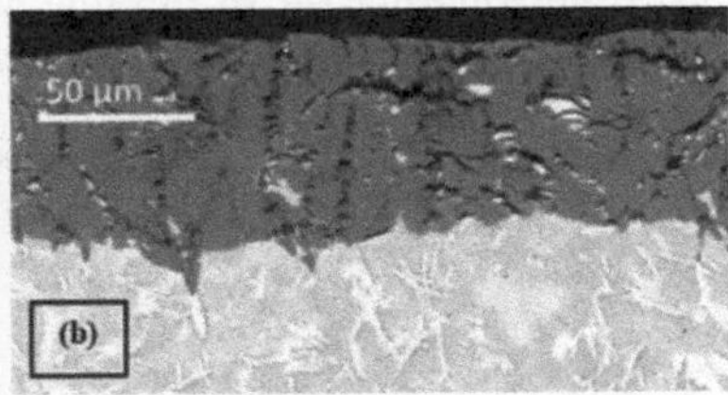


**Figure 5. (a) Free surface SEM of a hard-anodised coating, and (b) cross-sectional SEM of the coating.**

### 3.1.2 *Thermal properties measurements*

Laser flash measurements were performed on coupons from room temperature to 400°C. Thermal conductivity is measured to rise from ca. 0.45 W/mK to ca. 0.8 W/mK at 400°C, significantly below that target of 1 W/mK and also below those reported by Toyota for their "SiRPA" coatings (16). Good reproducibility has been attained between different samples, especially at lower temperatures.

### 3.1.3 *Piston coating*

The optimized coating was applied to the Ford Puma pistons used in this study. Since these were "series" pistons which were pre-treated with a phosphate layer of 1-2 µm, this was removed by an acid etch in dilute nitric acid (pH ~ 1.4) followed by light mechanical abrasion. Coatings were then applied to a nominal coating thickness of 65-70 µm. Following washing and drying a silicate rich sealer (as used on the coupons) was then applied to the coating and cured at 100°C.

Both PEO1 and PEO4 pistons were coated on the entire top face including the bowl area. Wax masking was used to ensure that no coating was present on the sides of the pistons. The as-coated surfaces had a Ra of ~2.7-2.9 µm. Following application of the sealer the Ra was considerably reduced to ~0.7-0.9 µm. For PEO-1 an extra smooth sealer layer was applied to the bowl which gave an Ra of ~0.4-0.5 µm. The higher smoothness of the sealer layer on PEO-1 is believed to be due to the sealer thickness being at the higher end of the range (~2 µm) versus being at the lower end of the range (~5 µm) on PEO4.

## 3.2 Engine test results

Due to space limitation, only results for engine testing at 1500 rpm and 3.8 bar IMEPn are presented in this section; other results exhibited fairly similar trends.

### 3.2.1 *Combustion and performance*

The in-cylinder pressure traces for different piston coatings, for injection timing A are shown in Figure 6(a); other injection timings were omitted due to space limitation, but exhibited similar behavior. It can be seen that the PEO coated pistons, particularly PEO1, demonstrate higher in-cylinder pressure levels relative to the uncoated piston; the increase in maximum cylinder pressure value with PEO1 relative to BSLN is as high as 3% (±1%). This is mainly attributed to the insulation effect of the TBC, where the reduced heat transfer to the cylinder wall increases the average gas temperature and hence pressure (31). With reduced heat transfer to the cylinder wall due to the TBC effect, the amount of useful heat that is converted into work acting on the top of the piston increases.

The ratio of the ultimate values of the cumulative HR for all piston coatings to the fuel energy supplied to the engine for injection timing A is shown in Figure 6(b). It can be seen that PEO coatings yeild better fuel conversion; a direct consequence of the reduced coolant loss. In comparison with BSLN, PEO1 offers more than 2% increase in the ratio of the cumulative net HR to the fuel energy supplied; signifying improved engine efficiency (11).

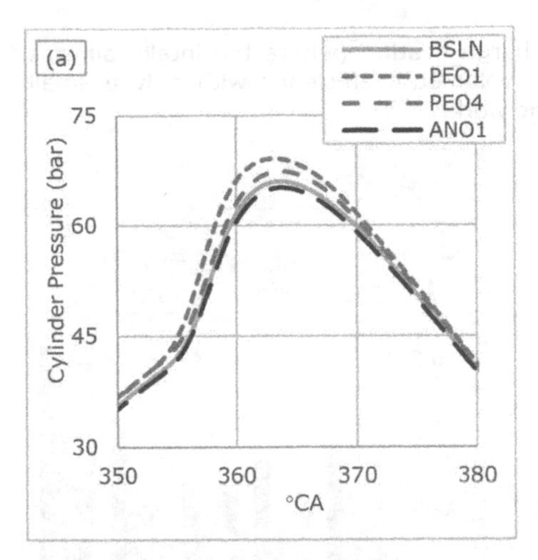


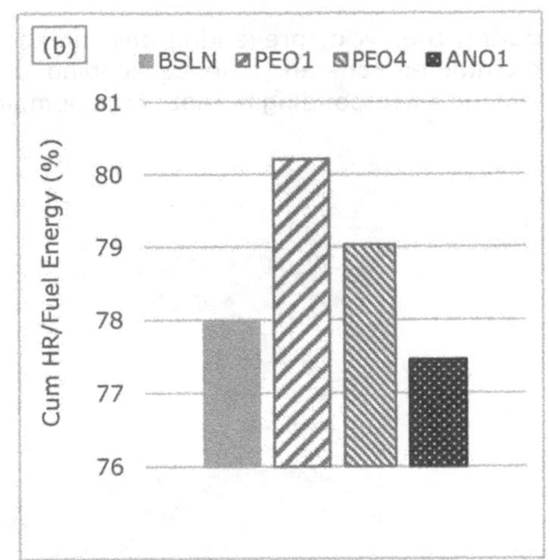


**Figure 6. (a) In-cylinder pressure, and (b) The ratio of cumulative HR to fuel energy supplied. Data for 1500rpm, 3.8 bar IMEPn, at injection timing A.**

Higher in-cylinder pressure values imply increased work per cycle transferred into the piston; i.e. indicated work. The gross indicated thermal efficiency trends versus the position of the CA50 dictated by the injection timing, are presented in Figure 7(a) for different piston coatings. It can be seen that for all coatings, the maximum value of the gross indicated thermal efficiency is achieved where the CA50 is attained at about 8° CA ATDC. Optimum injection timing, therefore, should be selected such that it achieves this point, so as to compromise between power, heat transfer and exhaust gas energy (26). As far as different coatings are considered, it can be seen from Figure 7(a) that the PEO1 piston provides the highest gross indicated thermal efficiency at all injection timings, with a maximum increase rising to 3% (absolute) relative to the baseline data. PEO4 also exhibit higher gross indicated thermal efficiency than BSLN despite the comparable in-cylinder pressure values. This is attributed to the improved fuel economy, where less amount of fuel is used to generate the same load levels. The case is reversed with ANO1, where the increased fuel amount results in lower gross indicated thermal efficiency. The total diesel fuel injection amount (pilot + main) for different piston coatings and different injection timings is shown in Figure 7(b).

The improvement in gross indicated thermal efficiency with PEO coatings primarily results from the reduced heat loss hence increased work per cycle transferred to the piston. Energy balance for different piston coatings at injection timing A is presented in Figure 8(a). Yet, it can be seen that part of the improvement in the gross indicated work is attributed to the reduced exhaust gas losses. From one side, the earlier start of combustion with PEO coatings mean that the exhaust gas will have longer time to expand during the power stroke, hence leaving the cylinder at lower temperature. In addition, the increased charge mass with the BSLN and ANO1 pistons mean the total exhaust mass flow will be larger and at higher temperature, hence the engine suffers more exhaust energy losses. As far as the engine breathing is considered, pumping loss increases with PEO coatings, mainly due to the higher cylinder temperature with the reduced heat loss. The values for pumping mean effective pressure (PMEP) for different piston coatings at injection timing A is presented in Figure 8(b). It can be seen that the increased pumping work with PEO coating remains marginal (1-2%), where the thermo-swing effect of the coating allows its temperature to change very rapidly during the cycle, preventing any intake charge heating (where the intake air was maintained constant), hence keeping the volumetric efficiency with only a small change corresponding to that of the pumping work.

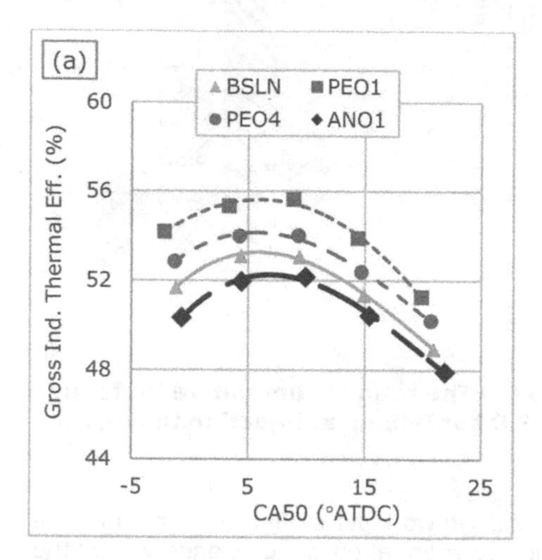


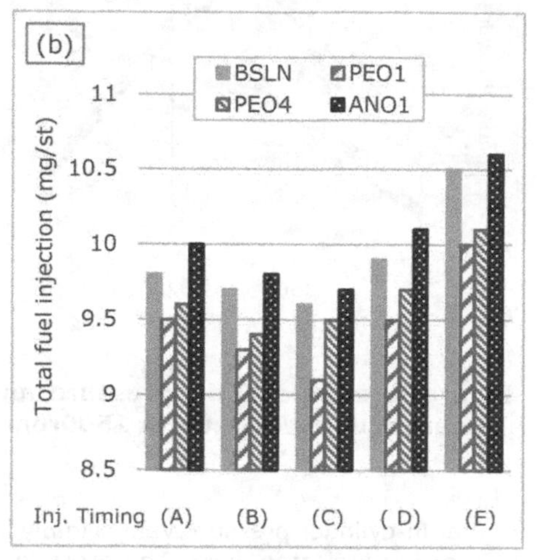


**Figure 7. (a) Gross indicated thermal efficiency (%), and (b) Total fuel injection (mg/st). Data for 1500rpm, 3.8 bar IMEPn, at different injection timings.**

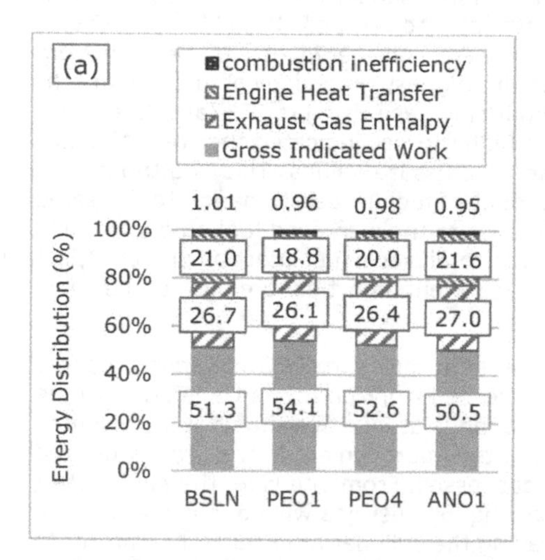


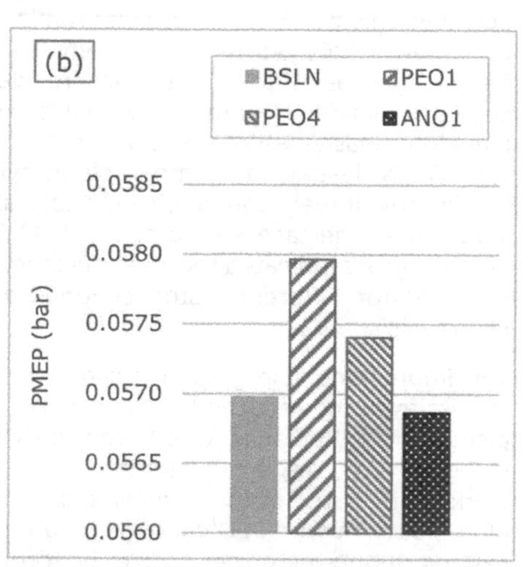


**Figure 8. (a) Engine energy balance, and (b) pumping mean effective pressure – PMEP (bar). Data for 1500rpm, 3.8 bar IMEPn, at injection timing A.**

### 3.2.2 *Exhaust emissions*

Brake specific NOx emissions with different piston coatings verses CA50 are shown in Figure 9(a). For all pistons, the highest NOx emissions are produced with advanced injection timing where peak gas temperature is higher; a more favourable condition for thermal NOx formation (32). Further, it can be seen that PEO1 demonstrates the highest NOx emissions, due to the higher cylinder temperature with the reduced heat loss. The increase of NOx emissions with PEO1 relative to BSLN exceeds 33%. Yet, PEO4 exhibits lower NOx emissions than BSLN despite the comparable cylinder pressure. This could be attributed to the reduced size of the spray cone of the main injection, where the majority of NOx is formed (33). The effect is reverse with ANO1, where the rich mixture zone in the fuel spray is larger hence it produces more NOx relative to BSLN, yet with only marginal increase.

The soot emissions with different piston coatings verses the CA50 are illustrated by Figure 9(b). The traditional trade-off between NOx and soot emissions is observed with advanced and optimum injection timings; the high temperature combustion with advanced injection timing leads to oxidation of the soot in the flame zone (7). With the increased diffusion combustion phase as the injection is retarded, soot formation increases. Still, late injection results in increased gas temperature in the late stages of the expansion stroke after the end of diffusion combustion phase, improving soot oxidation (34). Lowest soot emissions at all injection timings were generated with the PEO1 piston, owing to the high temperature and pressure while the earlier start of combustion offers more residence time for the soot in the high temperature zone, hence promoting oxidation (33). The reduction of soot emissions with PEO1 relative to BSLN at the optimum injection timing is almost 22%. ANO1 produces lower soot emissions than BSLN, potentially due to the larger volume of the yellow flame zone where the soot oxidation takes place (7).

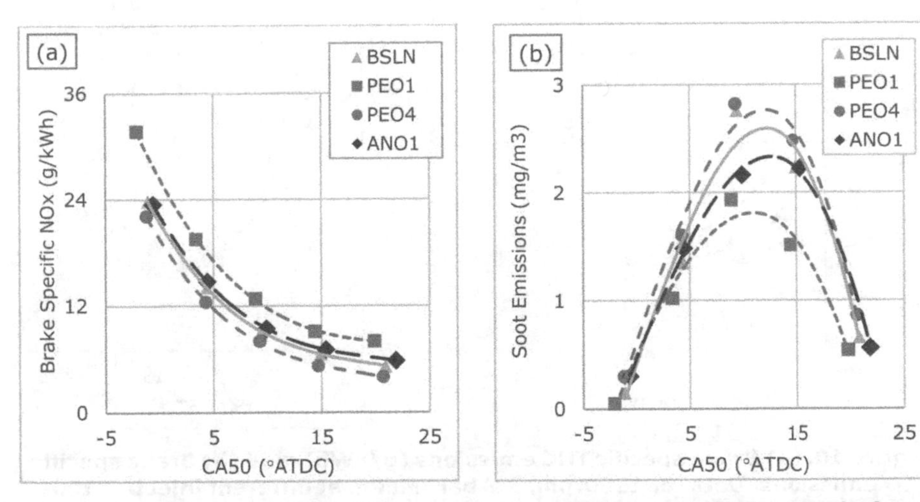


**Figure 9. (a) Brake specific NOx emissions (g/kWh), and (b) Soot emissions (mg/m$^3$). Data for 1500rpm, 3.8 bar IMEPn, at different injection timings (as represented by CA50).**

The brake specific THC emissions with different piston coatings verses CA50 are presented in Figure 10(a). It can be seen that lowest THC emissions are attained around the optimum injection timing. This is because advanced injection results in a higher cylinder pressure that could push the fuel vapour into the piston crevices hence increase the THC emissions, while retarded injection timing may lead to some of the fuel vapor escaping the combustion process and leave with the exhaust as unburned hydrocarbons (7). PEO1 demonstrates the lowest THC emissions, due to the high cylinder temperature and the associated better oxidation from one side, and the extra smooth surface finish of the sealed bowl hence no THCs are trapped inside the coating porosities. The reduction of THC emissions with PEO1 relative to BSLN at optimum injection timing exceeds 5%. PEO4, conversely, has regular coating on the bowl where the coating porosities are responsible for tapping some of the unburned hydrocarbon that do not experience combustion, hence increasing the THC emissions. This is analogous to the effect of the presence of porous deposits on the cylinder walls (7). With ANO1 coating, only the piston crown was coated, so the bowl was free of any porosities that could trap the unburned hydrocarbon hence THC emissions are comparable to those with PEO1.

The brake specific CO emissions with different piston coatings verses the CA50 are shown in Figure 10(b). For all piston coatings, the trends of CO emissions are fairly similar to those of THC emissions, where lowest values are mostly obtained with the optimum fuel injection timing. However, PEO1 exhibits comparable CO emissions to those with BSLN, despite the improved oxidation associated with the elevated cylinder temperature. One possible reason for that could partially be the dissociation that occurs in the high-temperature products, even with lean mixtures (7). As for PEO4, the effect of the nano pores of the bowl coating harbouring some of the partially burned products potentially hinders the oxidation of CO, contrary to ANO1 crown-only coating (35).

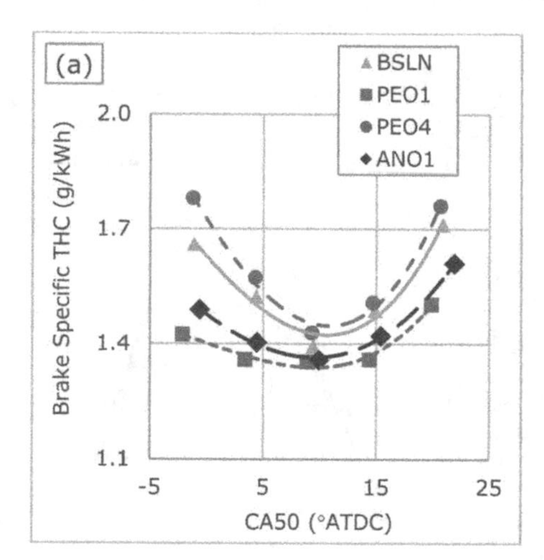


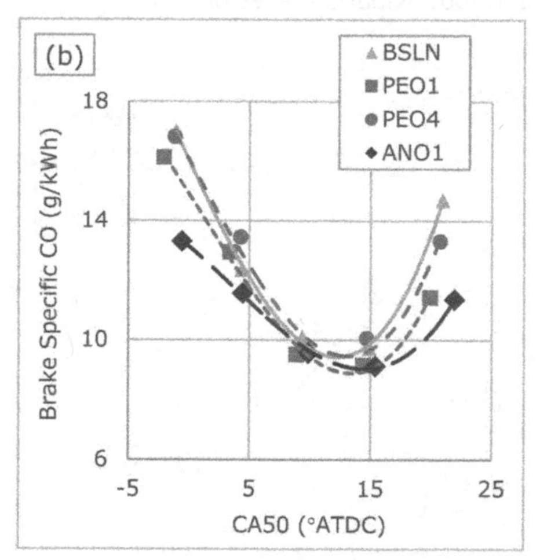


**Figure 10. (a) Brake specific THC emissions (g/kWh), and (b) Brake specific CO emissions. Data for 1500rpm, 3.8 bar IMEPn, at different injection timings (as represented by CA50).**

## 4 CONCLUSIONS

In this work, the potential benefits of applying a novel PEO coating to an aluminum alloy diesel engine piston were experimentally investigated. Four different pistons were used; a standard uncoated piston (BSLN), a PEO coated piston with an extra smooth surface finish in the bowl (PEO1), an ordinary PEO coated piston (PEO4) and a traditional anodised piston (ANO1). The main findings of the study are:

- PEO is an innovative surface coating technique that overcomes many of the traditional thermal barrier coating methods. It offers an attractive combination of hardness, wear resistance, corrosion resistance and interfacial adhesion, on top of effective insulation capabilities.
- PEO coatings have low thermal conductivities and low heat capacities as needed for thermo-swing coatings. Typical values were found to be around 0.45 w/m.K and 1,500 kJ/m3.K, respectively at room temperature. For the present study, the optimum coating thickness was found to be around 70 μm.
- Under the idealized part load conditions tested, PEO1 demonstrated the best reduction in wall heat transfer losses, with more than 3% gain (±1%) in the fuel useful energy and an equivalent improvement in the indicated thermal efficiency, due to the reduced heat loss to the coolant.
- PEO1 exhibits the highest NOx and lowest soot emissions owing to the higher gas temperature. the increase of NOx emissions with PEO1 relative to BSLN exceeds 33%, while the reduction of soot emissions at the optimum injection timing is almost 22%. Retarding the combustion phasing slightly from the optimum would lower NOx without largely sacrificing the fuel consumption benefit.
- PEO1 demonstrates the lowest THC emissions, due to the higher gas temperature and the extra smooth surface finish of the bowl that prevents THC trapping within the coating. The reduction of THC emissions with PEO1 relative to BSLN exceeds 5% at optimum injection timing. CO emissions with PEO1 remained comparable to those with BSLN despite the improved oxidation, potentially due to dissociation.
- For all piston coatings with the current setup, the optimum injection timing that brings about the best performance, where the CA50 occurs at around 8° CA ATDC, was found to be about -17° CA ATDC for the pilot and -2° CA ATDC for the main fuel injection. Nevertheless, these values are expected to change for different setups and test conditions, hence they should be taken as guidance only.

## 5 FUTURE WORK

Future work will extend to consider the thermal interactions of the new coatings with varying EGR effects (deactivation of EGR system have been deliberate in this work to establish relative piston coating effects without unwanted deviations in EGR rate and/or the thermodynamic state). Other engine surfaces (e.g. cylinder head, exhaust ports) will also be under investigation, along with examining the potential benefits of applying the PEO coating in advanced gasoline engines with novel fuel injection and combustion systems (e.g. jet ignition) that are applicable to future hybrid vehicles.

The application of PEO coating to the engine using carbon-free fuel (e.g. $H_2$, $NH_3$) also has a great potential for reducing engine heat loss and improving efficiency, especially with the high combustion temperature of $H_2$. Additional improvement of the power unit performance could also be attained by making the most of the increased exhaust gas temperatures via energy recovery by the use of thermoelectric generators (TEGs). Future work includes investigating the potentials of using a novel TEG in a heavy-duty truck engine using low/zero-carbon fuels, including $CH_4$, $NH_3$ and their mixture with $H_2$, with part of the work expected to be presented in the next IMechE conference.

## REFERENCES


[1] Department for Transport. Future of mobility: Urban strategy. March 2019.

[2] REUTERS. Electric cars win? Britain to ban new petrol and diesel cars from 2040. Available on: https://uk.reuters.com/article/us-britain-autos-idUKKBN1AB0U5. (accessed 20 May 2021).

[3] REUTERS. Electric dream: Britain to ban new petrol and hybrid cars from 2035. Available on: https://uk.reuters.com/article/us-climate-change-accord-idUKKBN1ZX2RY. (accessed 20 May 2021).

[4] International Energy Agency (IEA). Global EV outlook 2019 - Scaling-up the transition to electric mobility. May 2019.

[5] International Energy Agency (IEA). The Future of trucks – Implications for energy and the environment. 2017.

[6] Hegab A, La Rocca A, Shayler P. Towards keeping diesel fuel supply and demand in balance: Dual-fuelling of diesel engines with natural gas. Renewable Sustainable Energy Revs 2017;70:666–97. https://doi.org/10.1016/j.rser.2016.11.249.

[7] Heywood JB. Internal combustion engine fundamentals. New York, USA: McGraw-Hill; 1988.

[8] Kamo R. Adiabatic diesel-engine technology in future transportation. Energy 1987; 12 (10-11): 1073–80. https://doi.org/10.1016/0360-5442(87)90063-6.

[9] Kamo R. The adiabatic engine for advanced automotive applications. In: Evans RL, editor. Automotive Engine Alternatives, New York, USA: Plenum Press; 1987, p. 143–65.

[10] Woschni G, Spindler W, Kolesa K. Heat Insulation of Combustion Chamber Walls – A Measure to Decrease the Fuel Consumption of I.C. Engines? SAE Tech Pap 8703397; 1987. https://doi.org/10.4271/870339.

[11] Marr MA. An investigation of metal and ceramic thermal barrier coatings in a spark-ignition engine. MASc thesis, Totonto, Canada: Graduate Department of Mechanical and Industrial Engineering, University of Toronto; 2009.

[12] Arsie I, Cricchio A, Pianese C, Ricciardi V, De Cesare M. Modeling Analysis of Waste Heat Recovery via Thermo-Electric Generator and Electric Turbo-Compound for CO2 Reduction in Automotive SI Engines. Energy Procedia 2015;82:81–8. https://doi.org/10.1016/j.egypro.2015.11.886.

[13] Gao J, Tian G, Sorniotti A, Karci AE, Di Palo R. Review of thermal management of catalytic converters to decrease engine emissions during cold start and warm up. Appl Therm Eng 2019;147:177–87. https://doi.org/10.1016/j.applthermaleng.2018.10.037.

[14] Kumar V. Kandasubramanian B. Processing and design methodologies for advanced and novel thermal barrier coatings for engineering applications. Particuology 2016;27:1–28. http://doi.org/10.1016/j.partic.2016.01.007.

[15] Andrie M, Kokjohn S, Paliwal S, Kamo LS, Kamo A, Procknow D. Low heat capacitance thermal barrier coatings for internal combustion engines. SAE Tech Pap 2019-01-0228; 2019. https://doi.org/10.4271/2019-01-0228.

[16] Wakisaka Y, Inayoshi M, Fukui K, Kosaka H, Hotta Y, Kawaguchi A et al. Reduction of heat loss and improvement of thermal efficiency by application of "temperature swing" insulation to direct-injection diesel engines. SAE Tech Pap 2016-01-0661; 2016. https://doi.org/10.4271/2016-01-0661.

[17] Kosaka H, Wakisaka Y, Nomura Y, Hotta Y, Koike M, Nakakita K. Concept of "temperature swing heat insulation" in combustion chamber walls, and appropriate thermo-physical properties for heat insulation coat. SAE Tech Pap 2013-01-0274; 2013. https://doi.org/10.4271/2013-01-0274.

[18] Fukui K, Wakisaka Y, Nishikawa K, Hattori Y, Kosaka H, Kawaguchi A. Development of instantaneous temperature measurement technique for combustion chamber surface and verification of temperature swing concept. SAE Tech Pap 2016-01-0675; 2016. https://doi.org/10.4271/2016-01-0675.

[19] Kogo T, Hamamura Y, Nakatani K, Toda T, Kawaguchi A, Shoji A. High efficiency diesel engine with low heat loss combustion concept - Toyota's inline 4-cylinder 2.8-liter ESTEC 1GD-FTV engine -. SAE Tech Pap 2016-01-0658; 2016. https://doi.org/10.4271/2016-01-0658.
[20] Somhorst J, Oevermann M, Bovo M, Denbratt I. Evaluation of thermal barrier coatings and surface roughness in a single-cylinder light-duty diesel engine. Int J Engine Res 2019;1-21. https://doi.org/10.1177/1468087419875837.
[21] Mohedano M, Lu X, Matykina E, Blawert C, Arrabal R, Zheludkevich ML. Plasma electrolytic oxidation (PEO) of metals and alloys. Encycl Interfacial Chem 2018;423-38. https://doi.org/10.1016/B978-0-12-409547-2.13398-0.
[22] Clyne TW, Troughton SC. A review of recent work on discharge characteristics during plasma electrolytic oxidation of various metals. Int Mater Revs 2019;64 (3):127–62. https://doi.org/10.1080/09506608.2018.1466492.
[23] Hussein RO, Northwood DO, Nie X. The effect of processing parameters and substrate composition on the corrosion resistance of plasma electrolytic oxidation (PEO) coated magnesium alloys. Surf Coat Technol 2013;237:357–68. https://doi.org/10.1016/j.surfcoat.2013.09.021.
[24] Darband GB, Aliofkhazraei M, Hamghalam P, Valizade N. Plasma electrolytic oxidation of magnesium and its alloys: Mechanism, properties and applications. J Magnesium Alloys 2017;5(1):74–132. https://doi.org/10.1016/j.jma.2017.02.004.
[25] Ladommatos N, Abdelhalim S, Zhao H, The effects of exhaust gas recirculation on diesel combustion and emissions, Int J Engine Res 2000;1(1):107–26. https://doi.org/10.1243/1468087001545290.
[26] Abdelaal MM, Hegab AH. Combustion and emission characteristics of a natural gas-fueled diesel engine with EGR. Energy Convers Manage 2012;64:301–312. https://doi.org/10.1016/j.jma.2017.02.004.
[27] de O Carvalho L, de Melo TCC, de Azevedo Cruz Neto RM. Investigation on the fuel and engine parameters that affect the half mass fraction burned (CA50) optimum crank angle. SAE Tech Pap 2012-36-0498; 2012. https://doi.org/10.4271/2012-36-0498.
[28] Belgiorno G, Dimitrakopoulos N, Di Blasio G, Beatrice C, Tunestål P, Tunér M. Effect of the engine calibration parameters on gasoline partially premixed combustion performance and emissions compared to conventional diesel combustion in a light-duty Euro 6 engine. Appl Energy 2018;228:2221–34. https://doi.org/10.1016/j.apenergy.2018.07.098.
[29] Curran JA, Kalkanci H, Magurova Y, Clyne TW. Mullite-rich plasma electrolytic oxide coatings for thermal barrier applications. Surf Coat Technol 2007;201:8683–7. https://doi.org/10.1016/j.surfcoat.2006.06.050.
[30] Kawaguchi A, Tateno M, Yamashita H, Iguma H, Yamashita A, Takada N et al. Heat insulation by "temperature swing" in combustion chamber walls (third report). Trans Soc Automot Eng Jpn 2016;47(1):47–53. https://doi.org/10.11351/jsaeronbun.47.47.
[31] Borman G, Nishiwaki k. Internal-combustion engine heat transfer. Prog Energy Combust Sci 1987;13(1):1–46. https://doi.org/10.1016/0360-1285(87)90005-0.
[32] Abdelaal MM, Rabee BA, Hegab AH. Effect of adding oxygen to the intake air on a dual-fuel engine performance, emissions, and knock tendency. Energy 2013;61:612–620. https://doi.org/10.1016/j.energy.2013.09.022.
[33] Turns SR. An introduction to combustion: Concepts and applications. 2nd ed. New York, USA: McGraw-Hill; 2000.
[34] Stone R. Introduction to internal combustion engines. 3rded. Oxford, UK: Macmillan; 1999.
[35] Kawaguchi A, Iguma H, Yamashita H, Takada N, Nishikawa N, Yamashita C et al. Thermo-swing wall insulation technology; - A novel heat loss reduction approach on engine combustion chamber -. SAE Tech Pap 2016-01-2333; 2016. https://doi.org/10.4271/2016-01-2333.

# *Session 4: Simulation, modelling and experimental techniques*

# 'Modelling and simulation of transient clutch system dynamics'

**I. Minas[1], N. Morris[1], S. Theodossiades[1], M. O'Mahony[2], J. Voveris[2]**

[1]Wolfson School of Mechanical, Electrical and Manufacturing Engineering, Loughborough University, UK
[2]Ford Engineering Research Centre, Dunton, Laindon, SS15 6EE, Essex, UK

## ABSTRACT

The automotive industry faces a continuous challenge to design vehicles that meet increasingly stringent regulations, societal expectations and demands concerning tail pipe emissions, passenger safety and perceived quality. Frequently, any advancement towards meeting a single prescribed design objective may (without a fundamental understanding of the system-level behaviour) worsen the performance of the vehicle in another respect. For example, the reduction of vehicle weight, in order to improve fuel/ energy efficiency, while simultaneously meeting the anticipated Noise, Vibration and Harshness (NVH) behaviour by the discerning customer presents such a potential conflict. The aim of the current work is to formulate a numerical dry clutch system dynamics model that enables physical understanding of transient clutch oscillatory behaviour during the engagement process, enabling design optimisation and NVH improvement.

## 1 INTRODUCTION

Over the past decades, stricter regulation of automotive emissions has driven the need of introducing lighter powertrains. Furthermore, the growing awareness of the health impact due to vehicle noise and vibration is leading to ever tightening noise and vibration exposure regulations. Apart from the legal considerations, poor Noise, Vibration and Harshness (NVH) is also undesirable due to its association with poor quality by customers. Since the NVH performance of conventional panels and structures is mainly driven by their mass, NVH reduction often requires heavy or bulky palliatives, conflicting with the trend towards lightweight designs. Novel (reduced mass and volume) NVH solutions are required to face the challenging and often conflicting task of matching both NVH and lightweight design requirements (1).

Improvements in the NVH performance of automotive powertrains may also lead to improvements in powertrain efficiency. Several benefits are also attributed to less aggressive NVH, such as driver satisfaction and retention. Powertrain efficiency and NVH behaviour should be examined together to give the best result to the driver as well as for environmental purposes. For example, transmission efficiency improvements through reduction in oil viscosity and other assorted drag reductions, tend to have a negative impact on transmission NVH performance (gear rattle) (2). In conjunction with downspeeding of the Internal Combustion Engine (ICE) along with increased ICE boosting etc., ICE advancements tend to worsen transmission NVH (increased torsional signature). Reduced mass and consequentially often reduced rotating inertia (transmissions) may worsen aggressive NVH, and it seems likely that further improvements on NVH mitigation or palliative devices such as the Dual Mass Flywheel (DMF) (e.g. weight reduction) should also be investigated.

DOI: 10.1201/9781003219217-10

One source of NVH in automotive powertrains is the dry clutch system. Several modelling approaches have been presented in the literature. Wickramarachi et al. and Trinh et al (3,4) developed dry clutch models to simulate the wobbling motions of the pressure plate. It was observed that this motion can lead to aggressive NVH. Studies were conducted to define the most influential parameter on the system's stability and it was found that a thicker pressure plate led the system to be more stable.

Another modelling technique that was used for predicting clutch NVH is proposed by Aktir et al. (5) using a three dimensional finite element model to predict high frequency NVH in dry clutches. The natural frequencies of the pressure plate and clutch disc were identified using impact hammer testing. The dynamic response of the model was validated against measurements of the whole clutch assembly excited by an electromagnetic shaker attached to the flywheel. After conducting a stability analysis, it was found that high frequency NVH behaviour (up to 700 Hz) appears due to mode coupling of the input shaft bending motion and the clutch disc radial motion.

The main aim of the current research is to present a novel modelling approach for the dry clutch system, enabling NVH prediction and proposing an innovative approach for the optimization of the dry clutch system. Through the optimization the elimination of aggressive NVH can be achieved and new design guidelines are drafted.

## 2 METHODOLOGY

A complete dry clutch system model is presented in Figure 1, comprising 13 degrees of freedom. The main components are the flywheel (directly mounted to the crankshaft), the pressure plate cover (bolted to the flywheel and consequently rotating with the same angular velocity), the pressure plate (linked with the pressure plate cover via the diaphragm spring) and finally the clutch disc (directly connected to the gearbox input shaft via the clutch disc hub). The primary function of the clutch system is to transfer the engine torque to the rest of the drivetrain using the generated friction torque at the contact between the clutch disc and pressure plate/flywheel. All the above mentioned components are presented in the mathematical model shown in Figure 1.

The clutch model is studied using the following vector of generalized coordinates:

$$q = \{\theta_{fw}, \theta_{cd}, \theta_{in}, z_{pp}, \theta_{pp,x}, \theta_{pp,y}, y_{cd,y}, y_{cd,x}, z_{cd}, \theta_{cd,x}, \theta_{cd,y}, \theta_{insh,x}, \theta_{insh,y}\}$$

Where analytically the degrees of freedom are:

$\theta_{fw}$: Flywheel angular displacement
$\theta_{cd}$: Clutch disc angular displacement
$\theta_{in}$: Input shaft angular displacement
$z_{pp}$: Axial motion of the pressure plate
$\theta_{pp,x}$: Tilting motion of the pressure plate around x axis
$\theta_{pp,y}$: Tilting motion of the pressure plate around y axis
$y_{cd,y}$: Radial motion of the clutch disc at y axis
$y_{cd,x}$: Radial motion of the clutch disc at x axis
$z_{cd}$: Axial motion of the clutch disc
$\theta_{cd,x}$: Tilting motion of the clutch disc around x axis
$\theta_{cd,y}$: Tilting motion of the clutch disc around y axis
$\theta_{insh,x}$: Bending motion of the input shaft around x axis
$\theta_{insh,y}$: Bending motion of the input shaft around y axis

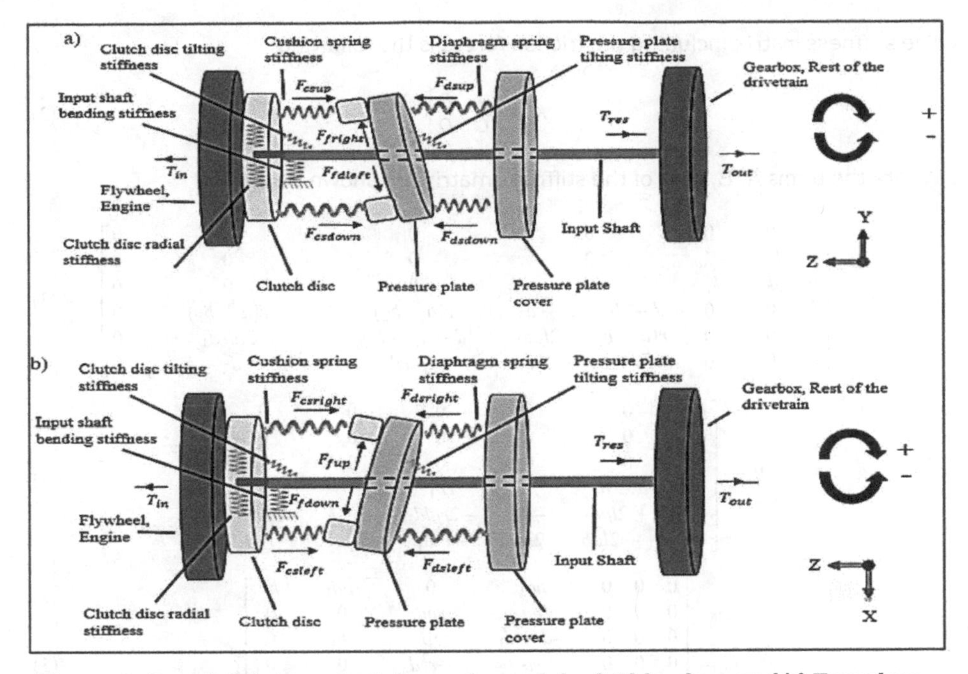


**Figure 1. Dry clutch transient dynamic model: a) Side view and b) Top view.**

The dry clutch model introduces local nonlinearities through the cushion spring stiffness variation ($k_{cs}$), which is a function of the cushion spring compression, $u$, and the diaphragm spring stiffness ($k_{ds}$) (6–9). The former connects the clutch disc with the pressure plate/flywheel and the latter links the pressure plate with the pressure plate cover. The tilting stiffness of the pressure plate ($k_{tiltpp}$) and the tilting stiffness of the clutch disc ($k_{tiltcd}$) are also considered. Finally, the bending stiffness of the input shaft ($k_{bend}$), is also considered. The clutch disc's connection to the gearbox input shaft is represented by radial stiffness ($k_r$) (10).

The equations of motion are presented in matrix form below. The mass (and inertia), the stiffness and the external load are each shown separately for convenience. The mass and inertia matrix is given by:

$$M = \begin{bmatrix} I_{fw}+I_{pp}+I_e & 0 & 0 & 0 & 0 & 0 & 0 & 0 & 0 & 0 & 0 & 0 & 0 \\ 0 & I_{cd} & 0 & 0 & 0 & 0 & 0 & 0 & 0 & 0 & 0 & 0 & 0 \\ 0 & 0 & I_{in}+I_g+I_{car} & 0 & 0 & 0 & 0 & 0 & 0 & 0 & 0 & 0 & 0 \\ 0 & 0 & 0 & m_{pp} & 0 & 0 & 0 & 0 & 0 & 0 & 0 & 0 & 0 \\ 0 & 0 & 0 & 0 & I_{pp,x} & 0 & 0 & 0 & 0 & 0 & 0 & 0 & 0 \\ 0 & 0 & 0 & 0 & 0 & I_{pp,y} & 0 & 0 & 0 & 0 & 0 & 0 & 0 \\ 0 & 0 & 0 & 0 & 0 & 0 & m_{cd,x} & 0 & 0 & 0 & 0 & 0 & 0 \\ 0 & 0 & 0 & 0 & 0 & 0 & 0 & m_{cd,y} & 0 & 0 & 0 & 0 & 0 \\ 0 & 0 & 0 & 0 & 0 & 0 & 0 & 0 & I_{cd,x} & 0 & 0 & 0 & 0 \\ 0 & 0 & 0 & 0 & 0 & 0 & 0 & 0 & 0 & I_{cd,y} & 0 & 0 & 0 \\ 0 & 0 & 0 & 0 & 0 & 0 & 0 & 0 & 0 & 0 & m_{cd} & 0 & 0 \\ 0 & 0 & 0 & 0 & 0 & 0 & 0 & 0 & 0 & 0 & 0 & I_{insh,x} & 0 \\ 0 & 0 & 0 & 0 & 0 & 0 & 0 & 0 & 0 & 0 & 0 & 0 & I_{insh,y} \end{bmatrix} \quad (1)$$

The stiffness matrix including the friction forces is the following:

$$K = \begin{bmatrix} A & B \\ C & D \end{bmatrix} \tag{2}$$

Where the terms A, B, C, D of the stiffness matrix are shown below:

$$A = \begin{bmatrix} 0 & 0 & 0 & 0 & 0 & 0 & 0 \\ 0 & k_{in} & -k_{in} & 0 & 0 & 0 & 0 \\ 0 & -k_{in} & k_{in} & 0 & 0 & 0 & 0 \\ 0 & 0 & 0 & d + d_1 - d_2 - d_3 & r(a + b_2) & r(a + b_2) & 0 \\ 0 & 0 & 0 & r(a + b_2) + 2l\mu a_1 & r^2\left[d - d_2 + k_{tiltpp}\right] & 2\mu lrd_1 & 0 \\ 0 & 0 & 0 & r(a + b_2) + 2l\mu a & 2\mu lrd & r^2\left[d_1 - d_2 + k_{tiltpp}\right] & 0 \end{bmatrix} \tag{3}$$

$$B = \begin{bmatrix} 0 & 0 & 0 & 0 & 0 & 0 \\ 0 & 0 & 0 & 0 & 0 & 0 \\ 0 & 0 & 0 & 0 & 0 & 0 \\ 0 & -d - d_1 & rb & rb_1 & rb & rb_1 \\ 0 & rb + 2l\mu b_1 & -r^2 d & -2r\mu ld_1 & -r^2 d & -2\mu lrd_1 \\ 0 & rb_1 + 2l\mu b & -2r\mu ld & -r^2 d_1 & -2\mu lrd & -r^2 d_1 \end{bmatrix} \tag{4}$$

$$C = \begin{bmatrix} 0 & 0 & 0 & \mu a_1 & 0 & r\mu d_1 & k_r \\ 0 & 0 & 0 & \mu a & r\mu d & 0 & 0 \\ 0 & 0 & 0 & -d - d_1 & rb & rb_1 & 0 \\ 0 & 0 & 0 & rb & -r^2 d & 0 & 0 \\ 0 & 0 & 0 & rb_1 & 0 & -r^2 d_1 & 0 \\ 0 & 0 & 0 & rb + 2L\mu b_1 & -r^2 d & -2\mu lrd_1 & k_r L \\ 0 & 0 & 0 & rb_1 + 2L\mu b & -2\mu lrd & -r^2 d_1 & 0 \end{bmatrix} \tag{5}$$

$$D = \begin{bmatrix} 0 & \mu b_1 & 0 & -\mu rd_1 & k_r L & -\mu rd_1 \\ k_r & \mu b & -\mu rd & 0 & -\mu rd & k_r L \\ 0 & d + d_1 & ra & ra_1 & ra & -rd_1 \\ 0 & ra & r^2 d + k_{tiltcd} & 0 & r^2 d & 0 \\ 0 & ra_1 & 0 & r^2 d_1 + k_{tiltcd} & 0 & r^2 d_1 \\ 0 & rb + 2L\mu a_1 & r^2 d & 2L\mu rd_1 & r^2 d_1 + L^2 k_{bend} + k_r L^2 & 2L\mu ra_1 \\ k_r L & ra_1 + 2L\mu a & 2L\mu rd & r^2 d_1 & 2L\mu rd & r^2 d_1 + L^2 k_{bend} + k_r L^2 \end{bmatrix} \tag{6}$$

In which the following terms are given by:

$$\begin{aligned} a &= k_{csup} - k_{csdown} \\ b &= k_{csdown} - k_{csup} \\ d &= k_{csdown} + k_{csup} \\ a1 &= k_{csleft} - k_{csright} \\ b1 &= k_{csright} - k_{csleft} \\ d1 &= k_{csright} + k_{csleft} \\ a2 &= k_{dsup} - k_{dsdown} \\ b2 &= k_{dsdown} - k_{dsup} \\ d2 &= k_{dsdown} + k_{dsup} \\ a3 &= k_{dsleft} - k_{dsright} \\ b3 &= k_{dsright} - k_{dsleft} \\ d3 &= k_{dsright} + k_{dsleft} \end{aligned}$$

Finally, the external load vector is given by:

$$f = [T_{in} - T_f, \quad T_f, \quad -T_{out} - T_{res}, \quad F_{ds}, \quad 0, \quad 0, \quad 0, \quad 0, \quad 0, \quad 0, \quad 0, \quad 0, \quad 0,] \tag{7}$$

The following assumptions are considered to simulate the tilting, radial and bending motions of the clutch components. The net force that is applied on the pressure plate must initially have a slight non uniform circumferential distribution in order for the aforementioned wobbling motions to be initiated. Such a distribution creates different cushion spring and diaphragm spring force actuations in each one of the four circumferential domains, depicted in Figure 2. As a result, the pressure plate starts tilting, and consequently the vector sum of the friction forces generated by the contact of the pressure plate and clutch disc in the perpendicular plane (YX plane) are non-zero. The schematic presentation of the model (Figure 1) reflects the pressure plate tilted and it can be observed e.g. in XZ plane (when simulating the cushion spring) that the upper side spring is compressed more when compared to the bottom spring and as a result, $F_{csup} > F_{csdown}$. For the current study an initial spring deflection mechanism was introduced, which initiates with infinitesimal bending of the input shaft. This enables the mechanism described above. If there is an initially circumferentially uniform applied forces, then the cushion spring stiffness is equally distributed in the four domains shown in Figure 2. Additionally, the direction of friction forces, due to the contact between the pressure plate and clutch disc are presented. Those forces are a function of the cushion spring stiffness and the kinetic friction coefficient. For the current research the following stiffness relations were considered to simulate the unbalancing of the forces applied on the pressure plate (3, 11-12):

$$k_{csup} + k_{csyleft} = 0.55k_{cs} \tag{8}$$

$$k_{csdown} + k_{csright} = 0.45k_{cs} \tag{9}$$

In which $k_{cs}$ is the total cushion spring stiffness and $k_{csup}$, $k_{csdown}$, $k_{csyleft}$, $k_{csright}$ are the cushion spring stiffness coefficients in each domain. Figure 2 shows the distribution of the cushion spring stiffness when equations 8, 9 are applied, compared to a uniform distribution in each domain.

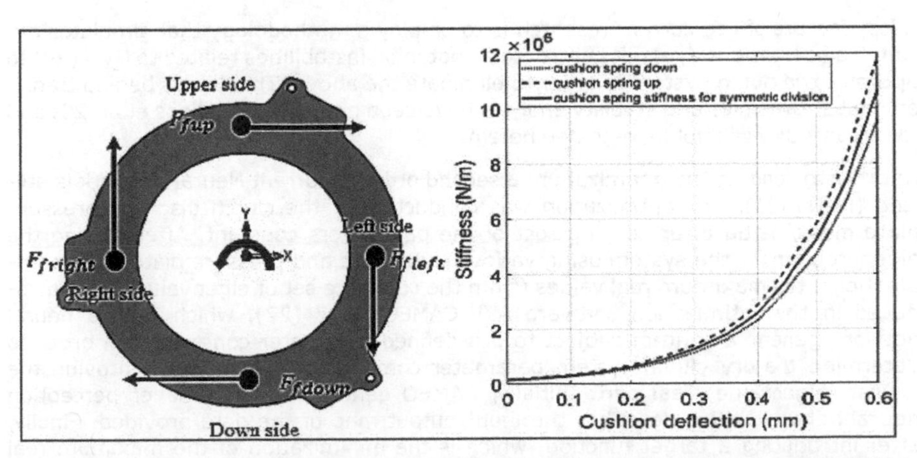


**Figure 2. Pressure plate schematic representation and cushion spring stiffness distribution. The dots show the cushion spring distribution domains.**

## 3 FRICTION COEFFICIENT AND FRICTION TORQUE

The transmission of the engine torque to the rest of the drivetrain is accomplished due to the generated friction at the contact of the pressure plate/flywheel and clutch disc. Assuming uniform contact pressure, the generated friction torque ($T_c$) is given by (13,14):

$$T_c = N_s R_{eff} \mu_{kin} F_{cushion} \tag{10}$$

For the clutch system under investigation a single plate clutch disc was examined, which corresponds to $N_s = 2$. The effective radius, $R_{eff}$, is calculated as a function of the internal and external radius of the clutch disc. In the current research a relative speed - related friction coefficient was introduced. The calculation of friction coefficient for an engagement manoeuvre is accomplished by (15–18):

$$\mu_{kin} = \mu_{st} + m_s(\dot{\theta}_{fw} - \dot{\theta}_{cd}) \tag{11}$$

In which:

$\mu_{st}$ : static friction coefficient, when there is no relative speed between clutch disc and flywheel
$m_s$ : gradient of the linearized measurements of the friction coefficient
$\dot{\theta}_{fw} - \dot{\theta}_{cd}$: relative velocity of the flywheel and clutch disc

Through the coefficient of friction, the system is coupled because the responses of the component motions are calculated as a function of the relative velocity of the clutch components. Moreover, in order to establish a more precise calculation of the friction coefficient variation throughout the engagement manoeuvre, the gradient as a function of the clamp load is also considered. The following equation fitted to the experimental measurements presented by Minas et al (19) is used for this purpose:

$$m_s = -1.01393e^{-14} F_{load}^3 - 1.26651e^{-10} F_{load}^2 + 3.74744e^{-7} F_{load} - 4.18368e^{-4} \tag{12}$$

## 4 OPTIMIZATION PROCESS

A key feature of the current research is to employ a methodology that simulates dry clutch disc transient dynamics and detects potential instabilities related to NVH (20) to optimize the clutch system in order to eliminate the above. The details behind transient clutch dynamics and stability analysis have been presented by Minas et al (20) and for brevity they will not be repeated herein.

In order to conduct the optimization, a second order Recurrent Neural Network is utilized (RNN) (21). The optimization was conducted for the clutch disc and pressure plate mass/inertia after keeping rest of the parameters constant. After solving the eigenproblem for the system using various clutch disc and pressure plate mass combinations, the maximum real values (from the complete set of eigenvalues) are introduced in the optimization software AVL CAMEO 4 R2 (22), which uses a neural network generic algorithm subject to pre-defined parameter constraints in order to determine the dry clutch system parameter combination which would provide the lowest (eigenvalue) real part. Initially, CAMEO generates a multi-layer perception neural network which describes the input/output map of the data provided. Finally, after introducing a target function, which is the minimization of the maximum real part (denoting the stability outcome of the simulations), the optimal combination was proposed. In Figure 3 the flowchart of the modelling and optimization process, followed in the current research, is presented.

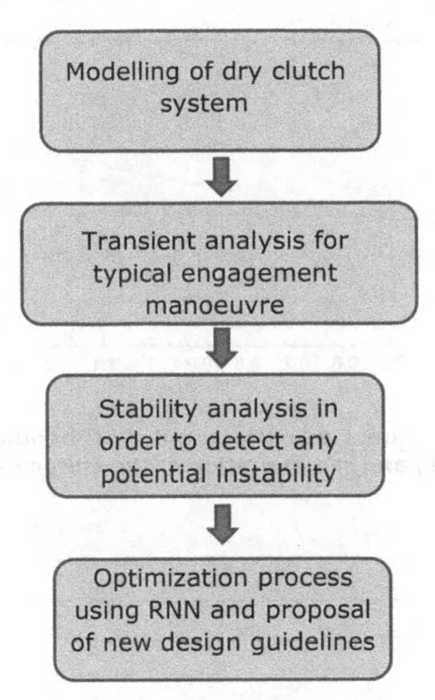


**Figure 3. Flowchart of the methodology followed.**

## 5 RESULTS AND DISCUSSION

The stability of the clutch system was investigated under a single engagement manoeuvre using the transient model presented above. For the time, $t = 0s$, that the pressure plate touches the clutch disc on the flywheel's surface, the cushion spring starts compressing and reaches its full compression, $u = 0.6mm$, at $t = 1.8s$. The cushion spring compression rate for the whole engagement process is steady at t/3. After simulating the engagement manoeuvre and the friction coefficient variation, stability analysis is conducted by solving the eigenproblem and projecting the real and imaginary parts of the eigenvalues to detect potential instabilities. When the real part of the eigenvalue is positive, instability occurs, with natural frequency being the corresponding imaginary part.

The real parts as well as the natural frequencies of the modes that cause unwanted instabilities are presented in Figures 4 and 5. It can be observed that from 0 to 0.4 s there are two converging natural frequencies. The natural frequency at 290 Hz corresponds to the bending motion of the input shaft around Y axis ($\theta_{insh,y}, mode$ 1) and the natural frequency which starts at 220 Hz corresponds to the clutch disc radial motion in the direction of the Y axis ($Y_{cd,y}, mode$ 2). Projecting the real parts of these two modes in Figure 5, it can be noted that both of them have negative real parts, thus until 0.4 s both modes are stable. However, when the natural frequencies meet and the two modes are coupled, system instability results. They reach a Hopf bifurcation point at 0.4s (3, 20, 23). After the coupling, mode 1 becomes unstable and mode 2 has negative real parts symmetric to those of mode 1. The maximum value of the positive real part is 40 and the mode with the negative real parts has value of -40. The natural frequency in the unstable region varies between 270Hz - 307Hz. After 0.6s the natural frequency of the input shaft bending increases at a greater rate than mode 2 and the modes uncouple. From the real values it can be observed that again both modes have negative real parts and become stable.

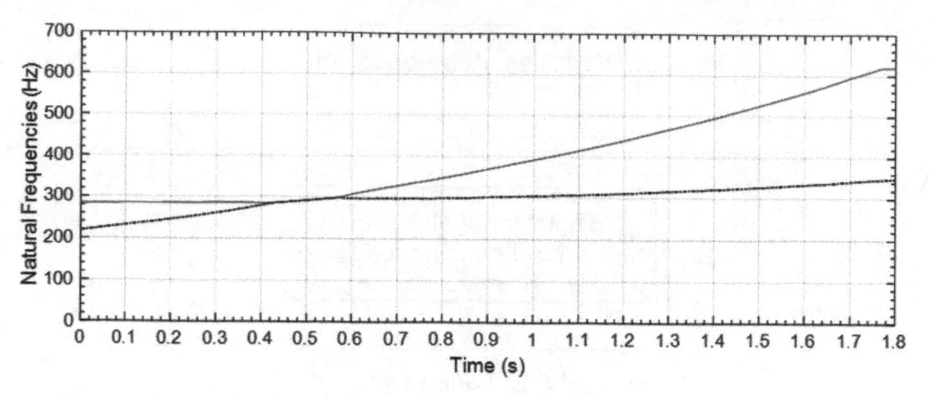


**Figure 4. Natural frequencies of the input shaft bending motion around Y axis (mode 1) and radial motion of the clutch disc (mode 2).**

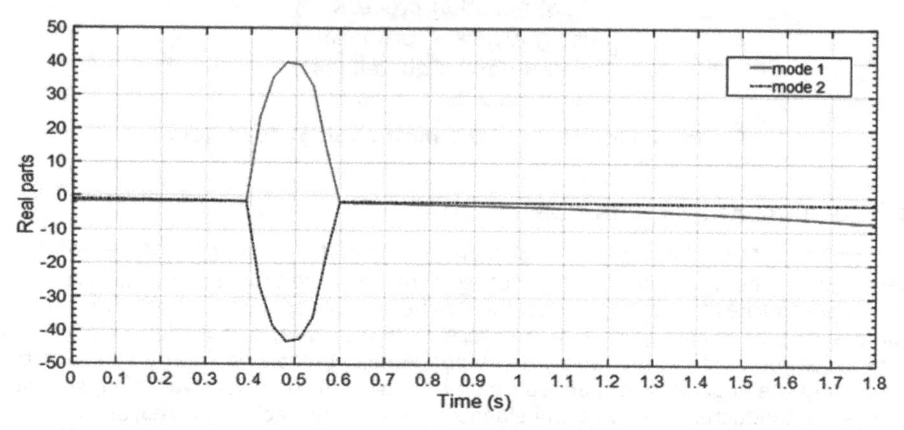


**Figure 5. Real parts of the input shaft bending motion around Y axis (mode 1) and radial motion of the clutch disc (mode 2).**

The last task is to conduct an optimization for two geometrical parameters of the clutch system (pressure plate mass and clutch disc mass) in order to reduce the real part which indicates instability in the clutch system. A parametric study was conducted after varying the pressure plate mass between 1.3 – 2.1 kg and the clutch disc mass between 0.8 - 2.2 kg (19). The nominal values for the clutch disc mass and pressure plate mass are 1.41 kg and 1.85 kg, respectively. The maximum value of the real part for the complete set of eigenvalues, for each combination of the masses, was isolated and inserted as an input to AVL CAMEO 4R2 for the optimization process. A recurrent Neural Network was deployed and the results of the optimization are presented in Figure 6. It is shown that the optimal combination consists of higher clutch disc mass and reduced pressure plate mass. Compared to the nominal case, the clutch disc mass should be increased by 26.65 % and the pressure plate mass should be decreased by 9.45 %.

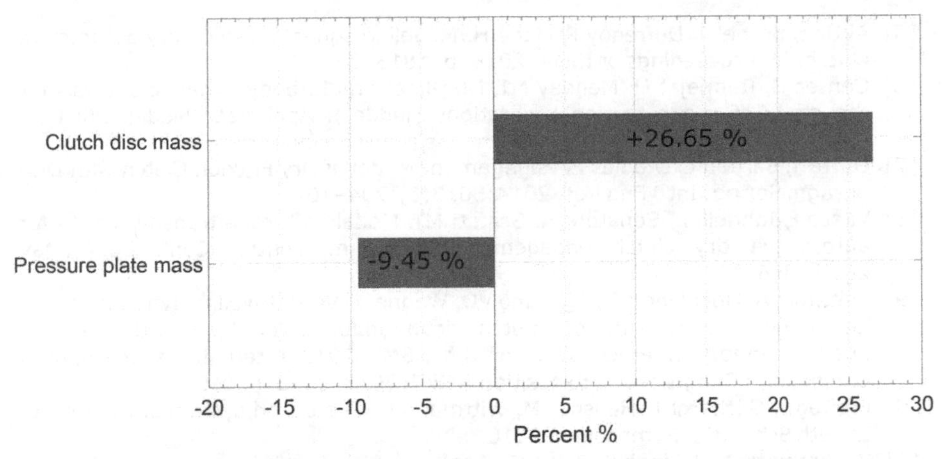


**Figure 6. Optimization results for the pressure plate mass and clutch disc mass.**

## 6 CONCLUSIONS AND FUTURE WORK

The main aim of the current research was to present a novel modelling methodology for the dry clutch system, to observe potential instabilities after simulating a typical engagement manoeuvre and to propose a methodology for eliminating those instabilities by optimizing key system parameters. Those instabilities can be related to aggressive NVH deriving from the automotive powertrain and the dry clutch system. In the current study a mode coupling instability at around 300 Hz was observed, and the modes causing this instability are the bending motion of the input shaft and the radial motion of the clutch disc. The optimization of the clutch system is conducted and the results showed that a clutch system with increased clutch disc mass and decreased pressure plate mass, can lead to the reduction of the real part of the instability as well as to the cushion deflection range that the instability exists. The future work will focus on the optimization of the whole dry clutch system and its link to powertrain efficiency improvement.


## ACKNOWLEDGEMENTS

The authors would like to express their gratitude to the Engineering and Physical Sciences Research Council (EP/N509516/1) and to Ford Motor Company for their financial support. We are also grateful to AVL for providing the software CAMEO 4 R2.


## REFERENCES


[1] Belle L Van, Claeys C, Deckers E, Pluymers B. Enhanced lightweight NVH solutions based on vibro-acoustic metamaterials. 2016;1–5.

[2] Birky AK, Ph D, Laughlin M, Lin Z, Ph D. Transportation Electrification Beyond Light Duty : Technology and Market Assessment. 2017.

[3] Trinh MH, Berger S, Aubry E. Stability analysis of a clutch system with multi-element generalized polynomial chaos. Mech Ind [Internet]. 2016;17 (2):205. Available from: http://www.mechanics-industry.org/10.1051/meca/2015061.

[4] Wickramarachi P, Singh R, Bailey G. Analysis of friction-induced vibration leading to "EEK" noise in a dry friction clutch. In: The 2002 International Congress and Exposition on Noise Control Engineering. Dearbon, Michigan; 2002.

[5] Aktir Y, Brunel J, Dufrenoy P, Mahé H. Modeling squeal noise on dry automotive clutch. In: Proceedings of ISMA. 2013. p. 1813–26.
[6] Centea D, Rahnejat H, Menday MT. Non-linear multi-body dynamic analysis for the study of clutch torsional vibrations (judder). Appl Math Model. 2001;25 (3):177–92.
[7] Dutta T, Baruah L, Studies E. Engagement Model of Dry Friction Clutch with Diaphragm Spring. Int J Eng Res. 2014;5013(3):704–10.
[8] Vasca F, Iannelli L, Senatore A, Scafati MT. Modeling torque transmissibility for automotive dry clutch engagement. Proc Am Control Conf. 2008;(May 2014):306–11.
[9] Senatore A, Hochlenert D, Agostino VD, Wagner U Von. Driveline dynamics simulation and analysis of the dry clutch friction induced vibrations in the eek frequency range. In: Proceedings of the ASME 2013 International Mechanical Engineering Congress and Exposition IMECE 2013. 2013. p. 1–9.
[10] Freitag J, Gerhardt F, Hausner M, Wittmann C. The clutch system of the future. In: 9th Schaeffler Symposium. 2010.
[11] Wickramarachi P, Singh R, Bailey G. Analysis of friction-induced vibration leading to "EEK" noise in a dry friction clutch. Noise Control Eng J [Internet]. 2005;53 (4):138. Available from: http://www.ingentaconnect.com/content/ince/ncej/2005/00000053/00000004/art00002.
[12] Miyasato HH. Modeling of the Clutch Squeal Phenomenon and Practical Possibilities for its Mitigation. 2015.
[13] Dolatabadi N, Rahmani R, Theodossiades S, Rahnejat H, Blundell G, Bernard G. Tribodynamics of a new de-clutch mechanism aimed for engine downsizing in off-road heavy-duty vehicles. SAE Tech Pap. 2017;1–5.
[14] Yang LK, Li HY, Ahmadian M, Ma B. Analysis of the influence of engine torque excitation on clutch judder. JVC/Journal Vib Control. 2017;23(4):645–55.
[15] Gkinis T, Rahmani R, Rahnejat H. Effect of clutch lining frictional characteristics on take-up judder. Proc Inst Mech Eng Part K J Multi-body Dyn. 2017;231 (3):493–503.
[16] Crowther A, Zhang N, Liu DK, Jeyakumaran JK. Analysis and simulation of clutch engagement judder and stick – slip in automotive powertrain systems. Proc Inst Mech Eng Part D J Automob Eng. 2004;218(12):1427–46.
[17] Naus GJL, Beenakkers MA, Huisman RGM, Van De Molengraft MJG, Steinbuch M. Robust control of a clutch system to prevent judder-induced driveline oscillations. Veh Syst Dyn. 2010;48(11):1379–94.
[18] Minas I, Morris N, Theodossiades S, O'Mahony M. Noise, vibration and harshness during dry clutch engagement oscillations. Proc Inst Mech Eng Part C, Mech Eng Sci. 2020;0(0):1–17.
[19] Minas I, Morris N, Theodossiades S, O'Mahony M, Voveris J. On the Effect of Clutch Dynamic Properties on Noise, Vibration and Harshness Phenomena. SAE Tech Pap. 2020.
[20] Minas, I, Morris, N, Theodossiades, S, O'Mahony M, Voveris J. Automotive dry clutch fully coupled transient tribodynamics, *Nonlinear Dynamics*, 105, 1213–1235 (2021).
[21] Sevilla J. Importance of input data normalization for the application of neural networks to complex industrial problems IMPORTANCE OF INPUT DATA NORMALIZATION FOR THE APPLICATION OF NEURAL NETWORKS TO COMPLEX. 2012; (August).
[22] Keuth N, Koegeler H, Fortuna T, Vitale G. DoE and beyond : the evolution of the model based develop-ment approach how legal trends are changing methodology. Conf Des Exp Powertrain Dev. 2015;(June).
[23] Sinou JJ, Jézéquel L. Mode coupling instability in friction-induced vibrations and its dependency on system parameters including damping. Eur J Mech A/Solids. 2007;26(1):106–22.

*International Conference on Powertrain Systems for Net-Zero Transport*
*Institution of Mechanical Engineers, ISBN 978-1-032-11281-7*

# Large-Eddy simulation of a Wankel rotary engine for range extender applications

**X. Shen[1], A.W. Costall[1]*, M. Turner[1], R. Islam[1], A. Ribnishki[1], J.W.G. Turner[2], G. Vorraro[2], N. Bailey[3], S. Addy[3]**

[1]Institute for Advanced Automotive Propulsion Systems (IAAPS), University of Bath, Claverton Down, UK
[2]Clean Combustion Research Center (CCRC), King Abdullah University of Science and Technology (KAUST), Thuwal, Saudi Arabia
[3]Advanced Innovative Engineering (UK) Limited, Lichfield, UK

## ABSTRACT

The Wankel rotary engine offers unrivalled power density as a consequence of having a combustion event every revolution, as well as lightness, compactness and vibrationless operation due to its perfect balance. These attributes have led to its success as a powerplant for unmanned aerial vehicles, and which should also make it an attractive proposition for range-extended electric vehicles. However, it is not currently in production for automotive applications having historically struggled with poor combustion efficiency, and high fuel consumption and hydrocarbon emissions. The purpose of this work is to study the in-chamber flow motion in order to better understand how such limitations may be overcome in future. A 225cc 30 kW Wankel rotary engine is modelled using Large-Eddy Simulation (LES) to ensure turbulent flow features are faithfully recreated, since Reynolds-averaged Navier-Stokes-based approaches are insufficient in this regard. The LES-predicted peak chamber pressure lies within 3.7% of the engine test data, demonstrating good model validation. Combustion simulation parameters are calibrated to match the measured heat release profile, for high-load engine operation at 4000 rpm. Simulation results provide insight into the generation of turbulent structures as the incoming flow interacts with the throttle, intake port, rotor and housing surfaces; how the turbulence breaks down as the combustion chamber is compressed; and how the flame propagates following ignition, leaving a pocket of reactants unburned. Indeed, the computational approach described here allows detailed understanding of the impact of design parameters on the detailed in-chamber flow phenomena, and consequently engine performance and emissions. This will enable the optimization of Wankel rotary engine geometry, port and ignition timing for maximum combustion efficiency and low emissions, reasserting its potential as an effective and efficient prime mover for hybrid and range extended electric vehicles.

## 1 INTRODUCTION

The electrification of transport is continuing apace. The consensus, laid down in the form of propulsion technology roadmaps (e.g., 1), is that decarbonization of urban mobility requires an entirely electric fleet, comprising battery electric vehicles (BEVs) and, later on, fuel cell electric vehicles (FCEVs). Yet the same roadmaps recognize that internal combustion engines (ICEs), in the guise of conventional and especially hybrid electric vehicles (HEVs), have an important supporting role in the transition to an electrified future. This is because the transition can only take place over decades, due to the low

*Corresponding author
DOI: 10.1201/9781003219217-11

baseline level of electric vehicle ownership and the slow rate of vehicle renewal. These factors may be illustrated by considering, for example, that of the 15.5M new passenger cars registered in the EU in 2019, 90% still employed a conventional gasoline or diesel ICE (98.1% if hybrids and alternative fuels are included) (2), while the average lifespan of a gasoline passenger car in Germany is around 18 years (3). Indeed, the Advanced Propulsion Centre Light Duty Vehicle Roadmap (1) uses a 30-year-plus timescale, out to beyond 2050.

So, although the electrified transition has begun, the automotive industry is still quite near the start. In the marketplace, consumer concerns over electric vehicle range and the high initial vehicle cost (largely due to the need to accommodate a sufficiently large battery for acceptable range) continues to inhibit uptake of BEVs. This is where HEVs, or range-extended electric vehicles (REEVs), can encourage fleet renewal by offering some of the advantages of electric propulsion (and while using a smaller, lighter, less expensive battery) without the issue of range anxiety. It is therefore essential that the automotive industry continues to advance engine technology in the interest of greater thermal efficiency, and thus lower $CO_2$ and pollutant emissions, because it is also the consensus that ICEs will persist, in one form or another, for at least the next three decades. Moreover, they are expected to be present beyond 2050 in long range, high power and performance vehicles, powered by dedicated hybrids running on net zero carbon fuels (1).

### 1.1 The Wankel rotary engine as a range extender

Although the Wankel rotary engine found mixed success as a prime mover for ground vehicles[1], it is currently in series production as the powerplant for unmanned aerial vehicles (UAVs), which take advantage of characteristics that differentiate it from reciprocating engines, e.g.,

- Perfect balance and hence vibrationless operation
- No noise, vibration, and harshness (NVH) caused by a conventional valve train
- Compactness and lightness, and hence high volumetric and mass power density
- Low friction
- Low rotational inertia

These same attributes make it a strong contender for the prime mover in a dedicated hybrid powertrain for REEVs, if its traditional weaknesses can be improved:

1. High fuel consumption at part- and especially low-load
2. High hydrocarbon (HC) emissions
3. High oil consumption
4. High combustion chamber surface area-to-volume ratio
5. High coolant and exhaust heat rejection
6. Poor combustion chamber shape

Item 1 in the above list can be avoided in a REEV application since the engine is not connected to the drivetrain and may, in the main, be operated at its best fuel efficiency point when running.

In response to item 2, though Wankel rotary engines have a reputation for high HC emissions, the authors' previous work (4) showed that the *side-ported* 13B-MSP

---

1. A brief historical overview may be found in ref. (4).

RENESIS Wankel rotary engine (5) used in the Mazda RX-8 (homologated at Euro 4) was able to meet more modern (Euro 5 & 6) HC emissions limits, highlighting the value of zero port overlap in inhibiting the escape of unburnt reactants into the exhaust. Follow-on work (6) extended the experimental investigation into the elimination of port overlap with dynamometer testing of the Mazda 13B-MSP, and further considered the benefits of zero port overlap applied to a *peripherally ported* Wankel rotary engine, namely the Advanced Innovative Engineering (AIE) 225CS (7), which was being developed as a range extender within the APC project ADAPT (8, 9). Intermediate conclusions were that, whether side- or peripherally ported, the Wankel engine should be considered for range extender applications because in this role it can exploit its aforementioned headline attributes (power density, lightness, compactness, vibration-free) while inefficient operating regions can be inherently avoided. Additionally, eliminating port overlap should expand the region of best fuel consumption while offering entirely satisfactory exhaust emissions.

Item 3 is, on the face of it, difficult to avoid since Wankel engines have an inherent lack of lubrication to the combustion chamber apex seals and require direct lubrication, which is consumed, leading to an additional source of HC emissions. This concern may be tempered by again considering the excellent HC measurements on the 13B-MSP engine (4), which included any contributions from lubricating oil. Furthermore, the latest generation of UAV Wankel rotary engines employ lubrication systems that can limit oil consumption to 10 cc/hour (6), comparable to automotive engines.

Item 4 relates to the inherent geometry of the combustion chamber in a Wankel rotary engine, where its higher surface area-to-volume ratio means that thermal losses will be greater, relative to a conventional engine. This characteristic also drives the high coolant losses mentioned in item 5, while high exhaust heat rejection is a function of the typical exhaust porting arrangement, which leads to under expansion of the working fluid when compared to a conventional engine.

Items 4 and 5 manifest themselves as lower thermal efficiency. To mitigate their impact, and to recover the performance lost due to the reduced compression and expansion ratios imposed by adopting zero port overlap, compounding the Wankel engine to extract some of the available exhaust enthalpy has also been studied. As previously reported at this conference (9), a rotary expander was designed and fitted to the exhaust of the AIE 225CS, further demonstrating the versatility of such devices. Test results demonstrated significant improvements in torque and brake specific fuel consumption (BSFC) of 30% and 12% respectively. Turbocompounding the AIE 225CS with a turbocharger turbine has also been simulated (6). This used the dedicated Wankel engine template available in AVL BOOST, which has been shown (10) to have important modelling advantages over the 'equivalent three-cylinder, four-stroke' approach that has been used elsewhere (e.g., 11).

Finally, item 6 refers to the difficulties imposed by the elongated shape of a Wankel engine combustion chamber, which does not lend itself to rapid and complete combustion of the fuel. Indeed, achieving a better understanding of the combustion process inside a Wankel rotary engine is the main objective of the current paper.

### 1.2 CFD modelling in Wankel rotary engines

Large-Eddy Simulation (LES) has been applied extensively for modelling turbulent flows in combustion engines, though has largely focussed on conventional reciprocating piston engines (e.g., 12). More recent work though has studied the deployment of different turbulence modelling approaches specifically for the in-chamber flow in Wankel rotary engines. A comparison of LES and a common Reynolds-averaged Navier-Stokes (RANS) technique using the standard k-F065 turbulence model recommended the former for better prediction of flow field structures (13). But while LES provides a more faithful representation of turbulent flow features, it imposes significantly greater computational expense than RANS modelling. A useful compromise is provided by hybrid LES-RANS techniques, such as the Detached-Eddy Simulation (DES) approach first introduced in ref. (14). Advancements since have led to the latest hybrid models such as the Improved Delayed Detached Eddy Simulation–Shear-Stress Transport (IDDES-SST) model, which aims to balance fidelity of turbulence modelling with reasonable computational expense, with its application to a small, peripherally ported rotary engine (15). Also notable in that work is the use of the Q-criterion for turbulent feature identification.

As will be described later, this paper employs a LES approach to turbulence modelling and uses the Q-criterion (defined in Section 3.2.3) to support tracking of turbulent flow structures. The novel contribution of the present work, however, is that the four 'strokes' (intake, compression, expansion, exhaust) of a complete cycle are simulated herein, which is a notable extension beyond recent LES simulations in rotary engines (e.g., 13, 15), which were limited to the intake phase of the cycle.

## 2 METHODOLOGY

This section provides a brief overview of Wankel rotary engine geometry and its set-up in the CFD simulation.

### 2.1 Wankel rotary engine geometry

Inside a Wankel rotary engine (Figure 1a), a triangular rotor, which has a shape similar to a Reuleaux triangle (16), rotates around an output shaft and inside a trochoidal housing. The particular design shown in Figure 1 is known as a 2:3 arrangement – two housing lobes and three rotor flanks. This creates three moving chambers separated from each other by the apices of the rotor, and the internal gearing is correspondingly configured such that the output shaft rotates three times for each rotor rotation. Each chamber experiences a complete four-stroke cycle during each full rotation of the rotor, i.e., there is a combustion event upon every rotation of the output shaft.

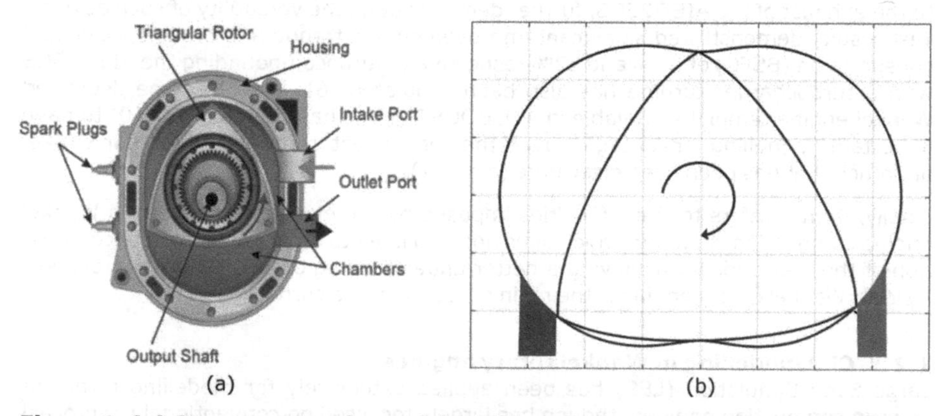


**Figure 1. (a) Components of a peripherally ported rotary engine (17); (b) 2D chamber geometry with respect to intake (blue) and exhaust (red) ports for the case of zero port overlap (4).**

The combustion chambers in a Wankel rotary engine are therefore continually moving and changing shape, with their boundaries defined by the eccentric motion of the rotor about the output shaft and the internal surface of the housing (Figure 1b).

## 2.2 CFD meshing

This subsection describes the moving and stationary mesh domains in 3D CFD. Ansys Fluent R19.2 was used for all CFD aspects described in this paper.

### 2.2.1 *Mesh motion generation and dynamic meshing*

In Fluent a moving mesh is described by a user defined function (UDF). The rotor boundary (red surface in Figure 2) is defined as a rigid wall, i.e., non-deformable. Its simultaneous translation and rotation are defined in the UDF by the velocity and angular velocity of its centre with respect to Cartesian (x,y,z) coordinates and time.

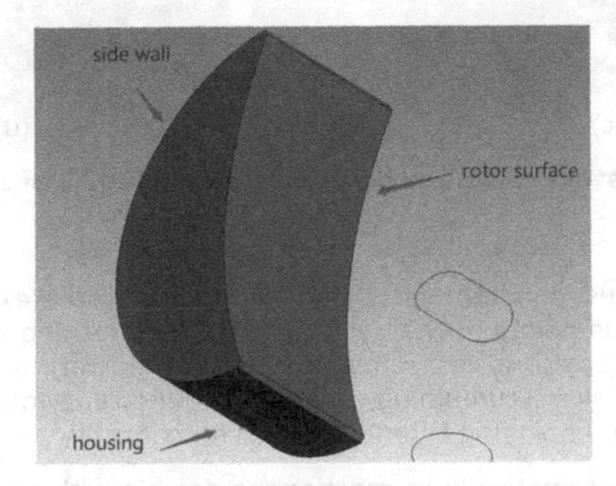


**Figure 2. Rotor domain and surface boundaries.**

On the other hand, the required variation in chamber geometry is enacted by setting up the housing and side wall boundaries (blue and green surfaces in Figure 2) as deformable walls, whose shape adapts according to the position of the rotor.

Due to the complexity of the resulting chamber geometry, a sixth-order polynomial fit is used to describe its variation within the UDF. This is an acceptable simplification, with the difference (in the *y*-coordinate) between the actual and approximated geometries quantified by a root mean squared error of $1.14 \times 10^{-5}$. Some examples of the approximated chamber geometry are shown in Figure 3.

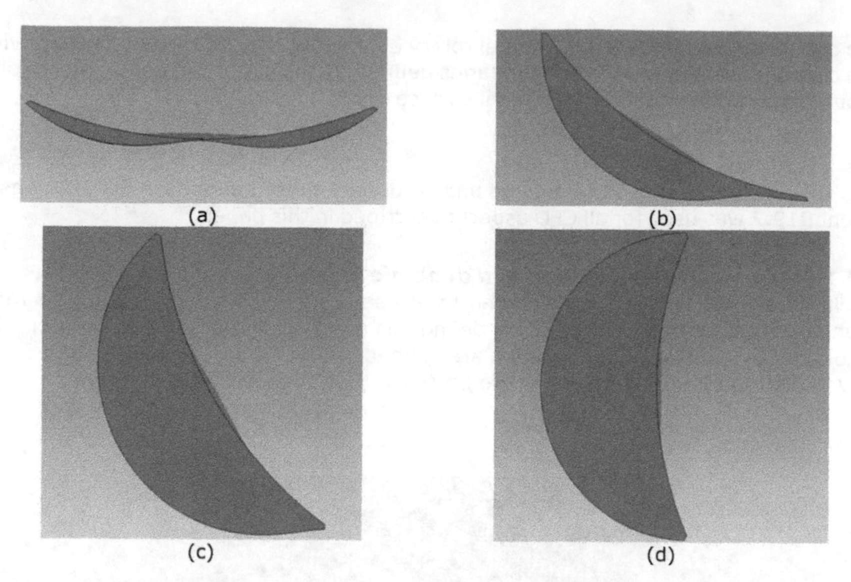

**Figure 3. Combustion chamber deformation during ¼ of a rotor rotation.**

Meshing of the combustion chamber is challenging due to the extreme levels of cell deformation at various points in the cycle. Indeed, re-meshing is unavoidable if an acceptable cell aspect ratio is to be maintained. In this work a hybrid approach to dynamic meshing is taken, combining mesh smoothing, and manual and automatic re-meshing, the latter employing Fluent's cut-cell functionality.

The cut-cell feature automatically re-meshes the domain with an unstructured hexahedral mesh, which offers a significantly lower cell count compared to a tetrahedral mesh, while permitting a level of flexibility unavailable in a fully structured mesh. The main drawback, however, is that automatic re-meshing does not recreate the intended geometry with complete accuracy. In addition, the more often automatic re-meshing is used, the less faithful the mesh becomes. Figure 4 shows this situation following multiple calls for automatic re-meshing. In Figure 4b, the curvature of the rotor surface has changed. Although only slight, this will eventually have a material effect on chamber volume at top dead centre (TDC) and therefore the simulated geometrical compression ratio, and hence the accuracy of the predicted performance. For these reasons, a sensible balance between user time and geometrical accuracy was found by invoking automatic re-meshing once every 1/30 of a cycle, with manual mesh replacement (which corrects any geometrical inaccuracy) performed once every 1/12 of cycle.

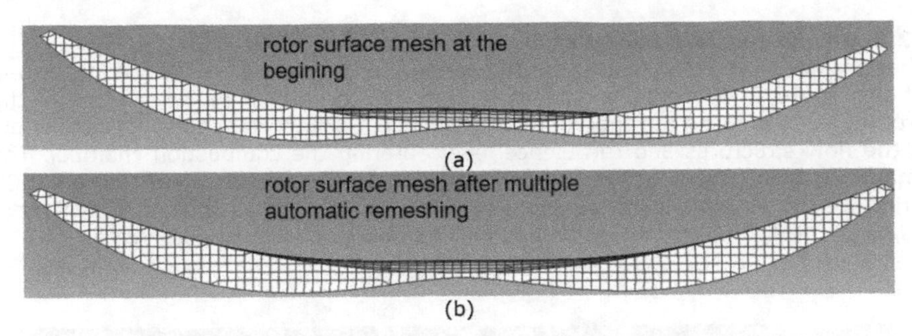


**Figure 4. Combustion chamber mesh at (a) beginning of simulation; (b) after multiple instances of automatic re-meshing.**

In between re-meshing events, the mesh deforms with Fluent's smoothing option. This seeks to retain mesh quality by diffusing the movement smoothly into every node, as Figure 5 shows (N.B. the boundary mesh is purposely refined due to high velocity and temperature gradients in these regions).

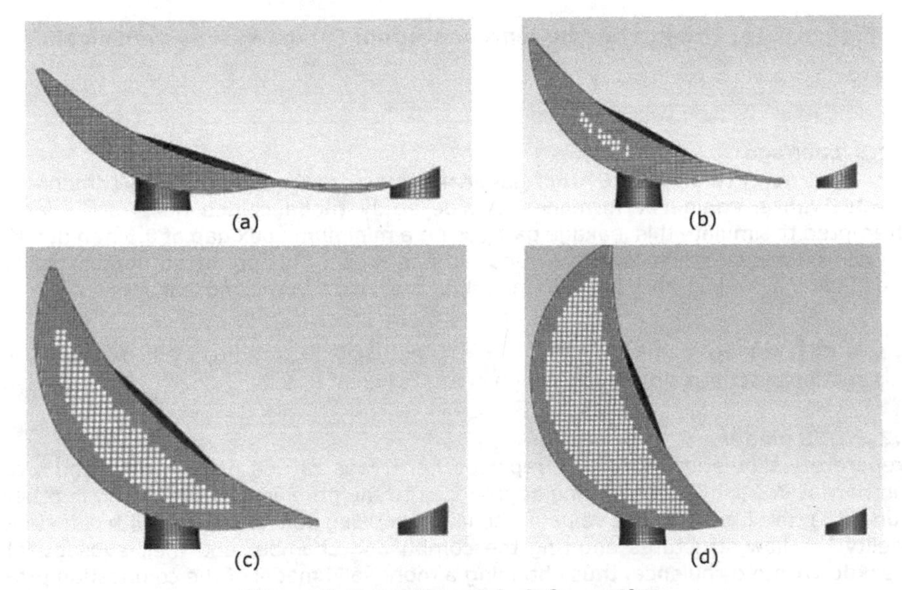


**Figure 5. Smooth mesh deformation.**

It is worth noting that the cell count in the chamber domain varies between approximately 60,000–100,000, depending on the complexity of the deforming domain and dictated by the dynamic meshing technique just described. For the sake of brevity, the mesh dependency study is not reported in this paper. Suffice to say, this order of mesh resolution is in line with the observation in ref. (13) that there is little to choose in the quality of CFD validation between cell counts of 100,000 and 1,000,000 elements, especially considering the computational expense of the latter.

### 2.2.2 *Intake and exhaust domains*

A hexahedral mesh is applied to the remaining stationary domains of the intake and exhaust systems. The intake domain (Figure 6a) includes a representation of the throttle, meshed using tetrahedral elements since it is important to capture its effect on the flow structures and turbulence level entering the combustion chamber. The exhaust domain (Figure 6b) is directly modelled on the one attached to the 225CS during the aforementioned engine testing, with a grid that becomes purposely more coarse as the distance from the combustion chamber and exhaust port increases.

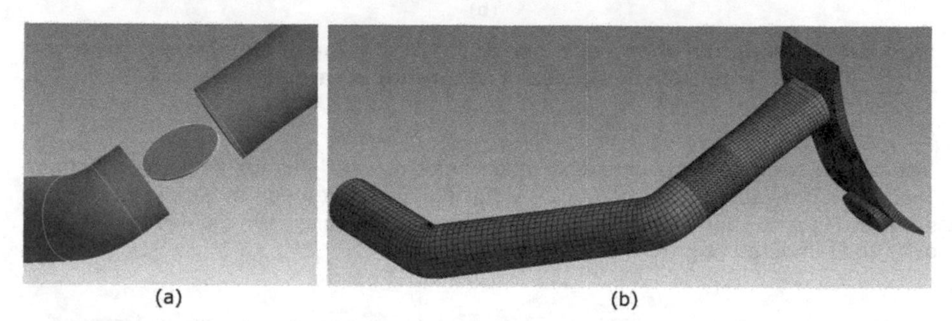

(a) (b)

**Figure 6. (a) Intake throttle representation; (b) exhaust system mesh.**

### 2.2.3 *Leakage*

It is well known (e.g., 18, 19) that gas leakage across the apex seals detrimentally affects Wankel engine performance. A side study (not included here for brevity) attempted to simulate this leakage path, using a minimum apex gap of 0.5 mm due to the computational expense of the required fine mesh. This predicted unrealistically large leakage levels. Hence the CFD model, at present, assumes no leakage.

## 2.2 CFD set-up

This section describes simulation set-up in Ansys Fluent R19.2.

### 2.2.2 *LES model*

Preparatory simulation work (not reported here) was carried out by the authors to compare RANS and LES modelling approaches to the problem. As suggested in other work (13), this identified the value of the latter in being able to capture with sufficient fidelity the flow structures entering the combustion chamber and their subsequent breakdown into turbulence, thus obtaining a more valid model of the combustion process occurring within a Wankel engine. Unlike the RANS approach, which purposely averages out all turbulence structures and cannot resolve at the eddy scale, LES is capable of resolving eddies larger than eight cells (i.e., a cube with side length equal to two cells). For turbulence structures smaller than eight cells, the simulation will switch to a sub-grid model that is similar to a RANS approach.

### 2.2.2 *Simplified combustion model*

The combustion modelling approach employed in this work used Ansys Fluent's built-in *Spark Model*. This creates a local reaction hotspot for user-defined values of position, point in time, and initial conditions for hotspot radius (in this case 2 mm) and energy release (in this case 0.05 J); these parameters were tuned to best match the experimental result. The combustion process itself is very much simplified, using a finite-rate reaction model and a single reaction for iso-octane fuel (Eq. (1)):

$$C_8H_{18} + 12.5O_2 \rightarrow 8\,CO_2 + 9H_2O \tag{1}$$

#### 2.2.3 *Wall heat transfer*

As explained in Section 1.1, compared to reciprocating engines, the Wankel engine suffers from higher heat losses through the wall due to its high surface area-to-volume ratio, particularly around TDC (e.g., 18, 19). Wall heat transfer is thus a key factor affecting both work output and thermal efficiency. In the current study, all walls are set to 90°C, in line with the coolant temperature measured during engine testing (and since no direct wall temperature measurements were made). However, a side study (not included here for conciseness but which applied temperatures of 60°C and 120°C) showed that the assumed wall temperature had little impact on simulation results. Although prescribing a single temperature value to all walls is an overly simplified approach, there are additional reasons that prohibit the simulated wall heat transfer being wholly representative of the real situation:

- Due to the way in which the combustion chamber moving mesh is generated, the internal wall of the housing effectively moves with the chamber, whereas it is stationary in reality. This reduces the relative velocity between the gas and the wall and hence the rate of convective heat transfer.
- To accurately simulate wall heat transfer, the near-wall mesh would need to be much finer than could be afforded here.

### 2.3 Engine testing

As previously mentioned, the subject engine of this paper is the production single-rotor 225CS Wankel rotary engine (7), designed and manufactured by Advanced Innovative Engineering. Its main geometric and timing details are listed in Table 1. The 225CS is port fuel injected, with the injector purposely angled to direct the fuel through the port and into the combustion chamber (Figure 7).

**Figure 7. The AIE 225CS Wankel rotary engine (7).**

**Table 1. AIE 225CS engine geometry and timing (9).**

| Parameter | Value | Units |
|---|---|---|
| Swept volume | 225 | cc |
| Generating radii | 69.5 | mm |
| Eccentricity | 11.6 | mm |
| Offset/equidistance | 2 | mm |
| Rotor number | 1 | |
| Width of rotor housing | 51.941 | mm |
| Geometric compression ratio | 9.6:1 | |
| Intake port opens | 71 BTDC | ° |
| Intake port closes | 60 ABDC | ° |
| Exhaust port opens | 69 BBDC | ° |
| Exhaust port closes | 57 ATDC | ° |
| Number of spark plugs | 2 | |
| Ignition timing | 18 BTDC | ° |

Data to be used for CFD validation was compiled from steady-state engine dynamometer testing conducted at the University of Bath as part of the Advanced Propulsion Centre and Innovate UK-funded project ADAPT (8). In a reciprocating piston engine, a single pressure sensor can be installed in the head to measure combustion chamber pressure. In a Wankel engine however, owing to its rotating chambers, multiple pressure sensors are required to capture a complete pressure trace over a full rotor rotation, and the corresponding data concatenated in post-processing. A live data acquisition system was designed and developed (20) for use with Wankel engines that could generate in-chamber pressure traces and *p-V* diagrams without post-processing. The details of this technique are outside the scope of the current paper, but a full explanation and analysis can be found in refs. (4, 6).

## 3 RESULTS & DISCUSSION

### 3.1 CFD validation

This section demonstrates validation of the 3D CFD model against both motored and fired engine test data, for high-load engine operation at 4000 rpm.

#### 3.1.1 *Comparison of chamber pressure: Motored operation*

A comparison of predicted and measured combustion chamber pressure for motored operation (i.e., without combustion) is shown in Figure 8, with the throttle position at approximately 93% open. It can be observed that there is very good correlation between the simulation and experimentally measured data.

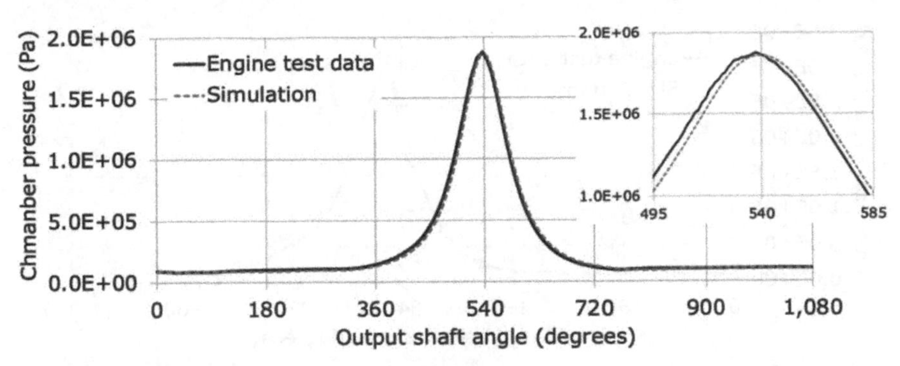


**Figure 8. Comparison of experimentally measured and simulated chamber pressure during motored operation at 4000 rpm and 93% throttle opening.**

In the engine test, *brake* torque is measured by the dynamometer, whereas the CFD simulation can only calculate an *indicated* torque quantity (i.e., that due to the pressure forces on the rotor over a cycle). By calculating the difference between these two values, and although they are taken from different sources (experiment and simulation), one may estimate the magnitude of mechanical friction torque present in the 225CS engine at this operating point. Table 2 shows this breakdown of torque values, resulting in an estimated mechanical friction torque of -3.69 Nm.

**Table 2. Estimate of mechanical friction torque using measured and simulated data for the motored case.**

| Type of torque | Data source | Value | Units |
|---|---|---|---|
| Dynamometer torque (i.e., accounts for mechanical friction and pumping losses) | Engine test | -4.0 | Nm |
| Indicated pumping torque | Simulation | -0.31 | Nm |
| Estimated mechanical friction torque | Combines engine test and simulation | -3.69 | Nm |

### 3.1.2 *Comparison of chamber pressure: Fired operation*

Next, Figure 9 compares engine test data and CFD predictions of the chamber pressure for a fired (i.e., with combustion) case, again at 4000 rpm and 93% throttle opening. There remains a good correlation overall, but a few regions of difference.

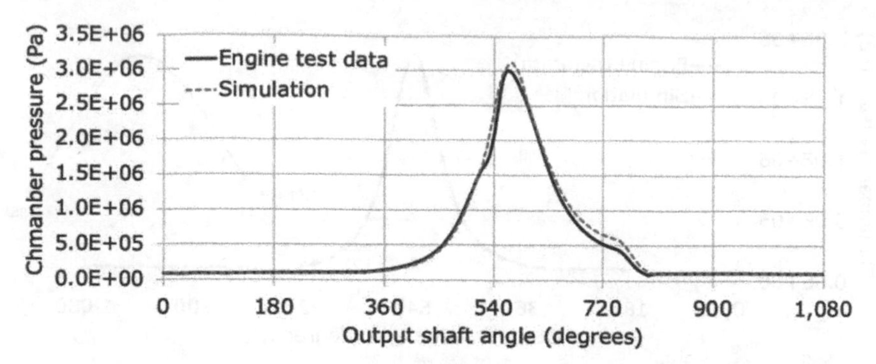


**Figure 9. Comparison of experimentally measured and simulated chamber pressure during fired operation at 4000 rpm and 93% throttle opening.**

The pressure rise in the simulation trace is smooth, but the engine test data shows a decrease in gradient at around 520° output shaft angle, which would coincide with the ignition timing (set at 18° BTDC, as per Table 1). It is surmised that this difference is caused by a mixture of fuel vaporization (not modelled in the simulation) and a transition to a much faster pressure rise rate due to the start of combustion.

The engine test and simulation pressure traces peak at 558° and 568° output shaft angle (18° and 28° ATDC) respectively, with the simulation showing a slightly greater peak pressure (by ~3.7%). This is likely due to the presence of greater thermal losses in the engine test data than are being accounted for in the simulation. The engine test data also shows a faster decrease in pressure from approximately 640°, leading to a pressure difference of approximately 36% by the point at which the exhaust port is uncovered, at about 750° output shaft angle.

### 3.2 Combustion chamber flow phenomena

Having shown that the 3D CFD simulations provide a faithful model of the 225CS engine in performance terms, the CFD-predicted flow structures are now examined by considering, e.g., the velocity field and turbulent viscosity levels.

#### 3.2.1 *Flow structures during intake stroke*

The flow structures during the intake 'stroke' are important for generating the turbulence that will later promote rapid and clean combustion.

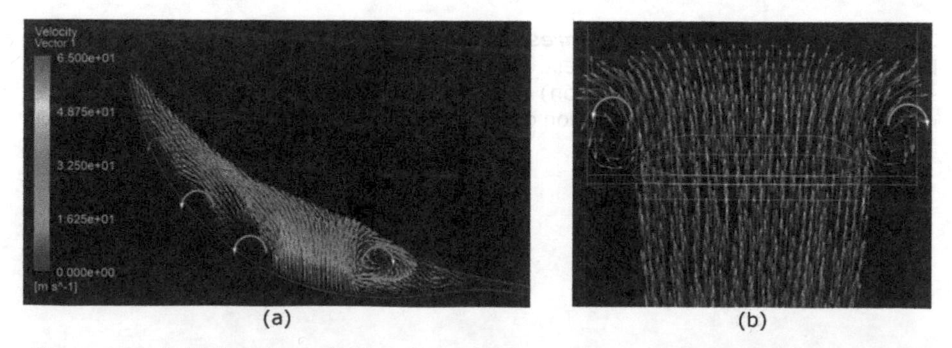


**Figure 10. (a) Chamber, (b) intake velocity vectors; 72° output shaft angle.**

Figure 10 shows velocity vectors for the point at which the output shaft is at 72° and the chamber is quite elongated. It shows formation of vortices as the gas enters the chamber (Figure 10a) from the intake (Figure 10b), before striking the rotor surface.

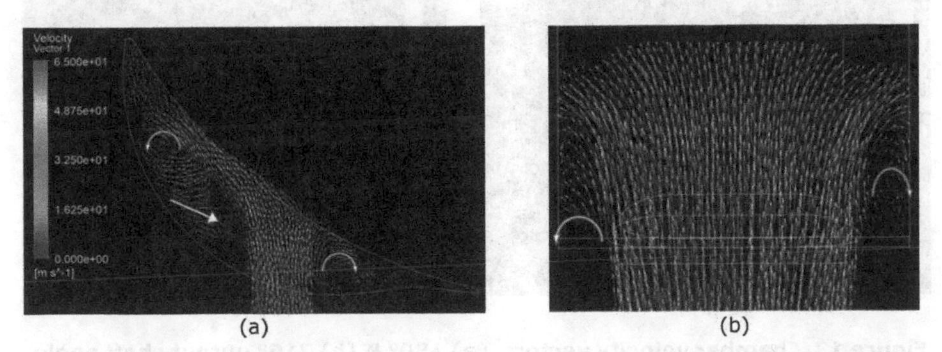


**Figure 11. (a) Chamber, (b) intake velocity vectors; 108° output shaft angle.**

As the chamber starts to expand in Figure 11 (108° output shaft angle), small eddies merge into larger ones near the rotor leading edge. But as the chamber expands further, and the intake flow starts to slow down (Figure 12), the large vortex near the leading edge starts to break into multiple smaller vortices. Meanwhile, the vortex near the trailing edge is sustained noticeably longer due to its relative isolation.

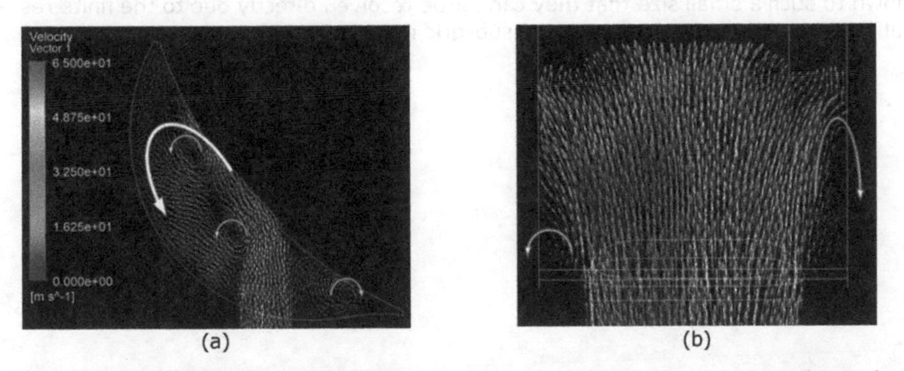


**Figure 12. (a) Chamber, (b) intake velocity vectors; 144° output shaft angle.**

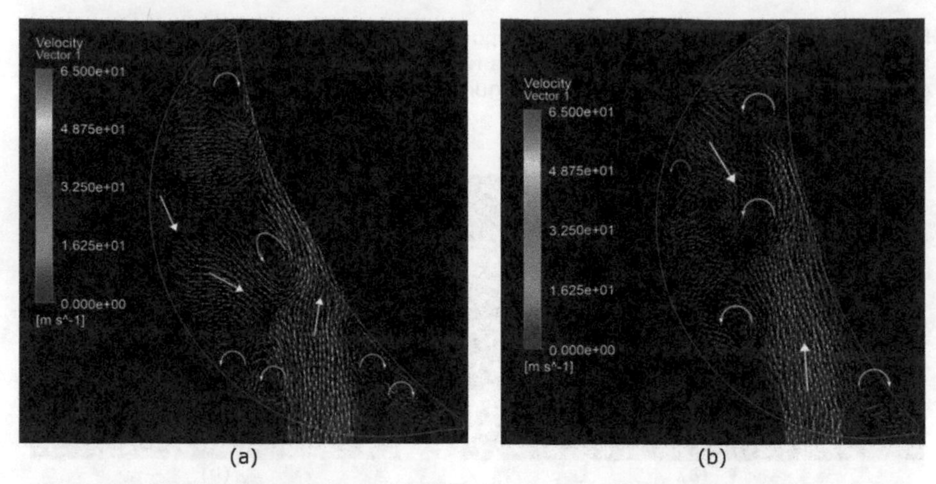


**Figure 13. Chamber velocity vectors, (a) 180° & (b) 216° output shaft angle.**

As the intake stroke ends (Figure 13), the flow quickly starts to decelerate, and the vortices start to break down.

### 3.2.2 *Flow structures during compression stroke*

As the chamber begins to shrink again during the compression stroke (Figure 14), the vortices quickly break down because there is insufficient space to sustain large scale rotation. Just before ignition occurs (Figure 14c), almost all vortices have broken down to such a small size that they cannot be resolved directly due to the finite resolution of the mesh, at which point the sub-grid model takes over.

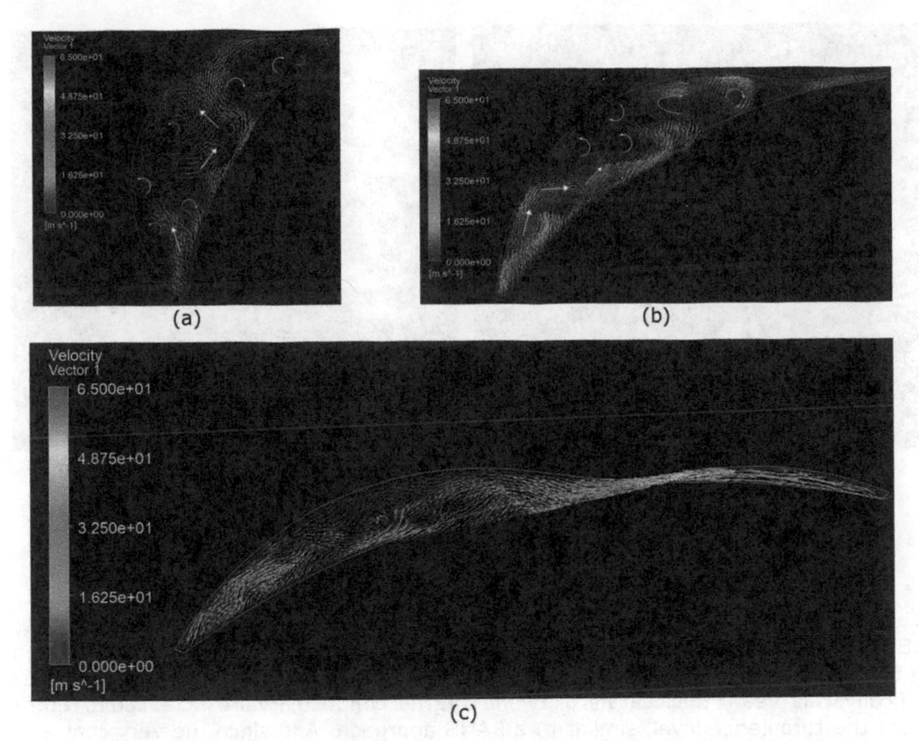


**Figure 14. Chamber velocity vectors at output shaft angles of (a) 360°, (b) 450°, & (c) 522.**

### 3.2.3 *Turbulence tracking*

To visualize the turbulence levels, the so-called Q-criterion is introduced (Eq. (2),

$$Q = \frac{1}{2}\left(\|\Omega\|^2 - \|\mathbf{S}\|^2\right) \tag{2}$$

where **S** is the symmetric part of the velocity gradient (better known as the rate of strain), and $\Omega$ is its antisymmetric part, otherwise known as the vorticity tensor. The Q-criterion is automatically available within the Ansys post-processing tool (CFD-Post). Further details of the Q-criterion can be found in the literature (e.g., 15, 21), but essentially, a higher Q-criterion indicates a region is dominated more by vorticity than strain. For example, the Q-criterion will increase towards the centre of a vortex - indeed, a local Q-criterion hotspot can aid vortex detection, as Figure 15 shows.

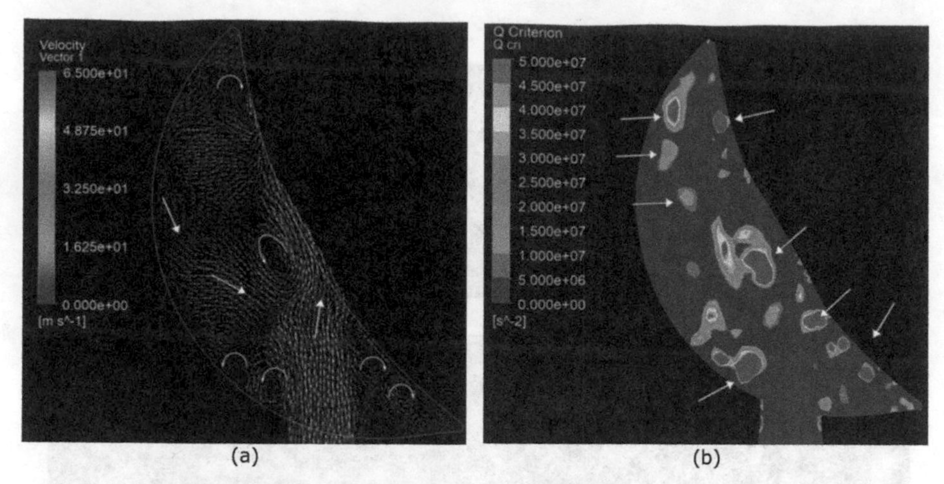


**Figure 15. Identifying vortical structures (a) as Q-criterion hotspots (b).**

The LES sub-grid model is also introduced for comparison. As previously explained, LES is able to resolve large scale eddies, but when they reduce to a scale smaller than a cube with a side length of two cells, the sub-grid model will be activated. More specifically, this means the local viscosity and thermal conductivity are increased to represent the turbulence level, similar to a RANS approach. And since the very centre of a vortex must be smaller than any finite mesh size, the sub-grid viscosity will increase at the vortex core, much like the Q-criterion.

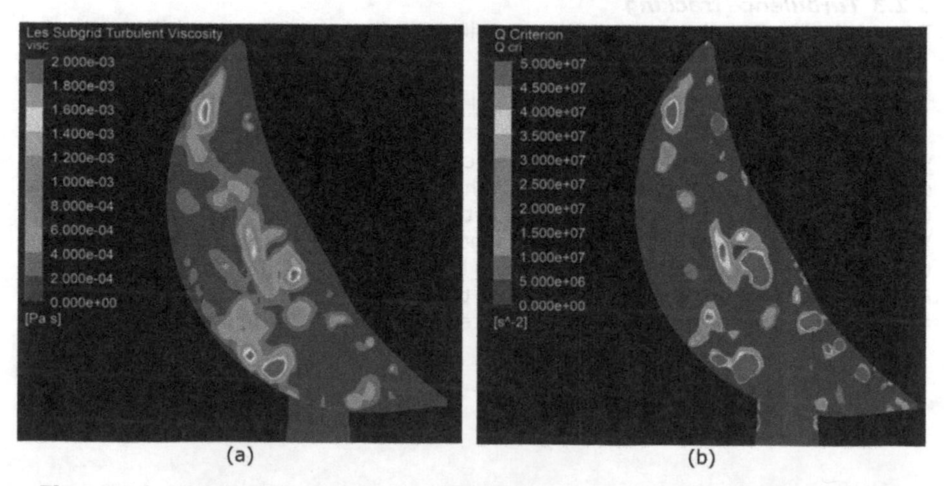


**Figure 16. Comparing contours of sub-grid viscosity (a) & Q-criterion (b).**

As Figure 16 shows, there is clear similarity between the contours of the LES sub-grid turbulence viscosity and Q-criterion. However, LES sub-grid turbulence viscosity is highly affected by mesh density, and since the sub-grid model only activates when the mesh size is too large relative to the turbulence, it is only comparable to itself when the mesh density is uniform – which is definitely not the case in this work. Indeed, it can be seen in Figure 16a that near the rotor surface, where the mesh is finer, the LES sub-grid turbulent viscosity is at a low level, whereas Figure 16b suggests small hot-spots of vorticity are, in fact, present. In conclusion, the Q-criterion is the better identifier of vortices for the purposes of the present work.

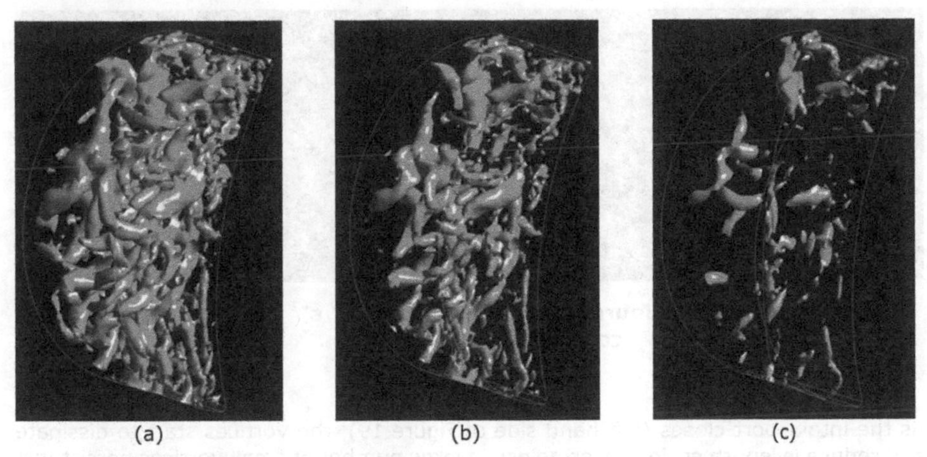

(a) (b) (c)

**Figure 17. Q-criterion iso-surfaces: (a) $1 \times 10^7$, (b) $2 \times 10^7$, & (c) $5 \times 10^7$ $s^{-2}$.**

Taking this a step further, Q-criterion iso-surfaces can help define the shape of vortices in 3D space, as Figure 17 shows. The higher the iso-surface setting, the tighter it envelopes a vortex core, and so eventually only the most intense vortices will remain. In this case, a Q-criterion iso-surface set to $2 \times 10^7$ $s^{-2}$ (as per Figure 17b) offers a balance between showing the quantity of coherent structures present while permitting individual vortices to be identified.

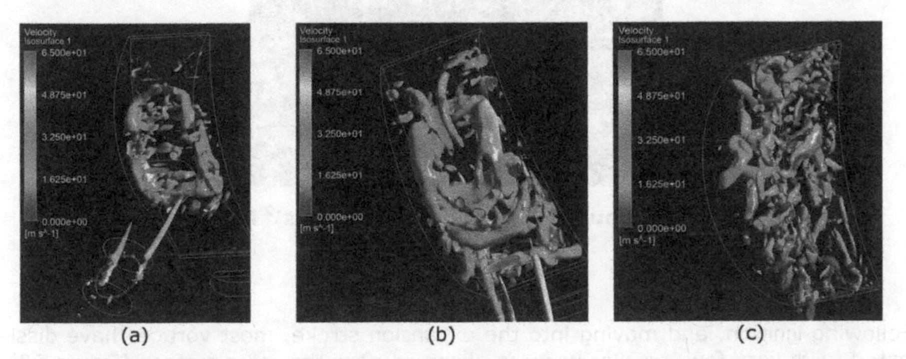


(a) (b) (c)

**Figure 18. Velocity-coloured Q-criterion = $2 \times 10^7$ $s^{-2}$ iso-surfaces during the intake stroke.**

The (velocity-coloured) 3D Q-criterion iso-surfaces in Figure 18 identify two strong vortices generated by the throttle plate (visible in Figure 18a), which are sustained through the intake port but dissipate upon entering the chamber (Figure 18b). As the intake stroke continues, the maximum vortex length scale is attained at an output shaft angle of approximately 160°. As the intake stroke proceeds, the flow starts to slow down, and the large vortices start to break up into multiple vortices of smaller scale (Figure 18c).

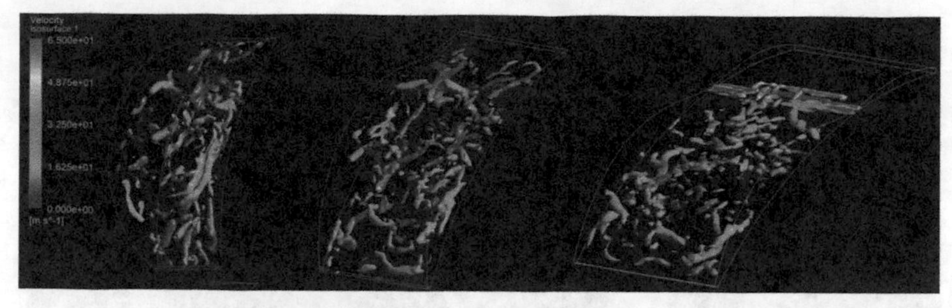


**Figure 19. Velocity-coloured Q-criterion = $2 \times 10^{7}$ $s^{-2}$ iso-surfaces between compression and ignition.**

As the intake port closes (left hand side of Figure 19), the vortices start to dissipate and reduce in length scale. Nevertheless, a large number of small vortices persist just in advance of ignition (right hand side of Figure 19) – which will support the development of the flame once the charge ignites.

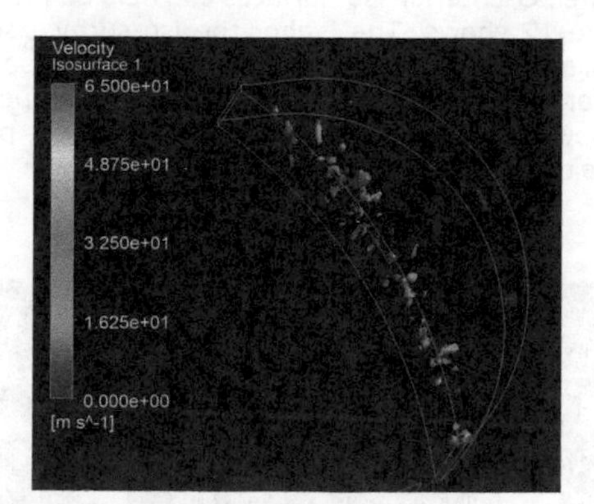


**Figure 20. Velocity-coloured Q-criterion = $2 \times 10^{7}$ $s^{-2}$ iso-surfaces during expansion.**

Following ignition, and moving into the expansion stroke, most vortices have dissipated, with very few, smaller vortices observed near the rotor surface (Figure 20), which could also be attributed to movement of the rotor.

### 3.2.4 *Flame structure*

The very simplified single-equation representation of combustion (Eq. (1)) prohibits tracking of intermediate species, which include those used as accepted markers of the flame front, e.g., OH, $C_2$, CH, $CH_2O$ (22, 23). The combustion model will be upgraded in future, but here visualization of the flame structure in 3D space is achieved by monitoring the mass fraction of $CO_2$ as an end product of combustion.

The flame is ignited once the two spark plugs strike at 18° BTDC (522° output shaft angle) and the two corresponding flame fronts spread out quickly (Figure 21a) and merge near TDC (Figure 21b). The flame reaches the leading edge of the rotor at approximately 36° ATDC (576° output shaft angle) in Figure 21c. However, even with the space afforded by the flank cut-out, the flame front has difficulty in passing into the narrow gap formed between the flank and housing. These results in an unburnt pocket near the trailing edge of the rotor, visible in in Figure 21d. This remains unburnt even after the exhaust stroke begins (Figure 21f) because the flame spreads at a rate slower than the chamber expands and is likely to be a major contributor to the engine's total HC emissions.

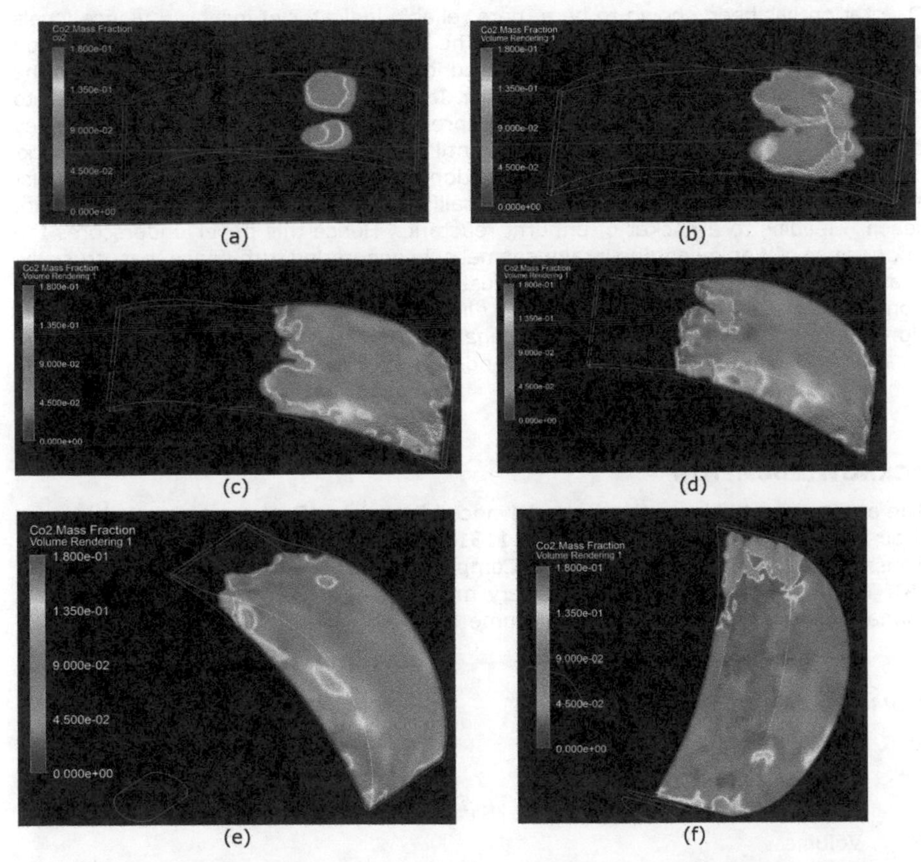


**Figure 21. $CO_2$ mass fraction contours at output shaft angles of (a) 530°, (b) 540°, (c) 576°, (d) 612°, (e) 648°, & (f) 900°.**

## 4 CONCLUSIONS

Despite the growing electrification of transport, internal combustion engines will persist for many decades to come. It is thus crucial to continue improving the thermal efficiency of engines, and reduce $CO_2$ and pollutant emissions, not least because they are an important part of the transition, in the form of range-extended electric vehicles. The Wankel rotary engine offers most of the characteristics of the ideal hybrid propulsion system prime mover (e.g., compactness, lightness, excellent NVH) but the continually changing chamber shape leads to combustion inefficiencies. To better understand their cause, the overall aim of this work was to develop a realistic model of the in-chamber turbulent flow and combustion processes.

A Large-Eddy Simulation (LES) model of the deforming combustion chamber inside a Wankel rotary engine has been successfully validated against engine performance test results. The complete intake-compression-expansion-exhaust cycle is simulated, an important extension beyond other LES simulations of rotary engines in the literature, which mainly focus on the intake phase. In this work, where mesh resolution is at a premium, a hybrid (manual/automatic) approach to dynamic meshing has been deployed in the 3D CFD simulations in order to keep the mesh count to reasonable levels, while maintaining resolvability of turbulent flow features. Furthermore, the Q-criterion has been shown to be a more reliable indicator of local turbulence levels than the LES sub-grid turbulent viscosity. The LES modelling approach, in conjunction with the Q-criterion, can be used to better understand the combustion process in the continually deforming combustion chamber. It allows the turbulent flow structures to be tracked throughout the intake and compression strokes, demonstrating that they persist, albeit at a reducing length scale, until at least the start of combustion. Tracking the extent of the flame following ignition, and into the expansion and exhaust strokes, identified regions at the rotor trailing edge into which the flame cannot reach – leading to a pocket of unburnt reactants. Hence this paper underscores the importance of Wankel engine intake geometry in generating turbulence that, crucially, is available at the beginning of the combustion process to promote rapid flame development. It also emphasizes the inherent difficulties posed by Wankel engine combustion chamber geometry on flame propagation. Indeed, the influence of spark plug location and ignition timing on the size and persistence of unburnt regions will be the subject of future work in this area.

## ACKNOWLEDGMENTS

The authors would like to thank the Advanced Propulsion Centre and Innovate UK for their financial support of APC6 Project 113127: ADAPT-IPT "Reducing Vehicle Carbon Emissions through Development of a Compact, Efficient, and Intelligent Powertrain" (8). The authors would also like to very much thank Kristina Burke for her project management of the ADAPT-IPT programme.

## NOMENCLATURE

$N$ Engine speed

$p$ Pressure

$Q$ Q-criterion, Q-criterion, $\frac{1}{2}\left(\|\Omega\|^2 - \|\mathbf{S}\|^2\right)$

$V$ Volume

## ABBREVIATIONS

| | |
|---|---|
| AIE | Advanced Innovative Engineering |
| APC | Advanced Propulsion Centre |
| ATDC | After Top Dead Centre |
| BEV | Battery Electric Vehicle |
| BSFC | Brake Specific Fuel Consumption |
| BTDC | Before Top Dead Centre |
| CFD | Computational Fluid Dynamics |
| DES | Detached-Eddy Simulation |
| FCEV | Fuel Cell Electric Vehicle |
| HC | Hydrocarbon |
| HEV | Hybrid Electric Vehicle |
| HPC | High Performance Computing |
| ICE | Internal Combustion Engine |
| LES | Large-Eddy Simulation |
| NVH | Noise, Vibration, Harshness |
| RANS | Reynolds-Averaged Navier-Stokes |
| REEV | Range-Extended Electric Vehicle |
| TDC | Top Dead Centre |
| UAV | Unmanned Aerial Vehicle |

## REFERENCES


[1] Advanced Propulsion Centre (2020) "Light Duty Vehicle <3.5t – Propulsion Technologies Roadmap", Automotive Council UK and Advanced Propulsion Centre UK 2020 Product Roadmaps, URL: https://www.apcuk.co.uk/product-roadmaps, last accessed 2021-08-21.

[2] Mock, P. & Diaz, S. (Eds) (2020) "European Vehicle Market Statistics — Pocketbook 2020/21", International Council on Clean Transportation Europe, URL: http://eupocketbook.theicct.org, last accessed 2021-08-21.

[3[ Weymar, E. & Finkbeiner, M. (2016) "Statistical analysis of empirical lifetime mileage data for automotive LCA", *Int. J. Life. Cycle. Assess.*, 21(2016), 215–223. doi:10.1007/s11367-015-1020-6

[4] Turner, J.W.G., Turner, M., Vorraro, G. & Thomas, T. (2020) "Initial Investigations into the Benefits and Challenges of Eliminating Port Overlap in Wankel Rotary Engines", SAE Technical Paper 2020-01-0280. doi:10.4271/2020-01-0280

[5] Ohkubo, M., Tashima, S., Shimizu, R., Fuse, S. & Ebino, H. (2004) "Developed Technologies of the New Rotary Engine (RENESIS)", SAE Technical Paper 2004-01-1790. doi:10.4271/2004-01-1790

[6] Turner, J.W.G., Turner, M., Islam, R., Shen, X. & Costall, A. (2020) "Further Investigations into the Benefits and Challenges of Eliminating Port Overlap in Wankel Rotary Engines", SAE Technical Paper 2021-01-0638. doi:10.4271/2020-01-0280

[7] Advanced Innovative Engineering (2021) "Wankel Rotary Engine 225CS 40BHP", Product Datasheet, URL: https://www.aieuk.com/225cs-40bhp-wankel-rotary-engine, last accessed 2021-08-21.

[8] Advanced Propulsion Centre (2018) "Westfield - ADAPT: Reducing Vehicle Carbon Emissions through Development of a Compact, Efficient, and Intelligent Powertrain", URL: https://www.apcuk.co.uk/portfolio/westfield-adapt, last accessed 2021-08-21.

[9] Vorraro, G., Islam, R., Turner, M. & Turner, J.W.G. (2019) "Application of a Rotary Expander as an Energy Recovery System for a Modern Wankel Engine", *In* Proc. Inst. Mech. Eng. Int. Conf. on Internal Combustion Engines and Powertrain Systems for Future Transport (ICEPSFT 2019), Birmingham, UK, 11–12 December, 260-282.

[10] Peden, M., Turner, M., Turner, J.W.G. & Bailey, N. (2018) "Comparison of 1-D Modelling Approaches for Wankel Engine Performance Simulation and Initial Study of the Direct Injection Limitations", SAE Technical Paper 2018-01-1452. doi:10.4271/2018-01-1452

[11] Tartakovsky, L., Baibikov, V., Gutman, M., Veinblat, M. & Reif, J. (2012) "Simulation of Wankel Engine Performance Using Commercial Software for Piston Engines", SAE Technical Paper 2012-32-0098. doi:10.4271/2012-32-0098

[12] Celik, I., Yavuz, I. & Smirnov, A. (2001) "Large eddy simulations of in-cylinder turbulence for internal combustion engines: A review", *Int. J. Engine Res.* 2(2), 119–148. doi:10.1243/1468087011545389

[13] Poojitganont, T., Sinchai, J., Watjatrakul, B. & Berg, H. P., (2019) "Numerical Investigation of In-chamber Flow inside a Wankel Rotary Engine", *IOP Conf. Ser.: Mater. Sci. Eng.* 501 (2019), 012043. doi:10.1088/1757-899X/501/1/012043

[14] Spalart, P. R., Jou, W.-H., Strelets, M. & Allmaras, S. R. (1997) "Comments on the Feasibility of LES for Wings, and on a Hybrid LES/RANS Approach", *In* Proc. First AFOSR Int. Conf. on DNS/LES, 137–147.

[15] Zhang, Y., Liu, J. & Zuo, Z. (2018) "The Study of Turbulent Fluctuation Characteristics in a Small Rotary Engine with a Peripheral Port Based on the Improved Delayed Detached Eddy Simulation Shear-Stress Transport (IDDES-SST) Method", *Energies* 11(3), 642. doi:10.3390/en11030642

[16] Martini, H., Montejano-Peimbert, L. & Braniff Oliveros, D. (2019) 'Bodies of Constant Width: An Introduction to Convex Geometry with Applications', Birkhäuser (Basel). doi:10.1007/978-3-030-03868-7

[17] Wankel Supertec GmbH (2015) "General Information about Rotary Engines", Web page, URL: https://www.wankelsupertec.de/en_rotary_engines.html, last accessed 2021-08-21.

[18] Feller, F. (1970) 'The 2-Stage Rotary Engine–A New Concept in Diesel Power', Proc. Inst. Mech. Eng. 185(1), 139–158. doi:10.1243/PIME_PROC_1970_185_022_02

[19] Yamamoto, K. (1981) 'Rotary Engine', Sankaido (Tokyo).

[20] Vorraro, G., Turner, M. & Turner, J.W.G. (2019) "Testing of a Modern Wankel Rotary Engine – Part I: Experimental Plan, Development of the Software Tools and Measurement Systems", SAE Technical Paper 2019-01-0075. doi:10.4271/2019-01-0075

[21] Zhan, J.-M., Li, Y.-T., Onyx Wai, W.-H. & Hu, W.-Q. (2019) "Comparison between the Q Criterion and Rortex in the Application of an In-stream Structure", *Phys. Fluids* 31, 121701. doi:10.1063/1.5124245

[22] Pfadler, S., Beyrau, F. & Leipertz, A. (2007) "Flame front detection and characterization using conditioned particle image velocimetry (CPIV)", *Opt. Express* 15 (23), 15445–15456. doi:10.1364/OE.15.015444

[23] Liu, X., Kokjohn, S., Wang, H. & Yao, M. (2019) "A comparative numerical investigation of reactivity controlled compression ignition combustion using Large Eddy Simulation and Reynolds-Averaged Navier-Stokes approaches", *Fuel* 257, 116023. doi:10.1016/j.fuel.2019.116023

## *Session 5: Real-world Driving Emission (RDE) and emissions analysis*

*International Conference on Powertrain Systems for Net-Zero Transport*
*Institution of Mechanical Engineers, ISBN 978-1-032-11281-7*

# RDE vehicle emissions improvements assessed on a London route

**J. Parnell[1], M Peckham[1], B. Mason[2], E. Winward[2]**

[1]Cambustion Ltd
[2]Department of Auto and Aero Engineering, Loughborough University

**ABSTRACT**

Euro 5 diesel, Euro 6 PHEV gasoline and Euro 6d-TEMP diesel passenger cars have been tested for tailpipe NOx on TfL's "West London Route". Fast response analysers were used to correlate transient emissions with ECU data, dashcam footage and GPS information to study the vehicles' emissions responses to stimuli from road lay-out, congestion and other real-world influences. The results show a whole route reduction of 94% total NOx for the Euro 6d-TEMP vehicle compared to the Euro 5 diesel suggesting that the turnover of the vehicle fleet and the growing proportion of improved vehicle emissions categories will, in itself, bring air quality benefits to urban areas. In addition, the particular events during the drive which produce (sometimes very short duration) spikes of NOx are analysed in terms of the engine and aftertreatment reasons for the tailpipe "event". The main reasons for the tailpipe NOx from the three vehicles mentioned were found to be respectively: harsh accelerations, high load engine restarts and accelerations immediately following congested slow driving.

## 1 INTRODUCTION

Vehicle manufacturers are facing increasing pressure by legislation and economics to reduce vehicle emissions and deliver improved fuel economy. By 2025 significant reductions in carbon dioxide ($CO_2$) emissions will need to be achieved to meet these requirements whilst at the same time satisfying the more stringent forthcoming legislation.

Air pollution is a major concern, in Europe it is estimated 400,000 people die prematurely every year as a consequence of short and long-term exposure to high levels of pollution in the air [1]. Short and long-term exposure to elevated concentrations of Particulate Matter (PM) and Nitrogen Dioxide ($NO_2$) in the air have been associated with cancer [2], respiratory [3] and cardiovascular diseases [4], hypertension and diabetes [5]. Road transport is a main source of $NO_2$ and ambient particle concentrations. In the European Union (EU), road transport contributes up to 39% of Nitrogen oxides (NOx) and 11% of PM emissions [1]. The contribution of PM2.5 and NOx emissions from road traffic increase significantly at the city level [6]. In London for example, PM2.5 increases by 30% and NOx by 49% [6], [7]. In 2017 the proportion of the European population exposed to concentrations exceeding the WHO (World Health Organization) air quality guideline annual mean PM2.5 (>10 μg/m3) and NO2 (>40 μg/m3) was 77% and 7%, respectively [6].

Many studies have shown that there are significant discrepancies between type approval test cycles such as New European Driving Cycle (NEDC) and real-world emissions [8]–[12]. The Worldwide harmonized Light vehicles Test Cycle and Procedure (WLTC and WLTP, respectively) were introduced into the certification process from 2017 to address this. However, real-world testing of vehicle emissions is more demanding of vehicle controller calibrations than laboratory-based testing, with

DOI: 10.1201/9781003219217-12

variable ambient conditions, road obstacles such a speed bumps, variable traffic conditions, etc. Therefore, the Real Driving Emissions (RDE) test has been introduced into the type approval procedure in Europe from September 2017. In the RDE the vehicle is driven on public roads in real traffic conditions and a Portable Emissions Measurement System (PEMS) continuously measures the tailpipe emissions.

The RDE NOx and Particle Number (PN) emissions must be below the respective Euro 6 not-to-exceed (NTE) limit multiplied by the Conformity Factor (CF). The test route must comply with a set of requirements including duration, ambient temperature and share of urban, rural and motorway operation [13], [14]. The CF is to account for the additional measurement uncertainties introduced by the PEMS compared to laboratory equipment. Vehicles homologated under the WLTP and the RDE procedures belong to the Euro 6d emissions standard. The Euro 6d-temp is applicable during the phasing in period 2017–2020) [14].

To date, four RDE packages have been released. The first package defines the RDE test procedure; the second package defines the NOx Conformity Factors and the introduction dates; the third package, adds a Particle Number (PN) Conformity Factor and includes cold-start emissions; the fourth package outlines conformity factors and the concept of surveillance, additionally it lowers the error margin of the 2020 NOx Conformity Factor from 0.5 to 0.43. The intent of the legislation is to ensure that the emissions generated under real driving conditions are suitably controlled, the differences between real and on-cycle emissions is well known [2].

The minimum trip duration requirement of RDE is 90 minutes, much longer than existing standardised cycles. For example, the New European Driving Cycle (NEDC) has a cycle duration of 20 minutes, and the Worldwide Harmonized Light Duty Test Cycle (WLTC) has a cycle duration of 30 minutes. The RDE sequence consists of urban driving following by rural and motorway driving. Unlike existing cycles which have a predefined vehicle speed profile, each RDE test has a unique vehicle speed profile. It is generated as a function of driver behaviour, road, traffic and vehicle conditions during the test. Hence, many dynamic manoeuvres will be undertaken, such as: gear change, traversing speed bumps, stopping for traffic lights, navigating roundabouts etc. Many of the resulting engine transients are not included in existing prescribed test cycles.

In terms of the engine the above will result in much greater coverage of the operating space, increasing the range of boundary conditions experienced and is much more representative of real-world driving. As a consequence of the less steady operation the dynamic behaviour of the system contributes significantly to performance.

Several studies report that dynamic driving can have a significant impact on NOx emissions. The study in [15] reports a 25% increase in the NOx emissions of Euro 6c diesel vehicles when driven dynamically compared to normal driving.

The study by Gallus et al. [16] investigated two Diesel test vehicles (Euro 5 and Euro 6) found that when driving outside the RDE boundary conditions, vehicle NOx emissions can be significantly increased whereas CO and HC emissions did not show a distinct separation from different driving styles.

It is reported in [17] that the greatest amount of NOx and $NO_2$ are emitted during the urban sections of a route that confirms to RDE requirements for a Euro 6 light-duty diesel vehicle, this study concludes that lower speeds with more accelerations and decelerations lead to higher NOx emissions levels compared to constant high-speed operation. They report that NOx emissions in real driving conditions are mainly a consequence of abrupt vehicle accelerations.

There is a more limited body of research for diesel Euro 6d-TEMP vehicles (homologated after September 2017), some recent studies include [14], [18], [19]. The study

[18] investigated two diesel Euro 6d-temp vehicles and found that the NOx emissions for both were below the Euro 6-temp requirements (80 mg/km × 2.1 CF) for the complete and urban sections. However, under dynamic driving conditions, the NOx emissions were ~ 3.5 times higher than in RDE compliant tests. The highest NOx emission factors were registered on the motorway section. They conclude that NOx emissions from Euro 6d-TEMP vehicles have improved overall but there are still some operating conditions that could be improved. They also observed that that use of different aftertreatment technologies led to the emissions of different, currently unregulated, pollutants.

The study [14] investigated six Euro 6d-TEMP passenger cars (two gasoline and four diesel vehicles). The NOx emissions were well below the applicable NTE limits and were significantly lower than pre-RDE vehicles with 10-12 times lower than Euro 6b diesel vehicles on the road. It was found that the NOx emissions of the only diesel vehicle not to use a SCR catalyst (it used a LNT) were particularly high on the motorway where the LNT showed limited deNOx capability. Under dynamic driving conditions when the RDE dynamicity criteria were exceeded, NOx emissions increased for most vehicles and were exceeded by two of the diesel vehicles. They conclude that the WLTP + RDE regulations have been very effective at reducing NOx emissions in real-world operation compared to pre-RDE vehicles.

Finally in [20] two Euro 6d-TEMP diesel vehicles, one a passenger car and the other a sports utility vehicle, were investigated along a route according to RDE procedure. Each vehicle was equipped with: Exhaust Gas Recirculation (EGR), Selective Catalytic Reduction (SCR), Diesel Oxidation Catalyst (DOC) and diesel particulate filter (DPF). The NOx emissions were 60% (SUV) and 75% lower (passenger car) lower than RDE regulations. The highest NOX emissions were measured for the urban section, with lower but similar NOx emissions both the rural and motorway sections. The study attributes the higher urban NOx emissions to the location of the test start and the effects of the characteristic driving and traffic conditions on the thermal state of the aftertreatment systems.

Hybrid Electric Vehicles (HEV) and plugin-HEV (PHEVs) are an alternative to conventional diesel and gasoline vehicles. A HEV/PHEV can potentially offer improved fuel economy and reduced emissions. For a HEV/PHEV such as the Toyota Prius, the engine can switch on and off whilst driving and the engine can be assisted by an electric motor. This behaviour however can lead to highly variable emissions rates compared to an equivalent conventional gasoline or petrol vehicle. The combination of circumstances such as driving conditions, vehicle speed, road grade, ambient temperature, driver aggressiveness, and HEV system design all influence the emissions behaviour of such vehicles [21], [22]. There is also potential for a high frequency of cold starts [23] leading to longer warm-up times for the aftertreatment system [21], [24]

It is reported in [21] that large NOx spikes can occur for a HEV during low-speed, high-load acceleration events where the engine is prone to operate lean for short periods. These NOx spikes could be between 16-55 times higher than the average values, other NOx spikes were also observed when the engine restarts.

In the study [25], the emissions of a PHEV are compared to a range extended battery electric vehicle and also Euro 6 conventional gasoline and diesel vehicles for -7°C and 23°C ambient temperatures. They report that the NOx emissions of the PHEV at -7°C were significantly higher than the conventional gasoline and diesel vehicles. They found that the NOx emissions of the PHEV in the first few seconds of use of the Internal Combustion Engine (ICE) could in some scenarios be greater than Euro 6 conventional gasoline vehicles during the cold-start or even during an entire WLTP test.

Measurement of the tailpipe NOx for a NEDC-compliant (prior to Euro 6d TEMP) PHEV is reported in [26] for an RDE compliant route. In most cases the tailpipe NOx emissions were very good however some short transients were observed to cause significant NOx emissions. These were typically harsh accelerations from speed bumps and when in congested traffic. It was observed that the Port Fuel Injected (PFI) gasoline engine when restarting and acerating to high load could have a short period of lean combustion which would cause high tailpipe NOx. It was observed that the engine would stop to recover energy when approaching the speed bump and then restart under high load as the driver accelerates after the speed bump, for a particular section of the route with successive speed bumps, this caused repeated tailpipe NOx emissions which exceed he NEDC limit.

This paper discusses the emissions from three different passenger vehicles (Euro 5 diesel, Euro 6 PHEV gasoline and Euro 6d-TEMP diesel) for a route around west London. GPS monitoring, camera and vehicle on board diagnostics information is combined to identify key features of real-world NOx emissions between the three different vehicles. The fast response analysers combined with the GPS and video enable the effects of local urban road to be evaluated for the three different vehicles.

## 2 EXPERIMENTAL SET-UP

Given that the focus of this paper is NOx emissions, the experimental measures concentrate on this aspect.

### 2.1 Vehicles

Three vehicles of differing emissions Euro classification have been compared for this work and their summary details are shown in Table 1. The main NOx abatement technology is briefly mentioned but some OEM calibration techniques involving boost pressure and fuel injection optimisation will also have been deployed though are not discussed in this work.

All vehicles were driven to the start of the route and switched off for approximately 1 hour while the emissions analysers were being warmed-up prior to the journey. The start-up transient of the vehicles can then be considered "warm starts" therefore and would tend to favour slightly better emissions characteristics compared to a typical cold start. For the PHEV vehicle, the drive from Cambridge to the start of the route depleted its traction battery to the point where it was acting more like a standard HEV; the engine stopping and starting to maintain the traction battery at ~15% charge and when high power was demanded.

The emissions from all vehicles were measured at the tailpipe location (after the final stage of any emissions aftertreatment) but before the vehicle silencer/muffler.

**Table 1. Passenger car summary details.**

| Vehicle type | Emissions class | Year of first registration | Main NOx abatement techniques |
|---|---|---|---|
| 7-seater MPV, 2.3litre diesel, manual | Euro 5 | 2011 | EGR |
| PHEV saloon, 1.8 litre gasoline, auto | Euro 6 | 2016 | TWC and EGR |
| Saloon, 1.5 litre diesel, manual with stop/start | Euro 6d-TEMP | 2018 | SDPF, SCR and EGR |

### 2.1.1 *Euro 5 diesel*

This 2011 vehicle was a well-maintained vehicle with 70k miles completed, after-treatment consisted of separate oxidation catalyst and Diesel Particulate Filter (DPF).

### 2.1.2 *Euro 6 PHEV*

This 2017 vehicle was brand new with only 200 miles service at the time of this study. More details of its aftertreatment system are available in [1]. The engine was 1.8 litre naturally aspirated PFI gasoline and the electric motors consisted of a 53kW traction-only motor and a 22.5kW generating/traction motor.

### 2.1.3 *Euro 6d-TEMP diesel*

This 2018, 1.5 litre 100ps vehicle was also brand new with only 260 miles service at the time of this study. Its aftertreatment consisted of a passive NOx adsorber followed by a SCR system with urea injection and a "SDPF" (SCR-coated DPF) and feedback control of the emissions via a solid-state tailpipe NOx sensor.

## 2.2 Instrumentation

A fast response chemiluminescence analyser (model CLD500 shown in Figure 1) was used for measurement of the real time tailpipe NOx emissions. The two channel analyser ($T_{10\text{-}90\%}$ approx. 5 milliseconds) was modified for on-board applications using an RDE Accessory kit and both NO and NOx were measured for the diesel powered vehicles (not considered necessary for the PHEV gasoline given that its main NOx content would be solely NO).

**Figure 1. Two-channel fast CLD with RDE accessory kit for real-time on-board measurements.**

## 2.2 Conversion of ppm to mg/km

Given that passenger car emissions limits are defined with unit mass per unit distance, there is a requirement to convert the volumetric "ppm" gas analyser output to mg/km units via measurement of the real time exhaust mass flow rate. The latter was recorded from the vehicle's own ECU data via the on-board diagnostics port and integrated with the analyser output in real time during the drive such that the instantaneous mg/km level could be viewed during the drive by the passenger. The fast RDE analysers' delay time in responding was of the order of 10 milliseconds and this was considered an insignificant correction to make before combining with the mass flow data. Similar or longer delay would be evident in the transit of exhaust gas from the engine to the sampling point. All vehicles except the Euro 5 diesel had been tested using this method and the resulting total cumulative NOx cycle results compared with a shorter RDE route on the Cambustion chassis dynamometer and CVS emissions system with resulting errors of less than 5%.

## 2.3 Route

The chosen route was originally derived by and obtained (with kind permission) from TfL and is known as their "West London route" shown in Figure 2. In short, its starting point is Brent Cross shopping centre where equipment can be warmed-up and stabilised on mains power before switching to battery power for the duration of the test. The route then proceeds in a clockwise direction along Finchley Road towards Central West London, passing through Swiss Cottage, Piccadilly Circus, Knightsbridge, Acton, Ealing, Neasden and joining the northbound M1 motorway at Brent Cross to the first junction where the direction is reversed at the service station before the return to the starting point at Brent Cross shopping centre. Prior to departure and after arrival at Brent Cross shopping centre, the analysers underwent a calibration check using certified gases.

**Figure 2. TfL's "West London Route" and the locations of NOx (ppm) from the PHEV.**

The route consists of some urban free-flowing dual carriageway driving (speed limited to 50mph), some inevitable congestion in central London, many sets of traffic lights, some residential areas including speed bumps, some congested dual carriageway (North Circular), acceleration on an up-ramp to join the motorway and some 70mph cruising. The route totals 55km and takes around 2 hours to complete.

The inherent and intended consequence of driving in the "real world" is that it is a somewhat "uncontrolled" environment but for the purposes of this comparison the length of the test and frequency of transients was, the authors supposed, sufficient to provide broad comparative results between the three different vehicles. Nominal control measures were deployed involving driving the route at the same time of day during a typical working weekday and during similar weather conditions.

A more repeatable way of performing a comparable and controlled set of tests would have been to record the speed vs time of the first drive and then repeat this drive using the visual aid or robot on the Cambustion chassis dynamometer thereby accurately replicating the same drive for all vehicles. However, access to the chassis dyno was limited and the portability of the fast response analysers meant that an actual drive was more practical

## 3 RESULTS AND DISCUSSION

The results and their discussion are confined to consideration of NOx given that this is a main concern for air quality (alongside particulates) and has for many years been one of the major challenges for engine researchers and calibrators to control. The dynamics of the transient NOx emissions and their correlation with driving conditions will be discussed in some detail; derived from data available from the vehicle ECU but also from additional thermocouples fitted by Cambustion.

### 3.1 Vehicle comparisons of NOx

A summary of the cumulative NOx results for all three vehicles is shown in Figure 3 along with the calculated g/km for each vehicle journey.

It is clear and obvious that the NOx abatement measures which the progressively stringent regulations have mandated on each vehicle class have shown direct benefits for newer vehicles. Although this testing did not consider the effects of vehicle aging (and in this respect, it may be considered that the Euro 5 with its higher mileage may be at a disadvantage).

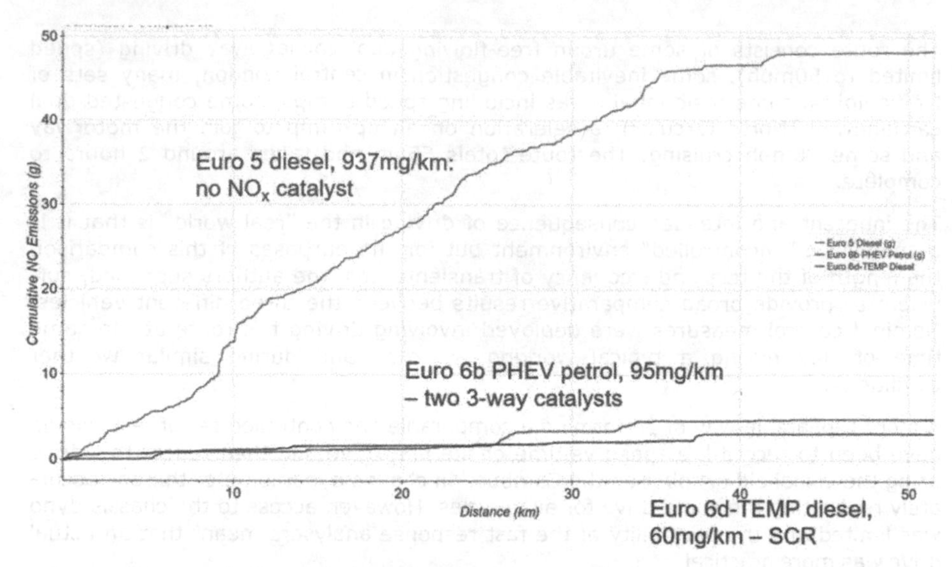


**Figure 3. Cumulative NOx results and the resulting mg/km over the entire route for all 3 vehicles.**

What is of more interest from this data is the evidence for the causes of the emissions steps and how much the short duration "spikes" of emissions contribute to the overall cumulative figure. This is especially true of the PHEV vehicle and the Euro 6d-TEMP vehicle whose NOx emissions are concentrated in to short-durations events with the potential for measuring, understanding and finally controlling the emissions. The events from these two vehicles will be studied in more detail in separate sections.

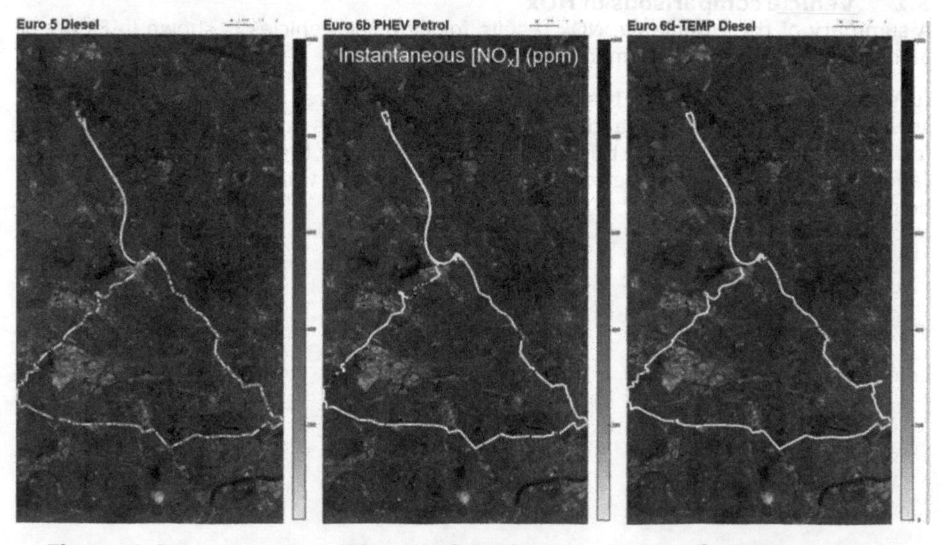


**Figure 4. Instantaneous NOx ppm location comparison for all 3 vehicles.**

### 3.2 Euro 5 diesel vehicle

The Euro 5 diesel (which has no NOx aftertreatment, but EGR for combustion control of NOx) tended to emit tailpipe NOx regularly and correlated with accelerations (Figure 4). Under harsh accelerations (often induced by urban driving conditions) the EGR is switched off to allow for high power. The combustion conditions are also ideal for the production of NOx (high temperature and pressure in the presence of oxygen). A remarkable example of this is shown during the rapid acceleration from the congested North Circular dual carriageway on to the Northbound M1 motorway on an "up-ramp". A screen shot of the real time emissions monitoring screen (Figure 5) shows the production of 1 gramme of NOx during only an 8s period of this acceleration. When this data was recorded the latest available published web data (2018), available from DVSA suggested that 68% of the UKs vehicle fleet was assumed to be Euro 5 or older. This blend of road layout, requirement for acceleration to motorway speeds and vehicle type is likely to made air quality down-wind of this point somewhat sub-optimal.

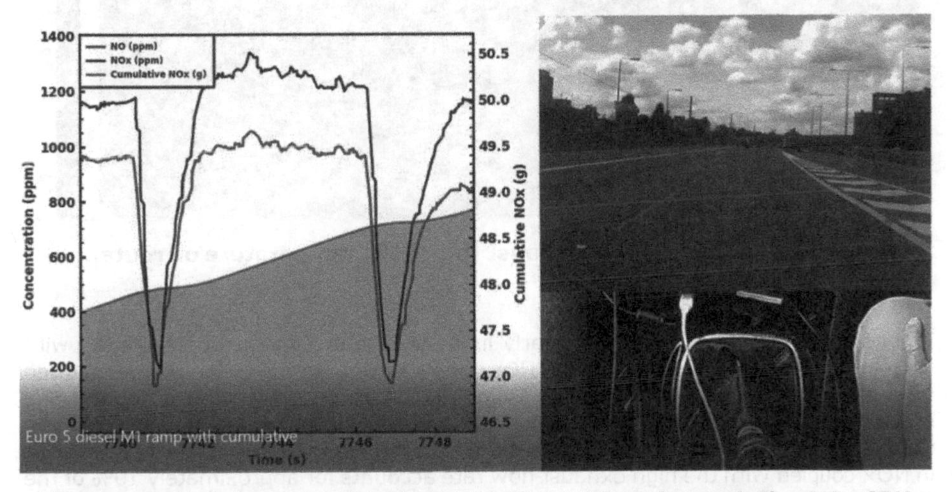


**Figure 5. Instantaneous NO & NOx ppm (LH axis) and cumulative NOx (RH axis) during acceleration on motorway up-ramp.**

### 3.2 Euro 6d-TEMP and its exhaust temperature for SCR

For the Euro 6d-TEMP vehicle equipped with SCR, the exhaust temperature is key to good conversion efficiency and a map of exhaust temperature on the route is shown in Figure 6.

**Figure 6. Euro 6d-TEMP exhaust (pre-SCR) temperature on route.**

The congested parts of the drive clearly influence the exhaust temperature but with very little adverse impact on tailpipe NOx. However, if a hard acceleration is required immediately following a slow-moving (cold) section of driving, a short-duration spike of NOx is evident at the tailpipe and is especially stark during the acceleration to join the M1 motorway following congested driving immediately before. The large increase in NOx coupled with the high exhaust flow rate accounts for approximately 10% of the total cycle NOx during this single manoeuvre (as shown in Figure 7).

Inspection of the cumulative emissions figures shows the significant contribution to the whole drive NOx from the motorway up-ramp.

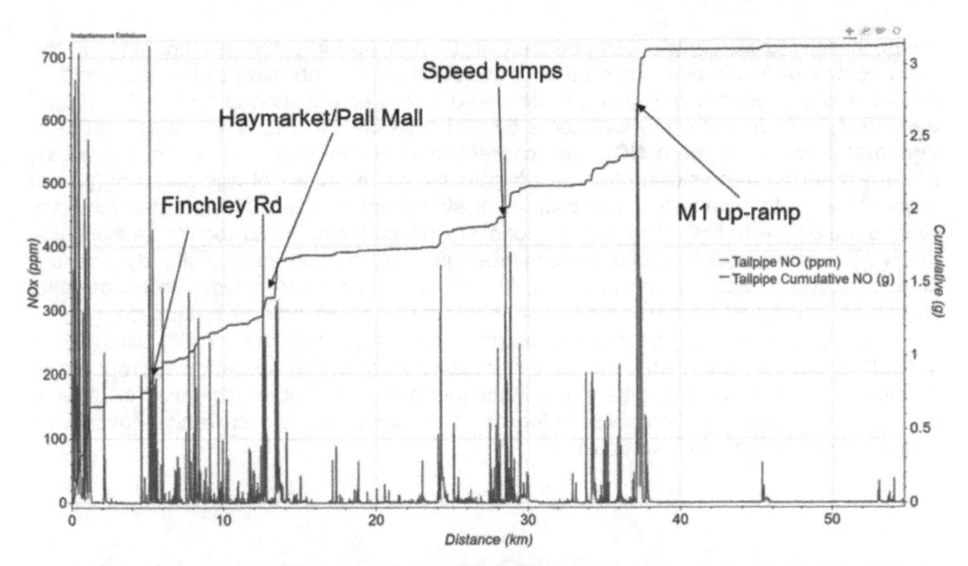


**Figure 7. Instantaneous and cumulative NOx emissions from Euro 6d-TEMP vehicles showing locations of major events.**

### 3.3 Transient NOx from PHEV

The cumulative NOx emissions from the PHEV vehicle are shown in Figure 8 and show that 70% of the total route's emissions are concentrated in to only 3 areas: Swiss Cottage (congested and lane-changing traffic interchange), Montpelier Park (speed bumps) and Neasden (more speed bumps).

The PHEV vehicle's TWCs require careful control to ensure balanced reduction and oxidation of all pollutants. This parameter was recorded and correlation between lean combustion and tailpipe NOx was clear as shown in Figure 9.

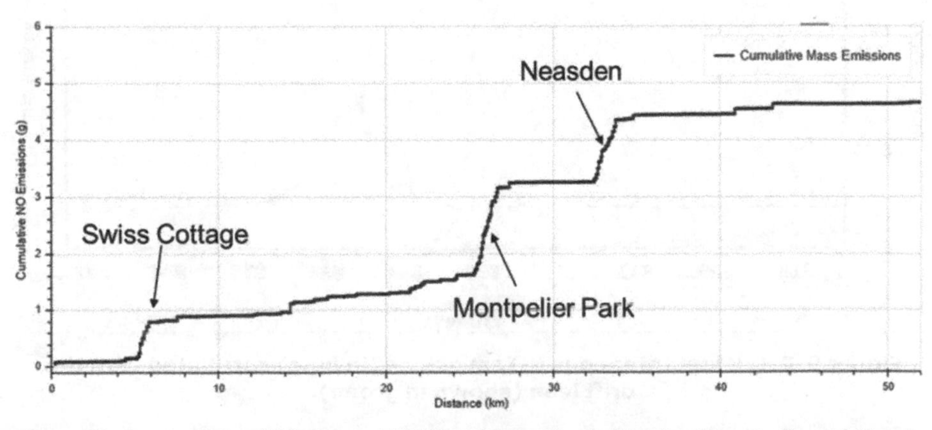


**Figure 8. Cumulative NOx from PHEV vehicle; both Montpelier Park and Neasden are sites of speed bumps.**

Given that the LHEV vehicle was powered with a port-fuel-injected gasoline engine, the rapid control of AFR is challenging due to the difficulty in controlling the intake port fuel puddle and subsequent fuel quantity delivered to the combustion chamber during fast transients; a problem largely overcome by GDI systems. For the PHEV vehicle, 60% of the total cumulative route NOx was concentrated within only 3 geographical areas, thereby meriting some closer study. During the transit of Swiss Cottage gyratory (partial roundabout traffic-controlled interchange) a significant number of rapid accelerations caused the engine to restart under high load. On inspection of the lambda trace alongside the NOx emissions reveals that the engine begins combustion initially slightly rich, but after a second or two, the combustion drifts lean with immediate adverse consequential effects on tailpipe NOx (see Figure 9). This effect is well-known in the industry (as the "PHEV cough") and although it is appreciated that the tight control of AFR using PFI is difficult, it appears clear that the post-start control from rich operation asymptotic to lambda=1 would have been likely to bring significant NOx abatement benefits. Indeed, the same effect appeared responsible for the other high emitting occurrences (over West London speed bumps, for example).

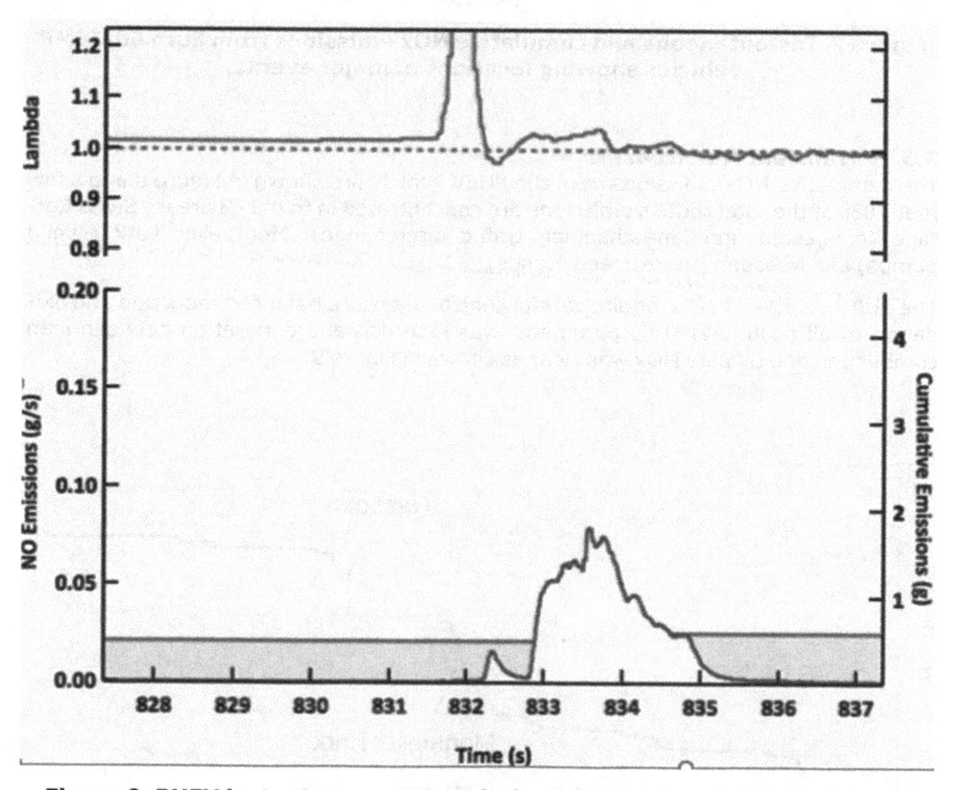


**Figure 9. PHEV instantaneous NO emissions (in blue) correlated with the drift lean (shown in green).**

The authors acknowledge that the vehicle need only be calibrated for the conditions inherent within the NEDC but it is also thereby a clear advantage of RDE testing that, within limits, these transients can then be identified and acknowledged.

The availability of GPS data alongside the real time tailpipe NOx allows, for each vehicle, some pollution position data which might be of use to air quality modellers of pollution dispersion. For example, the correlation of clouds of high NOx with the position of speed bumps indicating where the PHEV re-starts while accelerating away from speed bumps is clear in Figure 10.

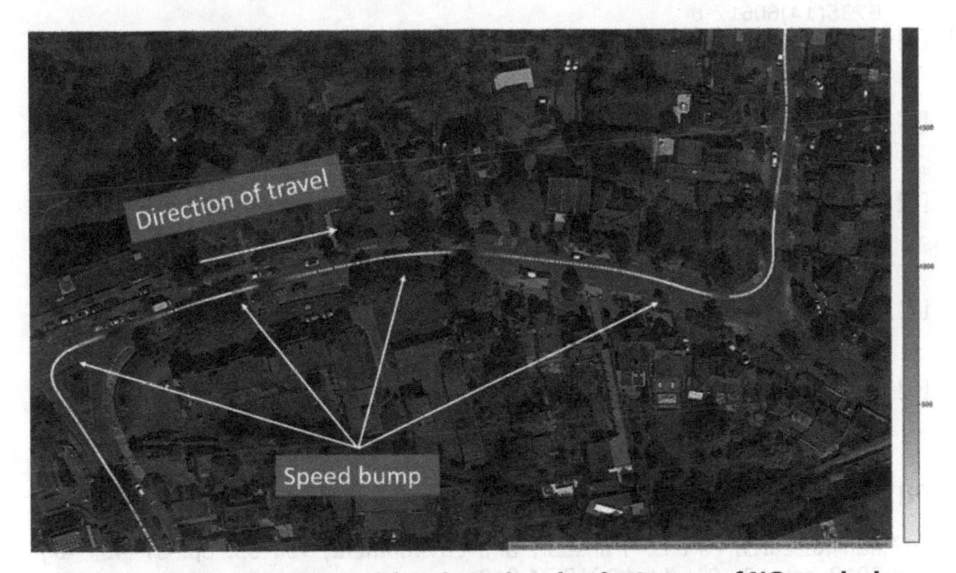


**Figure 10. Accurate GPS position data showing instances of NOx emissions correlated with individual speed bumps.**

## 4 CONCLUSIONS

The NOx emissions from Euro 5, Euro 6 and Euro 6d-TEMP vehicles vary greatly in their causes which are largely provoked by transient driving behaviour, but often as a consequence of road layout and imposed driving conditions. A 95% reduction in total NOx emissions from the Euro 5 to Euro 6d-TEMP vehicles on the route suggests an effective improvement in air quality is likely to be evident as a natural consequence of vehicle renewal and updating. The inclusion of hybrid and SCR systems in the Euro 6 vehicles are very effective but the NOx emissions that remain tend to be concentrated in very short duration events where the engine and aftertreatment control is temporarily undermined by driving conditions invoked by road geography.

## ABBREVIATIONS

DPF Diesel Particulate Filter

SCR Selective Catalytic Reduction

PFI Port Fuel Injected

## REFERENCES


[1] "Air quality in Europe - Publications Office of the EU." https://op.europa.eu/en/publication-detail/-/publication/7d42ac97-faca-11e9-8c1f-01aa75ed71a1/language-en (accessed May 25, 2021).

[2] D. Loomis et al., "The carcinogenicity of outdoor air pollution," The Lancet Oncology, vol. 14, no. 13, pp. 1262–1263, Dec. 2013, doi:10.1016/S1470-2045(13)70487-X.

[3] M. Guarnieri and J. R. Balmes, "Outdoor air pollution and asthma," The Lancet, vol. 383, no. 9928. Elsevier B.V., pp. 1581–1592, 2014. doi:10.1016/S0140-6736(14)60617-6.

[4] G. Hoek et al., "Long-term air pollution exposure and cardio-respiratory mortality: A review," Environmental Health: A Global Access Science Source, vol. 12, no. 1. BioMed Central, pp. 1–16, May 28, 2013. doi: 10.1186/1476-069X-12-43.

[5] L. Bai et al., "Exposure to ambient ultrafine particles and nitrogen dioxide and incident hypertension and diabetes," Epidemiology, vol. 29, no. 3, pp. 323–332, May 2018, doi: 10.1097/EDE.0000000000000798.

[6] V. Valverde and B. Giechaskiel, "Assessment of gaseous and particulate emissions of a euro 6d-temp diesel vehicle driven >1300 km including six diesel particulate filter regenerations," Atmosphere, vol. 11, no. 6, p. 645, Jun. 2020, doi: 10.3390/atmos11060645.

[7] "London Atmospheric Emissions (LAEI) 2016 - London Datastore ." https://data.london.gov.uk/dataset/london-atmospheric-emissions-inventory-laei-2016 (accessed May 25, 2021).

[8] S. Kwon, Y. Park, J. Park, J. Kim, K. H. Choi, and J. S. Cha, "Characteristics of on-road NOx emissions from Euro 6 light-duty diesel vehicles using a portable emissions measurement system," Science of the Total Environment, vol. 576, pp. 70–77, Jan. 2017, doi: 10.1016/j.scitotenv.2016.10.101.

[9] A. Ramos, J. Muñoz, F. Andrés, and O. Armas, "NOx emissions from diesel light duty vehicle tested under NEDC and real-word driving conditions," Transportation Research Part D: Transport and Environment, vol. 63, pp. 37–48, Aug. 2018, doi: 10.1016/j.trd.2018.04.018.

[10] G. Triantafyllopoulos, A. Dimaratos, L. Ntziachristos, Y. Bernard, J. Dornoff, and Z. Samaras, "A study on the CO 2 and NO x emissions performance of Euro 6 diesel vehicles under various chassis dynamometer and on-road conditions including latest regulatory provisions," Science of the Total Environment, vol. 666, pp. 337–346, May 2019, doi: 10.1016/j.scitotenv.2019.02.144.

[11] A. Dimaratos, Z. Toumasatos, S. Doulgeris, G. Triantafyllopoulos, A. Kontses, and Z. Samaras, "Assessment of CO2 and NOx Emissions of One Diesel and One Bi-Fuel Gasoline/CNG Euro 6 Vehicles During Real-World Driving and Laboratory Testing," Frontiers in Mechanical Engineering, vol. 5, p. 62, Dec. 2019, doi: 10.3389/fmech.2019.00062.

[12] N. Zacharof, U. Tietge, V. Franco, and P. Mock, "Type approval and real-world CO2 and NOx emissions from EU light commercial vehicles," Energy Policy, vol. 97, pp. 540–548, Oct. 2016, doi: 10.1016/j.enpol.2016.08.002.

[13] V. Valverde and B. Giechaskiel, "Assessment of gaseous and particulate emissions of a euro 6d-temp diesel vehicle driven >1300 km including six diesel particulate filter regenerations," Atmosphere, vol. 11, no. 6, p. 645, Jun. 2020, doi: 10.3390/atmos11060645.

[14] V. Valverde Morales, M. Clairotte, J. Pavlovic, B. Giechaskiel, and P. Bonnel, "On-Road Emissions of Euro 6d-TEMP Vehicles: Consequences of the Entry into Force of the RDE Regulation in Europe," in SAE Technical Papers, Sep. 2020, no. 2020. doi: 10.4271/2020-01-2219.

[15] R. A. Varella, M. v. Faria, P. Mendoza-Villafuerte, P. C. Baptista, L. Sousa, and G. O. Duarte, "Assessing the influence of boundary conditions, driving behavior

and data analysis methods on real driving CO2 and NOx emissions," Science of the Total Environment, vol. 658, pp. 879–894, Mar. 2019, doi: 10.1016/j.scitotenv.2018.12.053.

[16] J. Gallus, U. Kirchner, R. Vogt, and T. Benter, "Impact of driving style and road grade on gaseous exhaust emissions of passenger vehicles measured by a Portable Emission Measurement System (PEMS)," Transportation Research Part D: Transport and Environment, vol. 52, pp. 215–226, May 2017, doi: 10.1016/j.trd.2017.03.011.

[17] J. M. Luján, V. Bermúdez, V. Dolz, and J. Monsalve-Serrano, "An assessment of the real-world driving gaseous emissions from a Euro 6 light-duty diesel vehicle using a portable emissions measurement system (PEMS)," Atmospheric Environment, vol. 174, pp. 112–121, Feb. 2018, doi: 10.1016/j.atmosenv.2017.11.056.

[18] R. Suarez-Bertoa, M. Pechout, M. Vojtíšek, and C. Astorga, "Regulated and non-regulated emissions from euro 6 diesel, gasoline and CNG vehicles under real-world driving conditions," Atmosphere, vol. 11, no. 2, p. 204, Feb. 2020, doi: 10.3390/atmos11020204.

[19] R. García-Contreras et al., "Impact of regulated pollutant emissions of Euro 6d-Temp light-duty diesel vehicles under real driving conditions," Journal of Cleaner Production, vol. 286, p. 124927, Mar. 2021, doi: 10.1016/j.jclepro.2020.124927.

[20] R. García-Contreras et al., "Impact of regulated pollutant emissions of Euro 6d-Temp light-duty diesel vehicles under real driving conditions," Journal of Cleaner Production, vol. 286, p. 124927, Mar. 2021, doi: 10.1016/j.jclepro.2020.124927.

[21] P. Fernandes, R. Tomás, E. Ferreira, B. Bahmankhah, and M. C. Coelho, "Driving aggressiveness in hybrid electric vehicles: Assessing the impact of driving volatility on emission rates," Applied Energy, vol. 284, p. 116250, Feb. 2021, doi: 10.1016/j.apenergy.2020.116250.

[22] Y. Wang et al., "Fuel consumption and emission performance from light-duty conventional/hybrid-electric vehicles over different cycles and real driving tests," Fuel, vol. 278, p. 118340, Oct. 2020, doi: 10.1016/j.fuel.2020.118340.

[23] H. C. Frey, X. Zheng, and J. Hu, "Variability in measured real-world operational energy use and emission rates of a plug-in hybrid electric vehicle," Energies, vol. 13, no. 5, p. 1140, Mar. 2020, doi: 10.3390/en13051140.

[24] Y. Wang et al., "Fuel consumption and emission performance from light-duty conventional/hybrid-electric vehicles over different cycles and real driving tests," Fuel, vol. 278, p. 118340, Oct. 2020, doi: 10.1016/j.fuel.2020.118340.

[25] R. Suarez-Bertoa et al., "Effect of Low Ambient Temperature on Emissions and Electric Range of Plug-In Hybrid Electric Vehicles," ACS Omega, vol. 4, no. 2, pp. 3159–3168, Feb. 2019, doi: 10.1021/acsomega.8b02459.

[26] M. Peckham, H. Bradley, M. Duckhouse, M. Hammond, B. Mason, E. Winward and Z Yang, "Transient Emissions Measurement from a PHEV Vehicle During RDE Testing," Proceedings of the Fuel Systems Engines Conference 2018, ISBN: 9781510883482.

# The increasing importance of particles, volatile organic compounds, nitrous oxide, and ammonia in future real-world emissions regulation

**N. Molden**

Emissions Analytics Ltd, UK

## ABSTRACT

- Now that the immediate consequences of Dieselgate have been address through Real Driving Emissions, now is the time to reassess whether regulatory focus is in the correct place?
- Post-Euro-6 emissions regulation should not become an unnecessary burden, but rather should focus on the emerging environment threats.
- Beyond widening the measurement of particle number, what focus should be put on the composition of particles, and compounds carried on the surface of particles?
- Ammonia emissions from the tailpipe are a part of this, in so far as they lead to secondary particle formation of a particular chemical make-up.
- Nitrous oxide, a potent greenhouse gas, emitted at the tailpipe, may need to be regulated to avoid undermining carbon dioxide reductions, but evidence currently is limited.
- Volatile organic compounds are of interest from several angles: vehicle interior air quality and the off-gassing from materials; tailpipe speciation of hydrocarbons including formaldehyde; and off-gassing from tyres.
- Overall, a holistic view of pollutant emissions, carbon dioxide and fuel efficiency are needed.
- Emissions Analytics runs a large independent test programme, covering hundreds of vehicles across three continents each year.
- Data is analysed an available in a unique database that is accessed by a governments, industry, and others.
- Emissions Analytics' EQUA database contains a large amount of test data covering passenger cars, light commercial, heavy commercial and off-road.

To ensure environmental and health improvements, and to minimise unnecessary burdens on industry and consumers, it is necessary for regulation to be well focused and constructed. Therefore, new regulation should not just regulate what is easy measure or what has been addressed in the past. Rather, it is important constantly to reassess what new pollutants are growing in significance.

Emissions Analytics' presentation looks at several pollutant sources that may need to be considered, supported by data from its independent, real-world EQUA test programme. The innovation will be not just from the on-road measurement techniques, but the methods that allow comparative results between different vehicles.

DOI: 10.1201/9781003219217-13

## 1 INTRODUCTION

The downwards trend in regulated tailpipe pollutant emissions is marked in both the United States and Europe. The latest Real Driving Emissions (RDE) regulation in Europe appears in most respects to have brought down emissions of nitrogen oxides ($NO_x$), which were previously more than quadruple the certified limits in real-world driving. With particle filters on all modern diesel and gasoline direct injection vehicles in Europe, particle emissions are also generally well controlled. Combined with three-way catalysts, this has had the effect that, of the regulated pollutants, the dominant tailpipe emissions may now be just carbon dioxide ($CO_2$).

However, does this mean that the impact of vehicles on the environment is diminishing, or are there new pollutants emerging? This paper is intended as a review of additional, unregulated, pollutants that may prove an environmental risk. The issue is important to ensure that regulatory structures are continually verifying that improvements are real, and not just an artefact of existing regulations; and that what is being measured is not just being measured because it has been historically or because it is relatively easy to do so.

In short, the objective of this paper is to consider a more holistic view of the environmental impact of motor vehicles.

The paper adds to existing knowledge by looking at range of new pollutants in real-world conditions. Currently, little is published on ammonia ($NH_3$), nitrogen dioxide ($NO_2$) and speciated volatile organic compounds (VOCs) at the tailpipe. In-cabin air quality issues, in terms of particles and $CO_2$ build-up have been little-considered historically. Non-exhaust emissions have been the subject of some research, for example from the industry's Tire Industry Project, but a focus on the wear levels in real-world operation together with the chemical composition of the tyres are relatively new areas.

This paper is structured in four sections. First, it reviews the latest independent data from the EQUA Index test programme on regulated pollutants, to quantify the progress that has been made (1). Second, new data on real-world $NO_2$ and $NH_3$ will be presented, as these pollutants are most likely to be included in the new Euro 7 regulation being discussed, and potentially in a Euro 6e stage. Third, it will look beyond Euro 7 at a more detailed speciation of tailpipe VOCs. Fourth and finally, the paper takes a wider view beyond the tailpipe to present data on vehicle interior air quality and emissions ingress, and non-exhaust emissions from tyre wear.

## 2 EQUIPMENT SETUP

A range of measurement technologies were used to gather the data quoted in this paper. A common factor was that data was collected from light-duty vehicles operated in real-world conditions outside of the laboratory and on the public highway.

For regulated tailpipe emissions, Portable Emissions Measurement Systems (PEMS) were used, principally the SEMTECH-LDV from Sensors, Inc of Michigan, USA (http://www.sensors-inc.com). This equipment was also used to measure $NO_2$.

Ammonia emissions were measured on diesel vehicles using ECM's NH3CAN Module (http://www.ecm-co.com), which was integrated into the standard PEMS set-up by Emissions Analytics.

Tailpipe VOC speciation was measured using a Selected Ion Flow Tube Mass Spectrometry (SIFT-MS) from Anatune of the UK (http://www.anatune.co.uk) and built by the Christchurch, New Zealand-based company Syft Technologies. This equipment is not portable and therefore testing was conducted on stationary vehicles.

These vehicle interior air quality measurements comprise particle number (PN) using a Condensing Particle Counter (CPC) and $CO_2$ using a Non-dispersive Infra--red detector (NDIR) from National Air Quality Testing Services (NAQTS) of the UK (http://www.naqts.com). The lower particle size cut-off was approximately at 15nm.

Emissions Analytics has developed measurement techniques using the NAQTS system to analyse real vehicle interior air quality generically called PIMS (Pollution In-Cabin Measurement System). By taking measurements from two PIMS units, one placed inside the vehicle and the other outside, the relationship between the ambient environment and the interior can be measured.

Tyre wear emissions, both number and mass, were measured using an ELPI+ from Dekati of Finland (http://www.dekati.fi). Sampling was taken using a collection 'scoop' positioned behind one of the rear tyres, with a low-loss sample line leading to the ELPI+.

The chemical composition of tyres was measured using a Two-dimensional Gas Chromatograph with Time-of-Flight Mass Spectrometry (GCxGC-TOF-MS) from Markes International of the UK (http://www.markes.com). Samples were taken manually from tyres, heated in a microchamber and the gases captured on thermal desorption tubes prior to analysis.

## 3 METHODOLOGY

PEMS testing was conducted, in terms of equipment installation and operation, according to the best practice guidelines from the EU's Joint Research Centre.

Equipment calibration followed either regulatory requirements, where relevant, or equipment manufacturer recommendation.

The test route, dynamic conditions and vehicle conditioning followed the EQUA Index methodology employed by Emissions Analytics. It is approximately a four-hour test cycle comprising urban, rural, motorway and extended elements. Further details can be found at http://www.emissionsanalytics.com.

PIMS testing was conducted according to the SAE paper entitled *Development of a Standard Testing Method for Vehicle Cabin Air Quality Index* (https://saemobilus.sae.org/content/02-12-02-0012/).

## 4 ANALYSIS & DISCUSSION

### 4.1 Currently regulated pollutants

Considering regulated pollutants first, the following data is from the EQUA Index test programme on European vehicles, showing the average levels of $NO_x$, PN and $CO_2$ emissions.

Table 1 splits the $NO_x$ results between pre- and post-RDE homologated vehicles. Pre-RDE covers Euro 6 vehicles between sub-stages 6a to 6c, whereas post-RDE covers 6d-temp onwards. Roughly, the division falls in 2018.

**Table 1. $NO_x$ emissions pre- and post-RDE.**

| $NO_x$, mg/km | Pre-RDE | Post-RDE | Euro 6 limit | Post-RDE Conformity Factor |
|---|---|---|---|---|
| | | | | |
| Diesel | 347 | 49 | 80 | 0.61 |
| Diesel Hybrid | 53 | 33 | 80 | 0.42 |
| Diesel Plug-in Hybrid | n/a | 4 | 80 | 0.05 |
| Gasoline | 32 | 11 | 60 | 0.18 |
| Gasoline Hybrid | 7 | 3 | 60 | 0.05 |
| Gasoline Plug-in Hybrid | 27 | 2 | 60 | 0.03 |

It shows that $NO_x$ was poorly controlled for pre-RDE diesels, with an average exceedance factor of 4.3. Other powertrains were compliant. The effect of the introduction of RDE has been to reduce the diesel conformity factor to 0.6 – emissions well below the regulated limit, not taking into account the official "Conformity Factors" in the regulation that started at 2.1 times.

PN regulation was only introduced with Euro 6, validated on the road with RDE. As shown in Table 2, emissions were within the regulated limit even before RDE, although the conformity factors have reduced significantly with RDE. It should be noted that the PN regulation does not apply to non-direct-injection gasoline vehicles. Approximately a half of the gasoline vehicles (hybrid and non-hybrid) have direct fuel injection, all of which have gasoline particle filters installed.

**Table 2. PN emissions pre- and post-RDE.**

| PN, # x 10^11 | Pre-RDE | Post-RDE | Euro 6 limit | Post-RDE Conformity Factor |
|---|---|---|---|---|
| | | | | |
| Diesel | 0.640 | 0.106 | 6.000 | 0.02 |
| Diesel Hybrid | 1.117 | 0.021 | 6.000 | 0.00 |
| Diesel Plug-in Hybrid | n/a | 0.380 | 6.000 | 0.06 |
| Gasoline | 1.693 | 0.974 | 6.000 | 0.16 |
| Gasoline Hybrid | 0.267 | 0.630 | 6.000 | 0.11 |
| Gasoline Plug-in Hybrid | 4.943 | 0.000 | 6.000 | 0.00 |

Table 3 shows the equivalent $CO_2$ results. These emissions are not subject to the RDE regulation, but rather the fleet average targets. By 2021, the average fleet emissions target is 95g/km, and 130g/km prior to then. The table shows that real-world emissions are well above these limits, with gasoline hybrids getting the closest.

**Table 3. $CO_2$ emissions pre- and post-RDE.**

| $CO_2$, g/km | Pre-RDE | Post-RDE |
|---|---|---|
| | | |
| Diesel | 177 | 163 |
| Diesel Hybrid | 171 | 163 |
| Diesel Plug-in Hybrid | n/a | 167 |
| Gasoline | 198 | 180 |
| Gasoline Hybrid | 138 | 122 |
| Gasoline Plug-in Hybrid | 186 | 131 |

Therefore, considering these regulated pollutants, it can be seen that $NO_x$ and PN are now well-controlled. Although not presented here, this is also generally accepted as the case for carbon monoxide and total hydrocarbons as well. The remaining problem is $CO_2$, hence the push for electrification – this will not be discussed in more detail in this paper.

## 4.2 New pollutants to be regulated at Euro 7

Moving on to the second part of the paper, the discussions around future European regulation (dubbed "Euro 7") are considering adding new pollutants including $NH_3$ and $NO_2$. The latter is already included as part of $NO_x$, but may well become targeted explicitly.

Table 4 shows the real-world $NO_2$ results. Although there is no current regulatory standard, a limit of 20mg/km has been suggested by some (2). In this case, there would be general compliance with current vehicles.

**Table 4. $NO_2$ emissions pre- and post-RDE.**

| $NO_2$, mg/km | Pre-RDE | Post-RDE |
|---|---|---|
| | | |
| Diesel | 120 | 10 |
| Diesel Hybrid | 41 | 6 |
| Diesel Plug-in Hybrid | n/a | 2 |
| Gasoline | 3 | 1 |
| Gasoline Hybrid | 1 | 0 |
| Gasoline Plug-in Hybrid | 6 | 0 |

Table 5 shows some early real-world $NH_3$ emissions data, covering a diesel passenger car, and a fully-laden and unladen diesel van. A potential limit of 10mg/km has been suggested (2). If this were to be enacted, it would suggest that current vehicles would struggle for compliance, but the exceedance margin is relatively small. Therefore, relatively modest engineering improvements may be enough to achieve consistent compliance.

**Table 5. $NH_3$ emissions.**

| $NH_3$, mg/km | Kia Sportage | Citroen Relay – unladen | Citroen Relay – laden |
|---|---|---|---|
| | | | |
| Urban | 8.92 | 30.63 | 31.63 |
| Rural | 4.47 | 22.73 | 10.20 |
| Motorway | 4.85 | 2.62 | 1.55 |
| **Combined** | **6.08** | **18.66** | **14.46** |

Therefore, these additional pollutants may not be difficult to control at the future regulatory stages.

### 4.3 Beyond Euro 7 – additional pollutants

Going deeper, the third area considers what else may be in tailpipe emissions, and specifically in terms of species of VOCs that may be particularly noxious. While selected VOCs have been regulated in the US and Europe in the past, with enhanced measurement capability and understanding of the health effects of these compounds, there is an argument that a much wider range of compounds should now be considered.

Working with Anatune in the UK, a SIFT-MS was used to sample the tailpipes of a 2011 model year diesel Volkswagen Golf, a 2019 gasoline Peugeot 2008 and a 2019 diesel Renault Captur.

The cars were soaked in a controlled environment prior to testing. During the first 100 seconds the probe, positioned in the tailpipe, measured the prevailing ambient air, followed by 100-400 seconds ignition and idle; 400-800 seconds at 1,500rpm; 800-1,100 seconds at 3,000rpm; and finally back to idle for the last 100 seconds.

The analysis was conducted for hydrocarbons, sulphurs and oxygenates. Hydrocarbons are the product of combustion and include butadiene, heptane, styrene, benzene, hexane, toluene, butane, methane and xylenes/ethylbenzene. Concentrations were derived using the flow rate of the instrument.

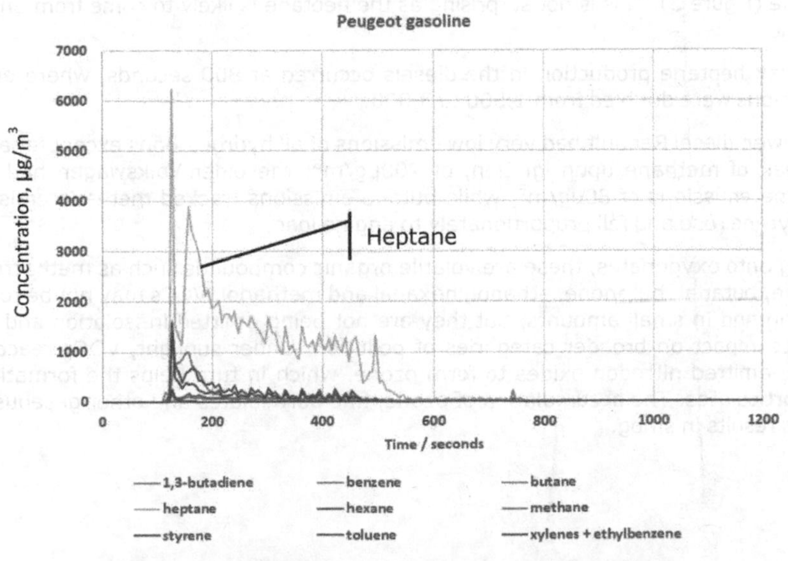


**Figure 1. Hydrocarbons from Peugeot gasoline.**

The notable result of the test was the initial spike in heptane as well as hydrocarbons methane, styrene, toluene and xylenes/ethylbenzene, observed in the gasoline vehicle tested, the Peugeot, as shown in Figure 2.

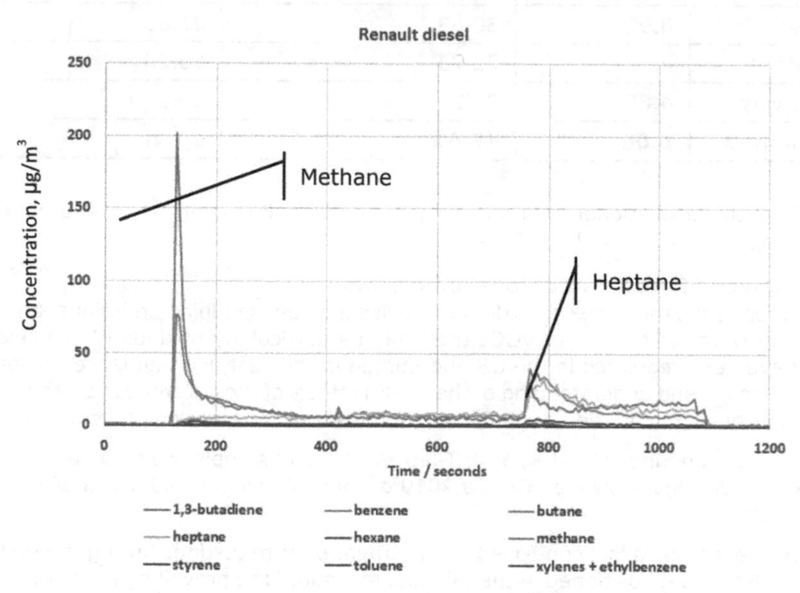


**Figure 2. Hydrocarbons from Renault diesel.**

In particular, the spike in heptane in the first few seconds after ignition reached a concentration of over 6,000µg/m$^3$. This was sixty times more than the highest reading for the older, diesel Volkswagen, while the Renault never produced more than 25µg of heptane (Figure 3). This is not surprising as the heptane is likely to come from unburnt gasoline.

The peak heptane production in the diesels occurred at 800 seconds, where engine revolutions were doubled from 1,500 to 3,000.

The newer, diesel Renault had very low emissions of all hydrocarbons except for an initial peak of methane upon ignition, of 200µg/m$^3$; the older Volkswagen had peak methane emissions of 80µg/m$^3$, while butane emissions tracked methane emissions and styrene rose and fell proportionately to engine load.

Moving onto oxygenates, these are volatile organic compounds such as methacrolein, acetone, butanal, butanone, ethanol, hexanal and methanol. VOCs may not be toxic in isolation and in small amounts, but they are not being emitted in isolation and have a direct impact on broader categories of pollution. Under sunlight, VOCs react with vehicle-emitted nitrogen oxides to form ozone, which in turn helps the formation of fine particulates. The accumulation of ozone, fine particulates and other gaseous pollutants results in smog.

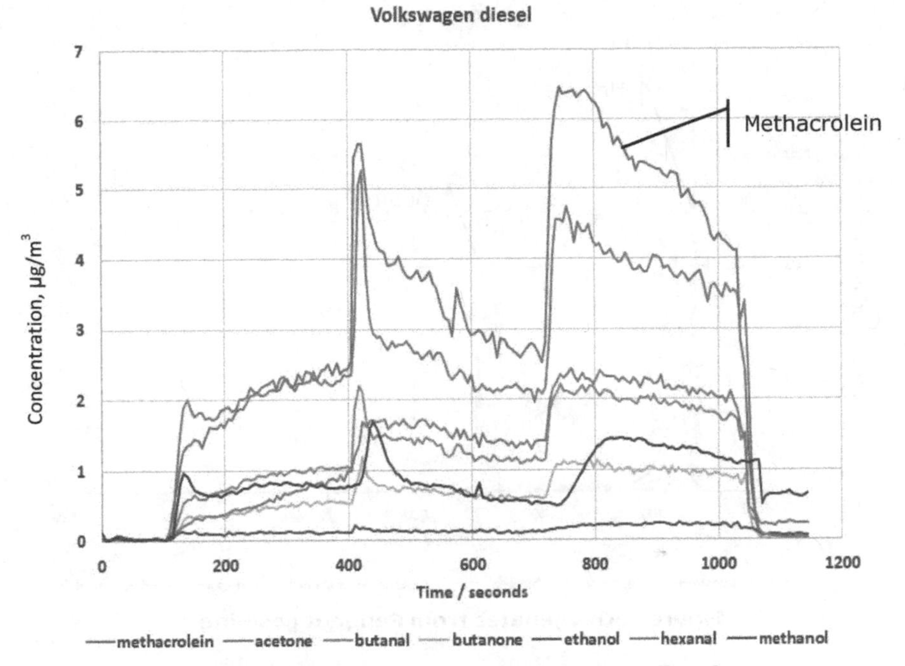


**Figure 3. Oxygenates from Volkswagen diesel.**

The Volkswagen produced methacrolein at over 6μg/m$^3$ when the engine was stepped up to 3,000rpm, as shown in Figure 5. Exposure to methacrolein is highly irritating to the eyes, nose, throat and lungs. The Volkswagen also produced just below 5μg/m$^3$ of acetone, less toxic than methacrolein but causing irritation to eyes and throat.

The Renault also produced noticeable amounts of acetone, peaking at almost 6μg/m$^3$, but never breached 3μg/m$^3$ for methacrolein.

The gasoline Peugeot offered a completely different pattern, as shown in Figure 6. Mirroring the hydrocarbon results, it showed a spike on first ignition in hexanal to nearly 140μg/m$^3$, and between 400 and 500 seconds two spikes in butanal of 80 and 50μg/m$^3$ respectively. Butanal (N-butyraldehyde) is an organic compound which is the aldehyde derivative of butane. It is judged to be of low toxicity to humans unless inhaled at high concentrations, causing chronic headaches and ataxia.

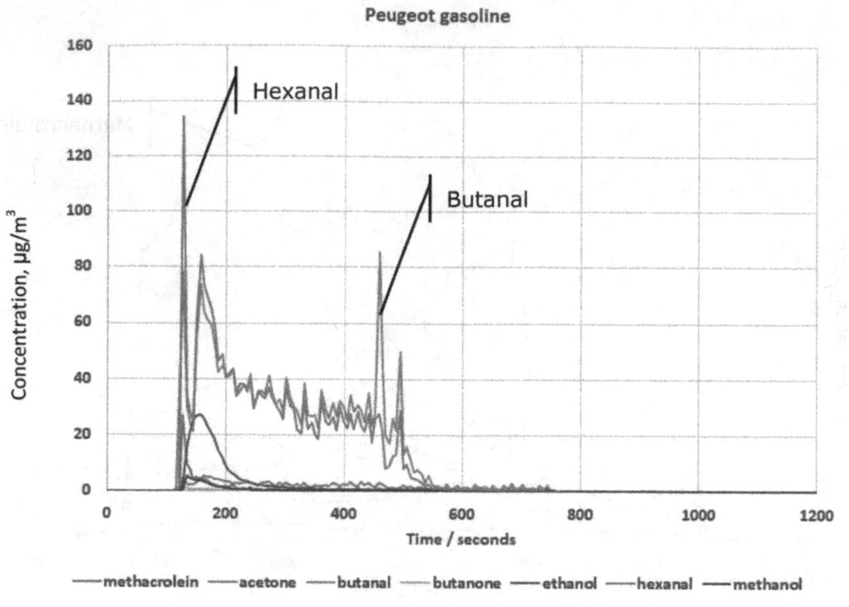


**Figure 4. Oxygenates from Peugeot gasoline.**

In summary, elevated emissions are seen primarily during cold start on the gasoline vehicle and during transient load events on both types of powertrain. The spike at the step-up in rpm represents the time gap between more fuel being sent to combust and control mechanisms responding, upon which a new equilibrium is reached.

## 4.4 Beyond the tailpipe – vehicle interior and non-exhaust pollutants

The fourth and final section moves beyond the tailpipe and considers non-exhaust emissions and health exposures.

The regulation of vehicle emissions has been concentrated historically on the tailpipe; vehicles are also regulated for other aspects such as safety and construction materials. A neglected area has the been the exposure of the occupants of vehicles to pollutants generated by the vehicle and from the outside. This exposure is largely governed by the effectiveness of the heating, ventilation and air conditioning system (HVAC).

There are three areas of exposure that are considered in this paper: PN infiltration into the cabin, $CO_2$ build-up during air recirculation mode of the HVAC due to human respiration, and off-gassing of VOCs from the cabin materials.

For PN and $CO_2$ there are no existing regulations, although the European CEN Workshop 103 is working on standardising measurement methods (3). Emissions Analytics has tested 97 different recent model year US light-duty vehicles across a wide range of manufacturers. Typical odometer readings were between 1,000 and 5,000 miles: the HVAC filters were therefore relatively new but not brand new. Results are shown in Figure 7.

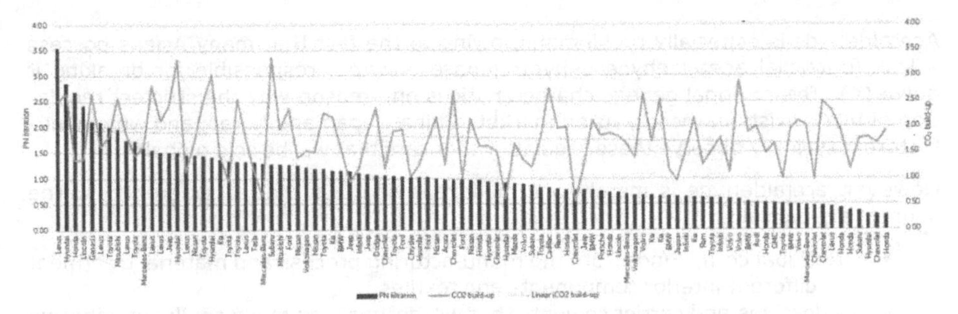


**Figure 5. PN infiltration and $CO_2$ build-up rates.**

Each test involved approximately one hour of driving on the public highway in Los Angeles, split between fresh air and recirculation modes. Measurements were taken simultaneously inside and outside the vehicle and the PN infiltration rate is the integrated ratio of inside to out, as described in the methodology section above. A value of one means the average PN concentration outside is equal to inside over the cycle on fresh air mode. A value of two means inside was twice as polluted as outside. For $CO_2$, only interior measurements are considered, with the average concentration being compared to the initial concentration. A value of two, for example, means the average is double the initial concentration.

Figure 7 shows that PN infiltration rates range from around 0.3 to over 3.5. For $CO_2$ build-up, the range is from around 1 to almost 3.5. There is no simple correlation between PN and $CO_2$. However, what this strongly shows is a wide variance in performance in both dimensions between different vehicles. This will be a function of the type of filter membrane together with the design and calibration of the HVAC system. Variances in performance directly affect the human exposures to PN and the cognitive effects of $CO_2$ levels, which impact upon driver safety.

Turning to the off-gassing of VOCs in the cabin, there are existing standards, but regulation is currently light. In the broad area there are existing ISO and SAE standards, and an active United Nations Economic Commission for Europe (UNECE) working group. Some countries have national standards, in particular Japan, Korea, China and Russia. There are 97 VOCs listed as hazardous air pollutants in Title III of the Clean Air Act Amendments of 1990. Overall, the arc of regulation is at an early stage and covers a limited number of pollutants.

What might be colloquially and informally referred to as 'new car smell' has typically been ignored, partly because it has been difficult to measure. Recent advances in instrumentation now allow the measurement of not only total, time-weighted average VOCs, but it can now distinguish between different species of VOCs in real time.

The subject has particular resonance in Asia. In China, 11.2% of buyers complained about the odours they found in their new cars, according to the 2019 JD Power China Initial Quality Study.

Car interiors, comprising dozens of separate materials ranging from natural textiles to synthetic polymers and adhesives, emit a wide range of VOCs, among them acetaldehyde. Symptoms that customers have cited range from sore eyes to nausea and headaches, and aggravated respiratory conditions.

Acetaldehyde is especially problematic, owing to the fact that many Asians possess a less functional acetaldehyde dehydrogenase enzyme, responsible for breaking it down (4). This regional genetic characteristic is one reason why the strictest regulation of VOCs exists in the key Asian markets China, Japan and Korea, and why manufacturers typically observe these regulations for cars that will be sold globally.

However, acetaldehyde is merely one of dozens of VOCs that a car produces. The sources are typically:

- Residual compounds from the manufacturing process and material treatment of different interior components and textiles
- Adhesives and carrier solvents that will de-gas – as much as 2kg of adhesive can be found in a modern car, much higher than in the past where mechanical riveting and bolting was more common
- Degradation of cabin materials over the longer term as a result of oxidation, ultra-violet light and heat.

The following figure, Table 6, sets out the regulated limits in key Asian countries, in micrograms per cubic metre, and the potential human symptoms from exposure.

**Table 6. Eight VOCs to be regulated.**

| Analyte | China | Japan | Korea | Symptoms |
|---|---|---|---|---|
| | | | | |
| Formaldehyde | 100 | 100 | 100 | Respiratory irritant and a contributory factor in asthmas and cancer |
| Acetaldehyde | 50 | 48 | No limit | Causes 'flush reaction' among some populations – itchiness or blotchiness of the skin and a flushed complexion |
| Acrolein | 50 | No limit | No limit | Highly toxic and severely irritating to the eyes, mucous membranes, respiratory tract, and skin |
| Benzene | 110 | No limit | 30 | Known carcinogen and declared as such by the US Environmental Protection Agency |
| Ethylbenzene | 1500 | 3800 | 1600 | Can cause throat irritation, and dizziness at higher concentrations |
| Xylene | 1500 | 870 | 870 | Causes headaches, dizziness, drowsiness, and nausea |
| Styrene | 260 | 220 | 300 | Causes headaches |
| Toluene | 1100 | 260 | 1000 | Solvent familiar as nail-polish remover, can cause headaches and nausea |
| Tetradecane | No limit | 330 | No limit | Irritating to the eyes, mucous membranes, and upper respiratory tract, and can cause skin irritation. |

In this testing, it was possible to identify a range of different VOCs and also quantify their mass using SIFT-MS, a type of direct mass spectrometry that uses precisely controlled soft ionisation to enable real-time, quantitative analysis of VOCs in air, typically at detection limits of parts-per-trillion level by volume (pptv).

A one-year-old gasoline Hyundai i10 was tested every 15 minutes for 60 seconds over five hours on an early summer's day, where temperatures rose to 20 degrees Celsius (68 degrees Fahrenheit). The measured concentrations were expressed as the mean across the 60-second duration of the sample. For the final 15-minute vent cycle, the car windows were opened, the car started and the air conditioning run at full power. The SIFT-MS then sampled continuously using the above conditions for the full 15 minutes. The sample point was at head height in the middle of the front seats.

The two principal outcomes of the test concern the steady accumulations of ten VOCs as temperatures rose, and the unexpected dynamic of emissions during the final fifteen minutes.

Most noticeably, the common solvents methanol and acetone rose from very low base points (18 and 12µg/m$^3$) to 935 and 576µg/m$^3$ respectively, as shown in Figure 8. The 52-fold rise in methanol is noteworthy. While it is a very common solvent and not directly regulated, it is toxic and could be an irritant at these levels.

The only exception to these across-the-board rises was benzene, which fell from 17 to 15µg/m$^3$. However, this is where the final fifteen minutes revealed unexpected results.

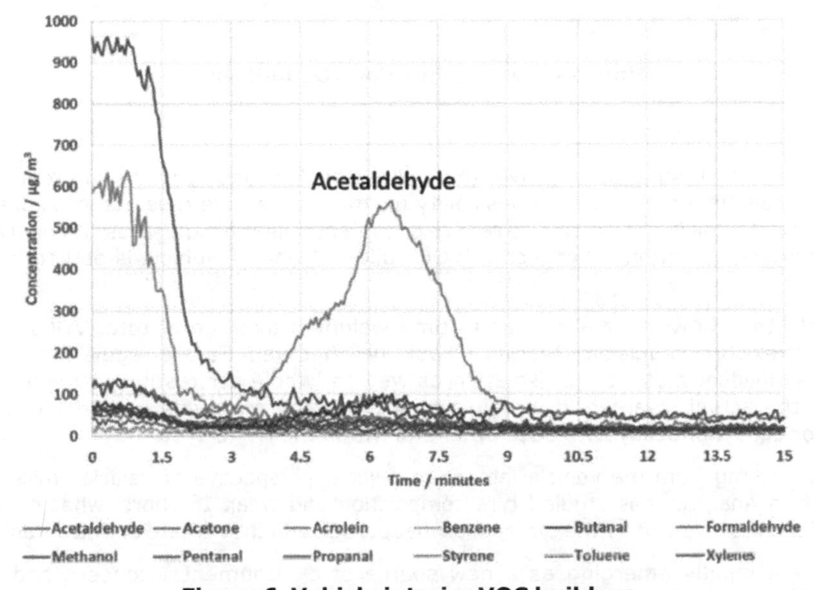


**Figure 6. Vehicle interior VOC build-up.**

Despite windows being open and the air conditioning turned on, some VOCs such as acetaldehyde rose steeply during the fourth to sixth minutes, as shown in Figure 9. During this phase acetaldehyde concentrations rose from an initial base of approximately 50 to 550µg/m$^3$, more than ten times the regulated limit in China and Japan.

A hypothesis for this is that the car's HVAC system may form a type of 'sink' for some VOCs. When the venting or AC are activated, the sink is flushed out into the cabin causing a pronounced spike. Three other analytes that rose at the same time frame included styrene, toluene and benzene, but by a much smaller degree.

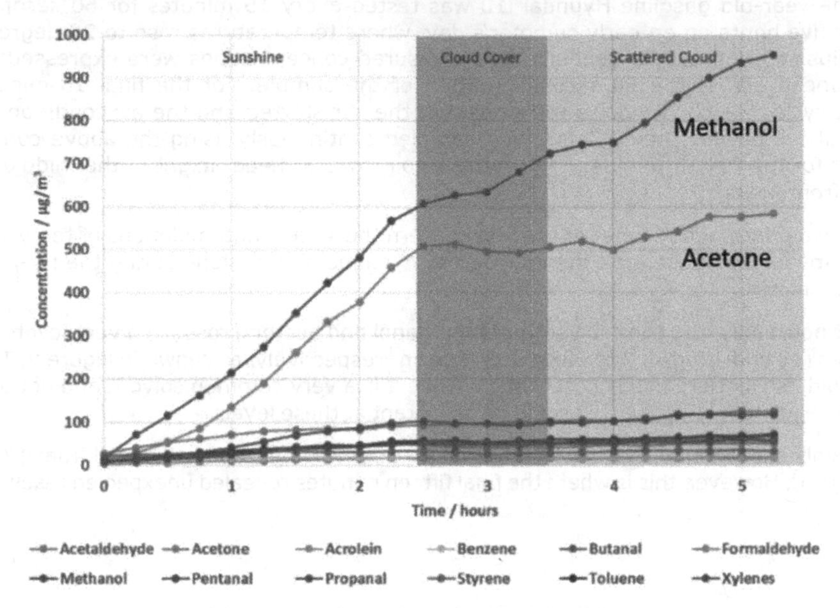


**Figure 7. Vehicle interior VOC build-up.**

From a vehicle testing perspective, the ability to detect and speciate different analytes in real time opens up the possibility for more extensive research of exposure and the potential for regulation to reduce detrimental health exposures. It could also assist driver education in respect of 'VOC build-up' when a vehicle is parked in hot weather.

Overall, this shows that a four-hour, time-weighted average of total VOCs – the basis of existing regulatory testing – could be improved. Future regulations could cover individual materials in isolation as well as 'whole car testing', by which we mean the actual, real-world way in which the many materials comprising a car interior act dynamically with each other and within the HVAC system.

Finally, moving from the vehicle interior to a wider perspective of vehicle emissions, Emissions Analytics has studied tyre composition and wear. In short: what noxious chemicals are given off by the tyres, and what particles do they create during abrasion?

Tyres are rapidly emerging as a new source of environmental concern and this will affect the automotive industry. In a recent BBC documentary, it was claimed that the world will have discarded three billion tyres in 2019, enough to fill a large football stadium 130 times (5). Beyond this broad issue of resource use and material waste, tyres impact on both air quality and microplastics in the marine environment.

Even when not in use, a tyre will give off a range of VOCs, especially when it is new. When in use, the surface of a tyre may reach towards 100 degrees Celsius, which will increase this off-gassing. Using a microchamber and a GCxGC-TOF-MS system from Markes International, it was possible to perform a detailed speciation of these VOCs on a relatively new, premium brand tyre, as shown in the two-dimensional chromatogram shown in Figure 10.

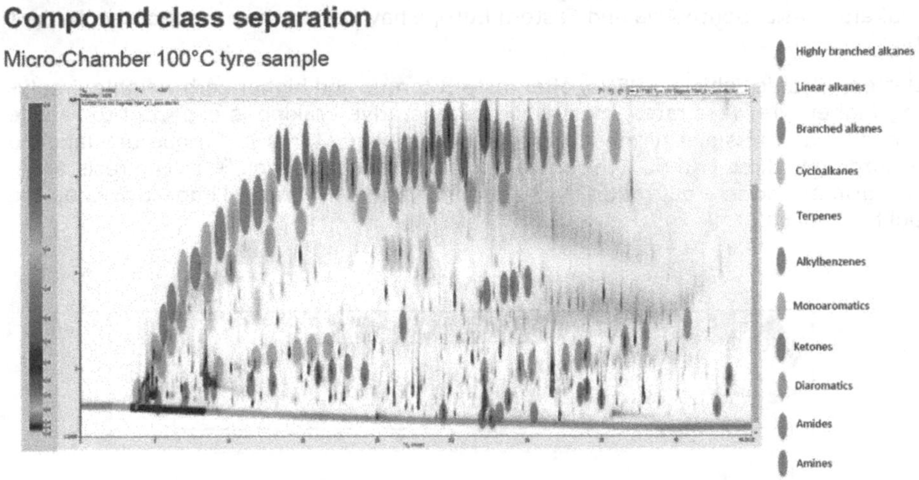

**Figure 8. VOC speciation from tyre off-gassing.**

The results show the large number of different species of VOCs present in the tyre material. These include aromatic hydrocarbons, benzenes, alkanes and ketones, each of which has health effects from causing minor irritation to cancer, and some of which are listed in Table 6. This information then puts the issue of tyre wear emissions into relief.

Over a lifetime of between 20-50,000km, a tyre will shed approximately 10-30% of its tread rubber into the environment, at least 1-2kgs (6). The wear factor (defined as the total amount of material lost per kilometre) varies significantly depending on tyre characteristics such as size – radius, width and depth – tread depth, construction, pressure and temperature. In one recent Emissions Analytics' test, conducted under real-world rather than laboratory conditions, the four tyres on a standard hatchback lost 1.8kg over just 300km of fast road speeds, far in excess of what had been anticipated.

A tyre abrades owing to the friction between its contact patch and the road surface. It 'emits' particles across a broad size spectrum, from coarse to fine to ultrafine to nanoscale. It may also emit other forms of aromatics such as benzopyrene and benzofluorene, the result of the incomplete combustion of organic matter resulting in evaporation of the volatile content of the tyres, which the EU has regulated to a degree (7).

Coarse particles typically fall rapidly to the ground. At the fine level and smaller, they are airborne for a certain duration, either being blown away from the carriageway before settling on the ground or falling to the carriageway where re-suspension may take place as other vehicles pass.

Particle dispersion and deposition eventually occurs, and the particles typically pass into the watershed through street drainage and are estimated to be a primary source of as much as 28% of microplastics found in the marine environment (8).

A typical car tyre comprises 45% oil-derived synthetic rubber (polymer), 40% oil-derived carbon-black (filler, 40%), and 15% various additives to aid production processes, some of which contain heavy metals and some of which are also oil-derived. Some tyres contain natural rubber, but as a minority component.

Tyres have become more disposable as their price has fallen in real terms, replacing an older tradition of re-treading carcasses for extended life. New entrant tyre-makers in Asia, South Asia and Eastern Europe have led to the advent of the 'budget tyre'.

Battery Electric Vehicles (BEVs) offer instant torque and higher kerb weights, implying higher tyre wear rates, even while regenerative braking is expected to reduce brake wear emissions. From a regulatory viewpoint, tyres in Europe are labelled according to three criteria, (the so-called 'performance triangle'): rolling resistance, wet grip and noise – but that may change as tyre environmental impact rises up the political agenda.

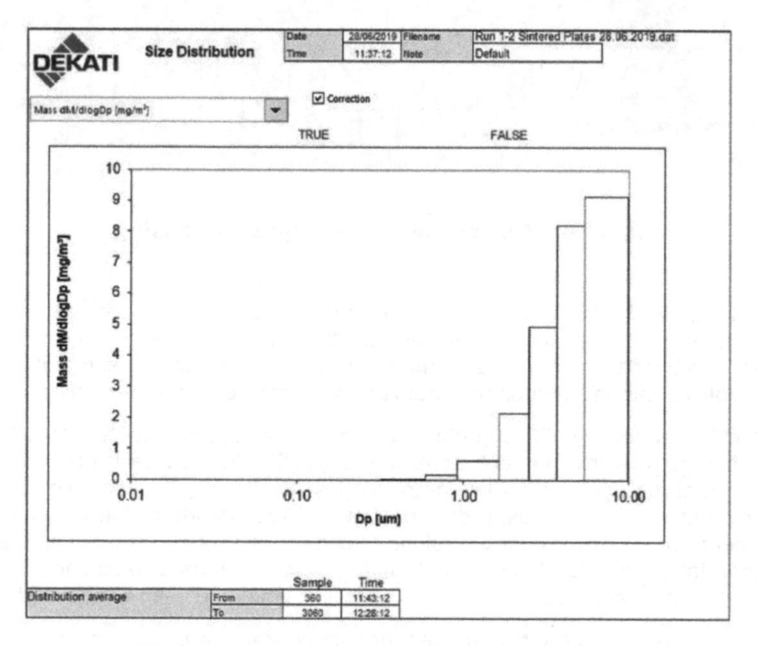


**Figure 9. Tyre particle mass distribution.**

In one recent test, Emissions Analytics used a Dekati ELPI+ to measure both particle mass and number at 1Hz. Figure 9 above shows the resulting mass distribution. It only corroborates what is broadly known, that a comparatively small number of coarser particles (up to and including the 10µm size shown in the far right column, familiar as PM10) account for most of the recorded mass.

Switching to the particle number distribution, as shown in Figure 10, there is a mirror of the first graph, with a tiny amount of mass expressed as a very high number of nanoscale particles right down to the 10nm level expressed in the first column (0.01µm).

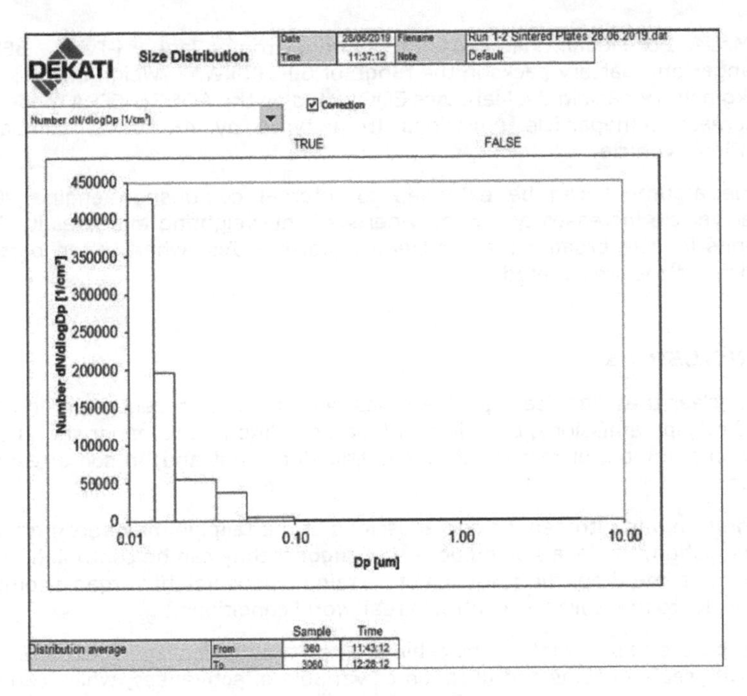


**Figure 10. Tyre particle number distribution.**

This is important because until now this high particle number count has typically either not been measurable or not been measured, owing to a regulatory preoccupation with mass and a lack of suitably sensitive real-time measurement instrumentation.

Regarding public health, there is growing concern among epidemiologists and other medical researchers that ultra-fine particles are potentially injurious to human health as a function of their number not just their mass, owing to their ability to translocate to the bloodstream through the lungs (9).

In a recent UK government report, *Non-Exhaust Emissions from Road Traffic*, authored by the UK Government's Air Quality Expert Group (AQEG), it recommended "as an immediate priority that non-exhaust emissions (NEEs) are recognised as a source of ambient concentrations of airborne PM, even for vehicles with zero exhaust emissions of particles."

This is particularly relevant as vehicles become heavier – even if they are zero emission at the tailpipe. The same UK government report noted that non-exhaust emissions are believed to constitute today the majority source of primary particulate matter from road transport: 60% of PM2.5 and 73% of PM10. While regenerative braking is expected to reduce brake wear emissions, the increased weight and torque characteristics of alternative drivetrains such as BEVs will likely be associated with increased tyre wear.

In the same report it is suggested that a 10kg increase in vehicle mass accounts for a 0.8-1.8% increase in nanoparticle emissions from tyres. This is particularly relevant as a whole generation of new BEVs is hitting the roads with considerably larger and heavier battery packs than in the past.

A Tesla Model S or Model X, Mercedes EQC, Audi e-tron or Jaguar i-Pace – BEVs with larger ranges and battery packs in the range of 60-100kWh – weigh 2.3-2.6 tonnes. The 600kg battery pack in the Mercedes EQC would on the AQEG/DEFRA model potentially increase nanoparticle emissions from tyres by 48-108%, compared to a conventional vehicle.

The same argument can be extended to internal combustion engine vehicles. A heavier vehicle increases tyre wear, whereas light-weighting mitigates it. This has implications for the broader market trend towards SUVs, where often particularly large rim tyre sizes are adopted.

## 5 CONCLUSIONS

While it is clear that significant progress has been made in recent years in reducing regulated tailpipe emissions, even in real-world conditions, this paper shows that the environmental impact of motor vehicles is still significant and, in some respects, is increasing.

While ammonia and nitrogen dioxide emissions at the tailpipe may become the focus of new regulation, the levels being observed suggest they can be controlled relatively easily. At the same time, the presence of certain noxious volatile organic compounds may be harder to measure and control in real-world conditions.

Human exposure to pollutants in the cabin is a new area of concern due to the combination of well-sealed cabins and filtration of variable effectiveness, which can lead to higher particle concentrations inside than outside of the vehicle.

Tyre wear emissions are of increasing concern as vehicles become heavier, both for the rate of wear but also the chemical composition of the material. The typical emissions contain both fine particles and coarser, which have consequences for both air and marine environments.

All these elements should help to inform and direct future surveillance monitoring and regulatory development. They should enable monitoring of this wider range of pollutants in real-world conditions.

## REFERENCES

[1] https://www.emissionsanalytics.com/equa-databases.
[2] Ricardo, "New emissions limits: the challenges and solutions for Euro VII and the US," https://mobex.io/webinars/euro-vii-new-emissions-limits-the-challenges-and-solutions/, accessed December 2020.
[3] European Committee for Standardization, "CEN/WS 103 – Real drive test method for collecting vehicle in-cabin pollutant data," https://www.cen.eu/news/workshops/Pages/WS-2019-017.aspx, accessed December 2020.
[4] Eng, M., Luczak, S., and Wall, T., "ALDH2, ADH1B, and ADH1C Genotypes in Asians: A Literature Review," Alcohol Research & Health, https://www.ncbi.nlm.nih.gov/pmc/articles/PMC3860439/, 2007.
[5] BBC Radio 4, "Costing the Earth: Tread Lightly," https://www.bbc.co.uk/programmes/m00035qh, 13 March 2019.
[6] Grigoratos, T., and Martini, G., "Non-exhaust traffic related emissions. Brake and tyre wear PM," JRC Science of Policy Reports, https://publications.jrc.ec.europa.eu/repository/bitstream/JRC89231/jrc89231-online%20final%20version%202.pdf, 2014.

[7] Since 2010 the EU has required the discontinuation of the use of extender oils which contain more than 1mg/kg Benzo(a)pyrene, or more than 10mg/kg of the sum of all listed polycyclic aromatic hydrocarbons in the manufacture procedure due to increased health concerns related to PAHs. Cited in European Chemicals Agency, "ANNEX XVII TO REACH – Conditions of restriction," https://echa.europa.eu/documents/10162/176064a8-0896-4124-87e1-75cdf2008d59, accessed December 2020.
[8] Microplastics are considered to be all plastic particles in the range of 0.1–5,000μm. A secondary source is when a larger plastic object breaks down once already in a marine environment. The figure of 28.3% was originally cited in Boucher, J., Friot, D., "Primary Microplastics in the Oceans: a Global Evaluation of Sources," *International Union for Conservation of Nature and Natural Resources*, 2017, http://dx.doi.org/10.2305/IUCN.CH.2017.01.en.
[9] C. Terzano, F. Di Stefano, V. Conti, E. Graziani, A. Petroianni, "Air pollution ultrafine particles: toxicity beyond the lung," *European Review for Medical and Pharmacological Sciences* 2010; 14: 809–821.

# *Session 6: Real-world Driving Emission (RDE) and emissions control systems*

# RDE Plus: Accurate replication of real-world vehicle testing within the chassis dynamometer laboratory using HORIBA RDE+ solutions

**Alex Mason[1]*, Richard Mumby[1], Phil Roberts[1], Yosuke Kondo[2], Yoji Komatsu[2], John Morgan[3], Steve Whelan[1]**
[1]HORIBA MIRA Ltd, UK
[2]HORIBA Ltd, JPN
[3]HORIBA UK Ltd, UK

## ABSTRACT

As legislative and market pressures increase on conventionally powered vehicles, new development tools will be needed to meet the demands and to deal with increasing systems complexity. To that end, this paper introduces a new technique for the replication of real-world driving in the laboratory environment. As part of HORIBA's RDE+ suite of development tools, HORIBA Torque Matching (HTM) improves on existing "road load simulation" techniques by being able to achieve very close replication of tailpipe emissions (with very low test-to-test variation), both in terms of mass and temporal evolution. Additionally, it can do so without the requirement for performing vehicle coast down testing or surveying road grades along test routes, in order to build an accurate, simulated road-load. Instead the exact road-load is generated implicitly by the technique.

Discussion of how the replication technique is performed is followed by demonstration using both a conventional ICE vehicle and a hybrid vehicle. Thus demonstrating the applicability of the HTM approach, not only to existing vehicles, but also to electrification technologies such as hybrid and BEVs.

Firstly, convectional vehicle replications were performed at sea-level and at high altitude (using altitude simulation equipment). $CO_2$ emissions replication is shown to be 2% or less of the respective road test and critical emissions like $NO_x$ shown to be within 5%. Finally the technique is applied to a hybrid vehicle, demonstrating not only very close emissions replication ($CO_2$, $NO_x$ < 2%) but also accurate replication of the hybrid systems including energy regeneration from braking, battery state of charge and critical system temperatures.

**Keywords:** RDE, RDE+, Chassis Test, real-world Driving, Replication, Hybrid, High Altitude

## 1 INTRODUCTION

In the next decade, vehicle manufacturers face tougher development challenges as tailpipe emissions from the ICE are subject to higher levels of scrutiny and stricter legislation. These emissions (alongside other unregulated emissions from tyre and brake wear) contribute to poor air quality metrics, especially in urban areas, where

*Corresponding author
DOI: 10.1201/9781003219217-14

the pollutants $NO_x$ and PM pose the most serious harms to health. According to the European Environment Agency, over 400,000 premature deaths in Europe can be linked to poor air quality [1]. Some estimate that there are more than 11,000 deaths per annum caused by excess $NO_x$ emissions from diesel vehicles alone in Europe [2]. Furthermore, many countries exceed the annual concentration for PM2.5 and $NO_x$ as suggested by the WHO. These excesses are often found in urban areas and areas with high vehicular traffic.

In the automotive industry, light duty emissions standards have been in place for over 25 years. With the introduction of each new regulation stage, allowable tailpipe emissions (on a per kilometer basis) have been gradually reduced. Whilst regulatory emissions tests were based on the NEDC, manufacturers were seen to be reducing tailpipe

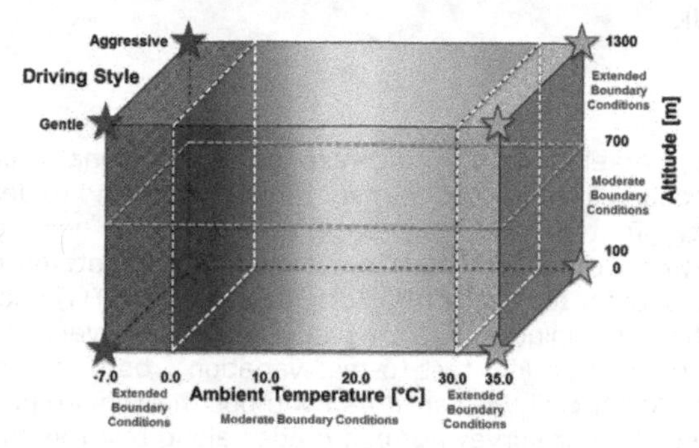


**Figure 1. Moderate and extended boundary conditions for current Euro 6d regulations.**

emissions in-line with the regulations. Unfortunately, real-world data has shown this not to be the case. Between Euro 2 (1996) and Euro 5 (in place until 2014) emissions standards, it is shown that there was virtually no real-world improvements in tailpipe $NO_x$ emissions [2,3]. Additionally, world wide consumer trends toward larger vehicles such as the SUV has worked against improvements in tailpipe emissions and fuel consumption [4].

The introduction of Euro 6b emissions regulations saw a step change in reducing allowable emissions from vehicles. Overtime, this regulation has been extended to include the more rigorous WLTP lab test (Euro 6c) and accompanying RDE test (Euro 6d-Temp/Euro 6d). Both elements are more representative of real-world driving, with the latter being undertaken outside the often favourable laboratory environment. These improvements to testing protocols have translated to an improved correlation between certified emissions and real-world performance. However, tailpipe out emissions can still be greater than the corresponding emissions limits when the test conditions stray away from the regulatory test boundaries.

There are numerous RDE boundary conditions scoped to cover 95% of real-world driving and environmental scenarios that could occur (some are shown in Figure 1). In addition, the vehicle is exercised to a much greater extent during RDE drives compared to laboratory test drive cycles, again aimed at representing real-world driving (Figure 2). However, since some scenarios involving various combinations of boundary conditions occur outside the 95th percentile, significantly lower than 95% of real-world driving scenarios would be covered by the RDE regulations [5].

As a direct consequence of the "Diesel-gate" scandal, numerous bodies have been active in identifying and testing common, real-world driving scenarios that are not

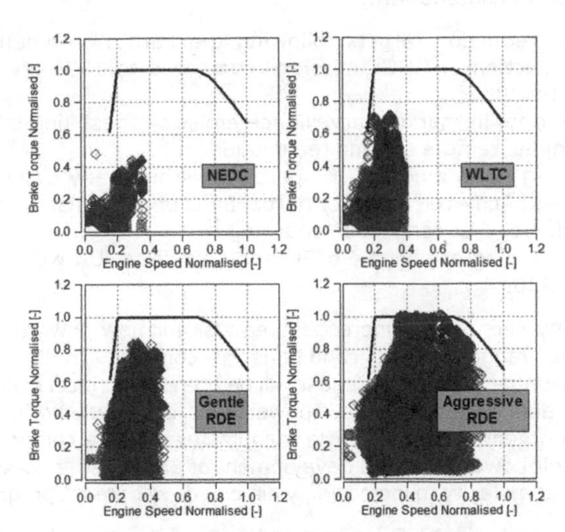


**Figure 2. NEDC,WLTC and typical gentle and aggressive RDE cycle residencies.**

covered by the standard RDE testing protocol. These include extended high-speed driving, heavily laden family vehicles, extra cold/extra hot soak periods, extended cold-start idling, snow and ice driving, towing and very hilly locales. In general, these studies, amongst others in [6–9] have been able to show that Euro 6 and even some of the latest Euro 6d TEMP vehicles can exceed $NO_x$ and PN limits when tested under such conditions. In particular earlier Euro 6 vehicles certified under the older NEDC testing can be seen exceeding limits by orders of magnitude.

Activity such as described above is pushing legislators world-wide to bring forward more stringent rules on tailpipe emissions. In Europe, the current Euro 6d standard coming into force at the start of 2021 includes In Service Conformity (ISC) testing up to 100,000 km, alongside a provision to measure evaporative emissions. In addition, RDE conformity factors for $NO_x$ emissions are reduced from 2.1 to 1.43[10].

Around 2025, new Euro 7 regulations will replace current RDE testing with "wide on-road testing". In brief, Euro 7 will effectively class almost any kind of driving over a 5 km distance as a valid real-world test [11]. Additionally:

- Temperature conditions will extend to cover driving in -10°C to 40°C conditions (including extended soak time at extremes of temperature).
- ISC will extend to cover 240,000 km of driving.
- Weighting adjustment of cold start emissions removed.
- Additional criteria emissions are likely to be moni- tored including: $NH_3$, $CH_4$, $N_2O$ and $NO_2$.
- PN monitoring will consider particle sizes down to 10 nm from the current 23 nm.

Other markets are not standing still either, with the 2021 US Tier 3/CARB LEV III posting the most stringent limits and China set to overtake Europe when China 6b is introduced in 2023. India and Brazil had also planned to adopt at least Euro 6 by the end of

2020[12]. In addition to more stringent emissions regulations, there will be further complications to navigate in some markets. For example, in the European Union manufacturers will also have to contend with:

- Fleet $CO_2$ reduction targets following the Paris Agreement (COP21) with a 40% reduction of GHG emissions compared to 1990 levels by 2030 and net- zero by 2050.
- Improvements in market surveillance emissions test- ing by use of "impossible to cheat" remote sensing techniques [13,14].
- The banning or limiting of ICE operation within many city centres or dense urban areas. Achieved by full or partial bans of particular types of vehicle and introduction of CAZ congestion charging [15].
- Outright banning of the sale of ICE-only vehicles. UK will ban such sales as early as 2030.

This level of scrutiny sets out an unprecedented task for new powertrain development. With such a wide range of test conditions to consider, traditional development approaches will no longer be suitable, and it will be therefore critical from a time and cost perspective to be able to evaluate new powertrains over real-world scenarios within a laboratory environment. Further, vehicle manufacturers have commenced aggressive electrification of their powertrains and development of BEV where necessary. All of which add time, cost and complexity to increasingly difficult development programs.

Of note, suggested by the AGVES [11] work on the Euro 7 approach, is the potential for replicating real-world driving tests within the lab in order to measure emissions that cannot be measured with sufficient accuracy by PEMS. This also applies for replicating real driving tests at extremes of temperature. This then puts onus on test labs and suppliers of automotive development tools to improve the capability of their offerings in order to meet this demand.

To address this situation HORIBA has been developing a whole vehicle and powertrain development, calibration and verification approach known as RDE+. This will connect real-world testing with chassis dynamometer, EiL and virtual test tools (Figure 3). This will allow manufacturers to front- load vehicle and powertrain development and deploy real- world scenarios much earlier in the development cycle, enabling high confidence that they will meet legislative requirements in their target markets. Aspects of the RDE+ offering have been introduced in detail previously [16–19]. This paper will expand on the work relating to replication of real- world driving within the chassis dynamometer lab (RDE+ Chassis, as per Figure 3.

In the previous work [16,19], chassis dynamometer based RDE replication covered a modified RLS type test with a robot driver. Termed RLR, the method used the dynamometer gradient function and significant robot driver controller tuning as a means of matching measured road- loads from previously conducted real-world testing. It also introduced an improved replication method, now called HORIBA Torque Matching (HTM), that gives the ability to re-create exactly, a real-world drive as it occurred on the road. The HTM methodology will be discussed in more detail in the following section. This paper focuses further on the HTM method and intends to demonstrate how the replication stage of technique can be applied with altitude simulation and extremes of temperature. Additionally it will show the technique remains valid when applied to hybrid- type vehicles.

## 2 METHODOLOGY

### 2.1 HORIBA torque matching process

The full HORIBA Torque Matching process comprises three main activities (shown in Figure 4). These are:

(1) **Road Test:** Perform any kind of road drive. Record vehicle velocity, pedal inputs and weather conditions.
(2) **Replication:** Chassis bench matches vehicle velocity, robot driver matches pedal inputs, MEDAS matches ambient conditions. Chassis bench produces and records **actual road-load**.
(3) **Emulation:** Chassis bench matches vehicle velocity, robot driver closed loop pedal control to match road- load recorded in (2).

As a minimum, road test data (or data output from simulation) will include the vehicle speed and the pedal inputs. Exact requirements are dependent on the type of vehicle. For example, a conventional ICE vehicle with manual gear box will require speed, accelerator and clutch pedal inputs and engine speed metrics (for gear ratio detection). A hybrid vehicle with automatic or CVT gearbox would not require engine speed, but would require brake pedal input in order to capture any regenerative effect if regenerative-friction brake blending is implemented. In some cases, a BEV with one-pedal driving would not need brake pedal monitoring as the accelerator pedal controls, entirely, the speed of the vehicle.

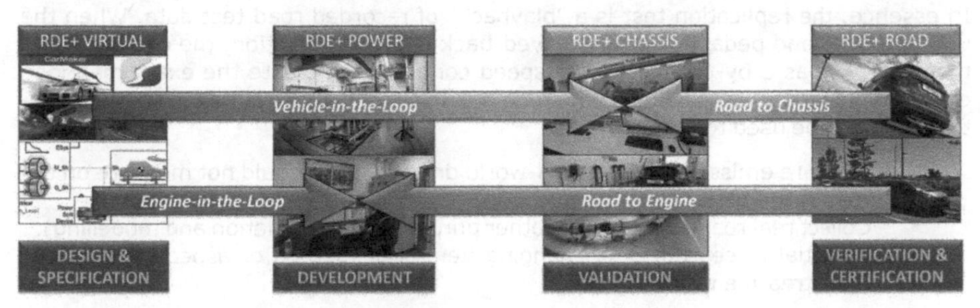


**Figure 3. HORIBA RDE+ development approach allows for bi-directional transition between four core areas.**

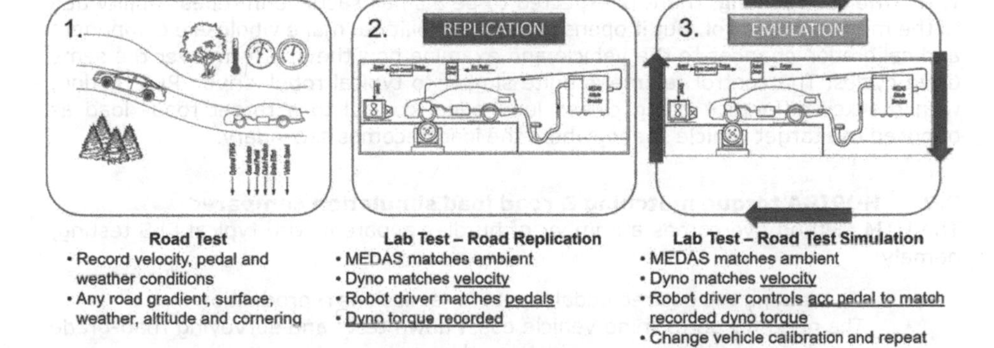


**Figure 4. HORIBA Torque Matching (HTM) method has three main stages; road testing, chassis replication and chassis emulation. Stages 2 and 3 can be used for different aspects of development.**

Coming into the lab there are a number of exercises to first perform before carrying out a replication. Mostly, these cover installation and teaching of the driving robot to the vehicle and mapping pedal positions to actuator positions. But it may also include modifications to items like the engine breathing system if altitude simulation is going to be used.

The replication test is orchestrated centrally from a test automation system (in this case, HORIBA STARS VETS). The following main actions are carried out during the test:

- Vehicle speed is communicated to HORIBA SPARC dynamometer controller.
- Pedal positions are communicated to driving robot (Stähle SAP2000, HORIBA ADS-Evo).
- Speed and pedal commands are synchronised with adjustment controls located within the test automation system.
- Ambient conditions are communicated to test chamber or altitude simulator if necessary.
- Emissions analysis equipment (exhaust gas analysers) is controlled.

In essence, the replication test is a "playback" of recorded road test data. When the vehicle speed and pedal inputs are played-back in synchronisation, the chassis dynamometer will, as a by-product of the speed controller, re-create the exact road load experienced by the vehicle during the road test in a highly repeatable fashion. Replication tests can be used to, for example:

- Measure emissions from a real-world drive that you could not measure on the road with PEMS.
- Collect real road-load data for other purposes (i.e. simulation and modelling).
- Potential to see effect of changing items like fuel, oil or aspects of exhaust after-treatment.

The emulation test is generally similar to the replication test with one key change:

- Pedal positions are now commanded by PID controller which performs a closed loop control on road-load force (target recorded during replication test, com- pared to feedback from the dynamometer controller).

With emulation testing, there is expected to be a small sacrifice in repeat-ability due to the method of control. But it opens up the possibility to make wholesale component and calibration changes to the vehicle and examine how they perform over the same drive cycles. The control regime is quite similar to typical robot-driven RLS testing, with the key difference being closed loop pedal control to a target road- load as opposed to a target vehicle speed, where the load becomes secondary.

## 2.2 HORIBA torque matching & road load simulation compared

The HTM method overcomes a number of hurdles apparent with typical RLS testing, namely:

- Re-creating the desired pedal inputs for emissions reproduction.
- The need for performing vehicle coast down tests and surveying road-grade in order to get the correct vehicle road load.
- Changes in load due to changes in road surface or environmental conditions such as wind

The first point is important if high quality replication of criteria emissions, particularly from a temporal perspective, is required. With the second and third points, HTM accounts for all of these by virtue of the way it works. The dynamometer matches the speed, the robot driver matches the pedal inputs and the result is that the dynamometer controller generates the exact road load experienced. So all these aspects are baked into the test.

It is certainly possible to achieve good cumulative mass emissions replication using the RLS method. But it will often be found that emissions are produced in different amounts and at different times during the test. This is particularly so for $NO_x$ and PN emissions which are more sensitive to pedal inputs. To demonstrate, HTM is compared with both robot- driven and human-driven RLS testing over the same driving route and with the same vehicle. (vehicle details located in supplementary material, Table 15).

Figure 1 shows mass emission results for HTM and the human and robot driven RLS methods. As shown, whilst HTM has generally performed the best, RLS is, in some instances like $CO_2$, able to replicate closely to the road test. However, examining the result by comparing how well a particular test correlates back to the original road test highlights the strength of HTM. This is shown in Table 2. Human/robot driven RLS and HTM were compared using linear fits ($y = m(x) + c$) back to the original road test. In

**Table 1. Mass emissions for HTM test compared with conventional RLS test (both human and robot driven) with the same vehicle over the same test route.**

| | $CO_2$ [g/km] | CO [g/km] | $NO_x$ [mg/km] | PN [#/km] |
|---|---|---|---|---|
| Road Test | 208.88 | 0.109 | 3.33 | 2.15E+11 |
| RLS (Human) | 205.44 | 0.055 | 6.40 | 4.30E+11 |
| RLS (Robot) | 195.71 | 0.067 | 4.41 | 3.82E+11 |
| HTM (Robot) | 214.02 | 0.105 | 3.53 | 3.96E+11 |
| Differences to Road Test | | | | |
| RLS (Human) | 2.46% | -4.17% | 5.96% | 84.15% |
| RLS (Robot) | -6.31% | -38.59% | 32.41% | 77.99% |
| HTM (Robot) | -1.65% | -49.31% | 91.87% | 100.26% |

**Table 2. RLS compared to HTM using linear fit characteristics. "M" is line fit coefficient and "$R^2$" is a measure of how closely correlated the respective lab test is to the original road test.**

| | Human Driver | | Robot Driver | | HTM | |
|---|---|---|---|---|---|---|
| Item | M | $R^2$ | M | $R^2$ | M | $R^2$ |
| CO | 0.47 | 0.33 | 0.66 | 0.25 | 0.77 | 0.39 |
| $CO_2$ | 1.03 | -0.29 | 1.01 | 0.56 | 1.00 | 0.99 |
| $NO_x$ | 1.13 | 0.14 | 1.09 | 0.48 | 1.12 | 0.89 |
| PN | 0.73 | -0.02 | 0.60 | -0.03 | 1.50 | 0.77 |
| Speed | 0.99 | 0.99 | 1.00 | 0.99 | 1.00 | 1.00 |
| Acc. Ped | 0.92 | 0.82 | 0.89 | 0.88 | 1.00 | 1.00 |
| Eng Torq | 0.92 | 0.83 | 0.91 | 0.89 | 1.05 | 0.99 |

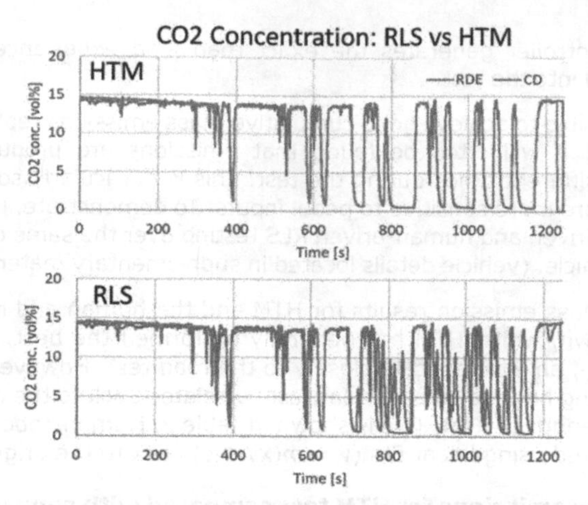


**Figure 5. $CO_2$ emissions rate for HTM (robot) compared with RLS (robot). Note production of emissions where none exist for road test.**

general HTM will provide fits with both linear fit coefficients and $R^2$ values close to or equal to 1. By contrast, RLS methods show much more variation with much poorer correlation on the emissions, even though the final mass results were similar. The emissions generation during the lab test is occurring at different times and by different amounts. Examples of this are shown in Figures 5 & 6.

Differences are caused, not only, by incorrect loading due to in-exact coast-down and altitude data (for gradient derivation) but also replication of pedal input (timing,

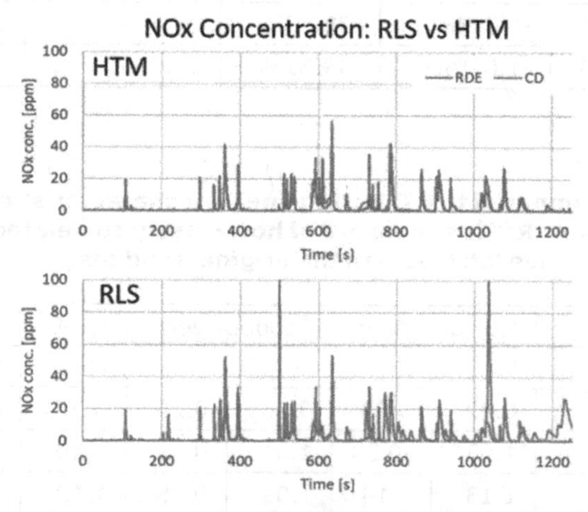


**Figure 6. $NO_x$ emissions rate for HTM (robot) compared with RLS (robot). Note production of emissions where none exist for road test.**

amount and feed rate). The authors have found during experimentation that robot drivers can be quite "nervous" with pedal inputs depending on how well PID control has been tuned [19]. On the human side, it differs driver to driver (which is already not ideal). Some are too smooth with pedal inputs, others react too late to faster changes in speed or gradient. An example of this is shown in Figure 7. They are both "reactionary" control methods controlling primarily to a speed target, rather than a load target.

### 2.3 Test vehicles

***Vehicle A:*** Vehicle A is a typical 5-door family car (C- segment) with small capacity turbo diesel engine and a manual six speed gearbox. Specifications can be found in Table 3.

**Table 3. Vehicle A Data.**

| Specification | Value |
|---|---|
| Vehicle Mass | 1500 kg |
| Engine Size | 1.5 Litre, 4 cyl |
| Engine Type | Turbo-Diesel |
| Engine Output | 88 kW, 270 Nm |
| Transmission | Manual, 6-speed |
| Aftertreatment | DOC, DPF, LNT |
| Emissions Cert | Euro 6b (NEDC) |

***Vehicle B:*** Vehicle B is a 5-door C-segment vehicle. It uses a highly efficient Atkinson-cycle combustion engine coupled to a hybrid drive system comprising two Motor- Generator (MG) units, planetary gearbox and a small high- voltage traction battery located in the rear. MG1 is used for turning over the ICE and also for regenerating energy into the battery pack. MG2 is used to drive the wheels in tandem with the engine. Specification can be found in Table 4.

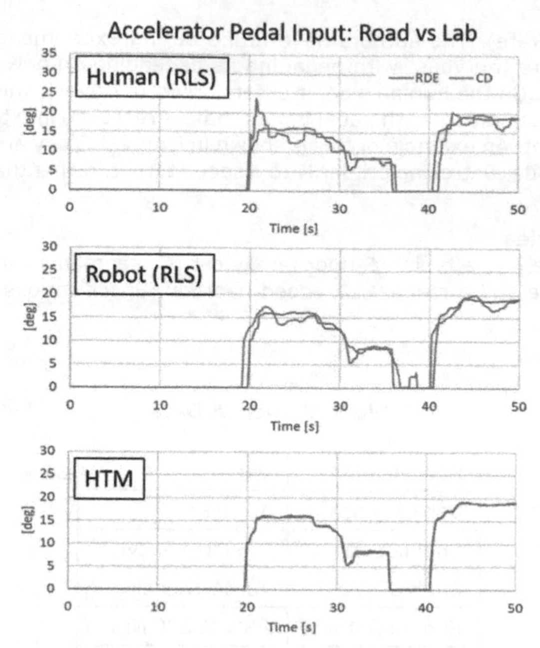


**Figure 7. Comparison of replicated pedal inputs for RLS (human and robot) and HTM. Note additional inputs and dynamics for RLS.**

**Table 4. Vehicle B Data.**

| Specification | Value |
|---|---|
| Vehicle Mass | 1435 kg |
| Engine Size | 1.8 Litre, 4 cyl |
| Engine Type | NA Gasoline |
| Engine Output | 73 kW, 142 Nm Electric |
| Motors | 42 kW (MG1), 60 kW (MG2) |
| Hybrid Battery | NiMH 1.3 kWh |
| Transmission | Automatic (cont.) |
| Aftertreatment | TWC Emissions |
| Cert | Euro 5 (NEDC |

## 2.4 Road testing & data collection

For on-road testing the test vehicles were instrumented differently according the requirements at the time. For example vehicle A (Table 3) was used to collect a large amount of RDE data from locations around Europe, taking in cold weather, hot weather and high altitude testing. This vehicle was fitted with additional instrumentation such as strain-gauges on the drive shafts for monitoring road- load and numerous thermo-couples for monitoring critical temperatures. Whereas Vehicle B had important hybrid system data collected entirely via dedicated smartphone application.

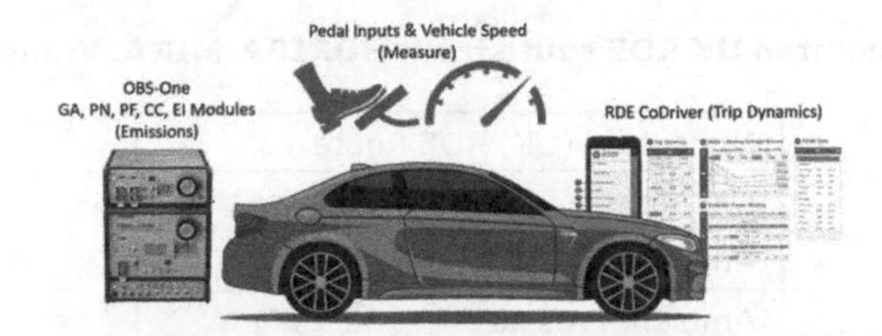


**Figure 8. Vehicle and data recording equipment.**

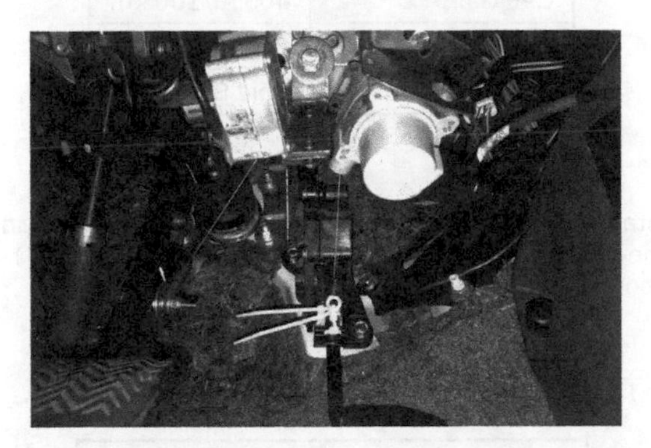

**Figure 9. Example of pedal position measurement using string potentiometer. Body of device is secured to hard points under the dashboard structure.**

In general, each vehicle had at least three main components shown in Figure 8: An OBS-One PEMS unit for emissions and environmental conditions, a method of recording pedal inputs (CAN bus data or direct measurement with string potentiometer, Figure 9) and a drivers-aid software for ensuring a compliant RDE test. All data was logged at a minimum of 10 Hz. Road tests were performed to the current RDE regulations whereby the routes take at least 90-minutes to complete and cover urban, rural and motorway sections in roughly equal proportion.

## 2.5 Road testing locations

***Nuneaton, UK:*** Data for the Nuneaton RDE routes are given in Tables 5 & 6. Map and elevation data can be found in the supplemental materials section (Figures 25 & 26). Nuneaton is approximately 100 m above sea level.

**Table 5. Nuneaton UK RDE route from HORIBA-MIRA. Vehicle A Test.**

| Nuneaton, (UK) RDE Route | |
|---|---|
| Distance | 85.09 km |
| Ambient Temp | 10°C |
| Ambient Pressure | 97.2 kPa |
| Ambient Humidity | 94.1 %RH |
| CPE Gain | 421 m/100 km |
| Drive Style | Gentle |
| Start Type | Cold Start |

**Table 6. Nuneaton UK RDE route from HORIBA-MIRA. Vehicle B Test.**

| Nuneaton, (UK) RDE Route | |
|---|---|
| Distance | 82.74 km |
| Ambient Temp | 23.6°C |
| Ambient Pressure | 100.7 kPa |
| Ambient Humidity | 34.5 %RH |
| CPE Gain | 403 m/100 km |
| Drive Style | Medium |
| Start Type | Cold Start |

***Avila, Spain:*** Data for the Avila RDE route is given in Table 7 Map and elevation data can be found in the supplemental materials section (Figures 27 & 28). Avila is approximately 1200 m above sea level.

**Table 7. Avila Spain, high altitude RDE Route.**

| Avila, (ESP) RDE Route | |
|---|---|
| Distance | 79.42 km |
| Ambient Temp | 32°C |
| Ambient Pressure | 89.1 kPa |
| Ambient Humidity | 31.0 %RH |
| CPE Gain | 935.9 m/100 km |
| Drive Style | Gentle |
| Start Type | Hot Start |

## 2.6 Test procedure creation

Conversion of road test data into a drive cycle for lab testing is dependent on the type of vehicle being used. The primary differentiation will be whether the vehicle uses a manual transmission or is automatic in some capacity. With a manual vehicle, the main consideration is matching the road drive to the performance of the driving robot. This requires manipulation of the road drive data to account for some critical performance metrics of the robot driver. The goal of these changes is for enabling the robot driver to complete the cycle safely. Mutli-speed automatic, continuously variable and single speed transmissions do not require such manipulation.

For example, the testing presented here utilised a Stähle SAP2000 driving robot. Installed into vehicle A, the robot has the following critical performance metrics:

- **Max clutch depress time** = 0.5 s
- **Max shifting time** = 0.8 s

These metrics will be different for different vehicles. For the clutch, the depressing rate depends on the length of pedal travel but also the stiffness of the pedal return spring. For the Stähle drive robot, the clutch actuator is itself spring loaded, so that the actuator works to pull the pedal upwards rather than pushing down. In the event of power-out, the default action is therefore the full depressing of the clutch pedal. The depressing rate of the clutch is therefore entirely dependent on the relative strength of the clutch return spring and the robot clutch actuator spring. Maximum shifting time depends on the geometry of the gear shift gate and linkages. The drive cycle manipulations then, are covered by the following and depicted in corresponding graphic in Figure 10:

1. **Adjusting clutch depress position:** For bringing the depressed position of the road data into the "safe region" of robot operation. Some drivers do not fully press the clutch to the floor. For example, they may stop around 70% of the travel. It is enough to change gear OK, but the robot requires greater than 80% to operate. Therefore the 70% position is stretched to meet the requirement of 80+%.
2. **Adjusting clutch rate:** For gear changes, the clutch depress profile is adjusted to match the robot performance if the human driver has pushed the pedal too fast.
3. **Adjusting gear placement:** The gear selection request is time-shifted to coincide with the point at which the clutch position enters the "safe region".
4. **Stretching clutch depress duration:** The duration for which the clutch is held in the "safe region" is extended to match the maximum shifting time. This is so the shift can be complete before the clutch is released.

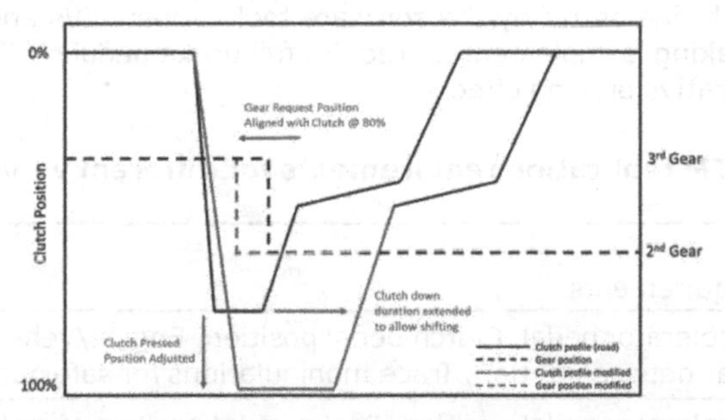


**Figure 10. Schematic representation of the drive trace manipulations necessary for safe robot driving operation.**

All of the manipulation tasks are performed by a software tool developed by HORIBA-MIRA to make it easy for the end user to jump from a road test into a chassis lab test, without having to spend a long time on preparing the drive cycles. As stated earlier, such manipulations are really only required by manual transmission vehicles. By dint of item number 4) in the above list, the resulting drive cycle is by necessity, of longer duration to the original road drive. Typically, 1 to 2 minutes of additional time can be added along with increasing driven distance slightly. However, all of the time adjustments come during gear changes where the engine is decoupled and throttle position is closed, so emissions are not significantly affected when comparing lab and road.

It would be possible to have zero cycle manipulation in two ways. The first requires the human driver performing the road test to drive according to the robot capability. This would not require a drastic change in driving style, but being mindful to not rush gear changes in particular, goes a long way toward minimising the need for adjustments. This is because the Stähle SAP2000 operating in manual driving mode does not allow for much overlap between clutch actuation and shifter actuation. A human driver can overlap these controls very well, whereas the robot requires a more sequential operation.

The second method would be to remove some of the robot driver safety limits. For example, in the manual mode, the default robot operation requires the clutch to be pressed by at least 80% before allowing a gear change under manual control. Requesting a gear change before this rule is met results in the robot controller assuming control of the clutch pedal. It will perform a complete automatic gear shift process according to its own control logic and will relinquish control back to the remote controller when complete. This is safe operation, but also not ideal for replication. Removing these safety features would allow for full control by the remote controller, but would shift the onus for safe operation. This approach is planned to be implemented, but for the first iteration of the drive cycle tool, it was decided to work around the in-built, proven safety features first rather than remove them.

In Table 8, the requirements by different types of vehicles is given. The conventional, combustion-only manual vehicle has the most requirements, with effort focused around ensuring safe gear changes. However it does not require any brake input; the dynamometer handles this. Vehicles such as BEVs with one-pedal driving functions require the least, needing only the accelerator pedal. A conventional automatic may require brake pedal input, if only for stationary periods to allow the ECU to disengage any clutches in the transmission. This can be recorded on the road test or automatically generated as brake on/off by the software tool. Other BEVs and hybrid vehicles where blended braking is implemented, require full brake pedal replication in order to match any regenerative braking effects.

**Table 8. HTM replication requirements for different vehicle types.**

| Vehicle Type | Requirements |
|---|---|
| Manual | Accelerator pedal, Clutch pedal position, Engine/vehicle speed for gear determina- tion, Trace manipulations for safe operation |
| Automatic | Accelerator pedal position, Brake pedal position (for stationary moments) |
| Hybrid | Accelerator pedal position, Brake pedal position (full, for regenerative braking) |
| Electric | Accelerator pedal position, Brake pedal position (only for blended brakes) |

## 2.7 Chassis lab test centre

All lab test work was carried out at the Advanced Emissions Test Centre located at HORIBA-MIRA in Nuneaton, UK (Figure 29). The centre features a climatic test cell with 4-wheel chassis dynamometer and full emissions analysis suite. Additional equipment includes robot drivers and an altitude simulator. There are also three temperature controlled soak chambers for conditioning vehicles for tests at high/low temperature. Where possible, the same PEMS kit as used on the road testing is used for the lab testing in addition to the lab analysers where possible.

Figure 11, shows the high level layout of the control system that performs the HTM test. The test automation system (STARS-VETS) controls all aspects of the test centrally.

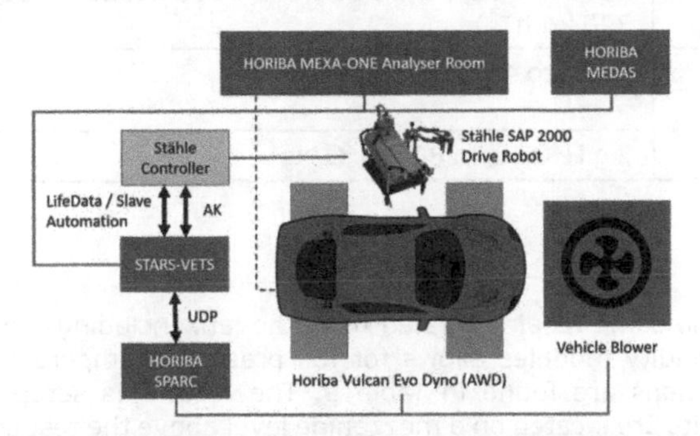


**Figure 11. Control layout for HTM in AETC. STARS-VETS test automation system orchestrates the test, controlling both dynamometer and robot driver.**

More generally, each replication of an RDE follows the same process as below. Replications are run a minimum of three times to obtain an average result.

1. Vehicle is soaked for 8 hours at starting engine oil temperature (for cold and hot start).
2. Test chamber is set to test average temperature and humidity.
3. If hot-starting, vehicle engine is warmed to starting engine oil temperature by holding 2-3000 rpm in idle (same as pre-warm up carried out on RDE road test).
4. If MEDAS is used it will follow pressure, temperature and humidity profiles recorded on test. Otherwise, cell will maintain average ambient conditions at 100 m altitude.
5. If performing "Nuneaton level" test, PEMS unit is used in conjunction with lab analysers via CVS emissions tunnel (tailpipe including PN & bag measurement).
6. If performing altitude test, PEMS unit is used but CVS tunnel is not. Tailpipe analyser is used in conjunction with PEMS pitot flow measurement to get a mass emission result for the lab. No lab PN or bag measurement in this case.
7. Vehicle goes back to soak for next test.

It should be noted that Vehicle A is operated by the robot driver using throttle-by-wire input rather than a pedal actuator. Vehicle B used the pedal actuator for both accelerator and brake pedals. Pedal positions recorded on the road are translated into robot driver actuator positions via a pedal mapping exercise performed as part of the robot driver installation.

***Altitude Simulator:*** HORIBA Multi-function Efficient Dynamic Altitude Simulator (MEDAS), is designed to simulate the environmental conditions at the engine air intake and exhaust (separate to the conditions set in the test chamber). In standard form, MEDAS can alter the engine intake air pressure, with air temperature and

**Table 9. MEDAS.**

| MEDAS 5012 VO (with MHM & MTM) | |
|---|---|
| Mass Flow | up to 1200 $kg.h^{-1}$ |
| Altitude Range | 2000 m below chamber level 5000 m above sea level (at 720 $kg.h^{-1}$) |
| Temperature Range | -10°C to 40°C |
| Humidity Range | 2 g $H^2O$/kg to 80 g $H^2O$/kg |

humidity being the same as air ingested from the lab. Including the additional temperature and humidity modules allows for full pressure, temperature and humidity control. Specifications are found in Table 9. The MEDAS is setup in AETC as per Figure 12. The units are located on a mezzanine level above the test chamber. There is a cell air intake in the ceiling for using MEDAS stand-alone (cell air ingestion) and the bypass channel has been extended into the test chamber to bring the conditioned air source closer to the vehicle.

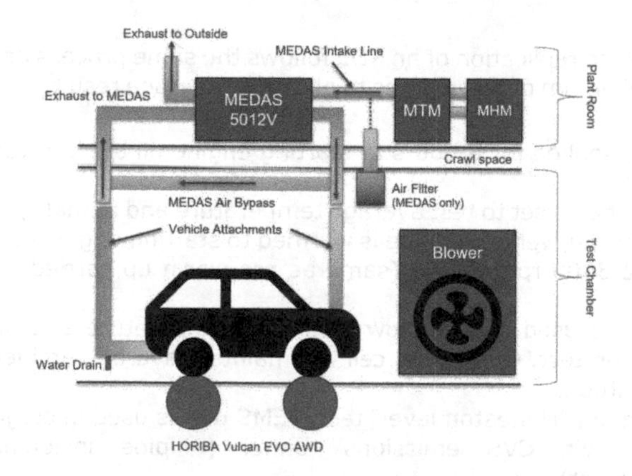


**Figure 12. Schematic diagram of MEDAS installation in AETC facility. The air bypass pipe was a post-installation modification, for bringing the conditioned air source closer to the vehicle.**

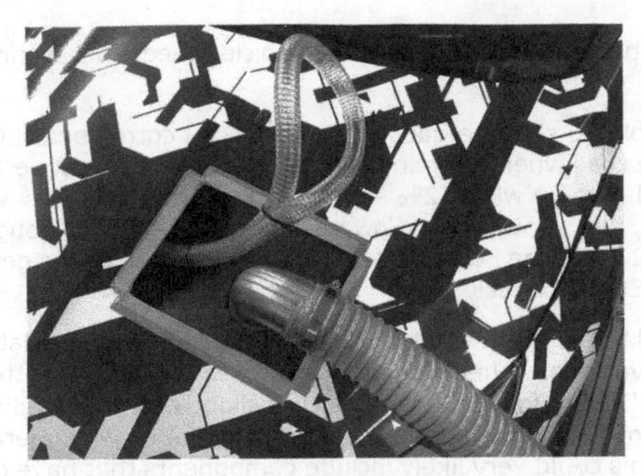

**Figure 13. Detail photo of example vehicle modification for MEDAS. Breather extension to check for oil carry over and modified air-box to connect to the system.**

Vehicle A modification for MEDAS is shown in Figure 13 MEDAS engine intake pipe is connected to the top of the vehicle air-box. The air-box was made air-tight and a hole was cut into the bonnet to facilitate this. The loop of clear pipe is an extended breather connection that allows for keeping an eye on any oil carry over. However, this vehicle exhibited none. Further images are found in supplemental material (Figures 30,31 & 32).

## 3 RESULTS & DISCUSSION

### 3.1 Results notes

1. Vehicle A is diesel vehicle with Euro6b certification, therefore $NO_x$ emissions are expected to be very high in real-world testing protocols.
2. Where possible, PEMS emissions results will be the main road to lab comparison tool as it is equivalent equipment for both test types. Lab emissions will also be included however.
3. PEMS units are not expected to match laboratory analysers exactly. Below are the allowable differences between lab emissions and portable emissions for acceptable correlation. This is based on current RDE regulation.
   (a) $CO_2$ +/-10% allowable
   (b) $NO_x$ +/- 15% allowable
   (c) PN +/- 50% allowable

### 3.2 Sea level RDE replication (Vehicle A)

Presented are the results of HTM replication test using the Vehicle A road test over the Nuneaton test route (Table 5). The test was replicated three times and the results will compare the road test to the averaged lab test data. The test cell was set to the road test average ambient temperature of 10°C and 90% relative humidity (this is the maximum achievable by the equipment and it was able to achieve at least 80% humidity at all times).

Figure 14 and Table 10 show the main emissions result. Figures 15 & 16 show further detail of the test replication. As shown characteristics such as engine speed and torque demand are very closely replicated, resulting in very close reproduction of total work done by the engine over the test. Figure 16 shows how well vehicle speed is matched along with accelerator pedal and clutch pedal inputs. Important to note is the synchronisation between pedal inputs and the vehicle speed. This alignment is critical for a good result.

Examination of Table 10 shows a good mass emissions correlation. $CO_2$ is a little over 2% different to the road when comparing PEMS results. The target performance for this replication method is to be within 2% so this is close. $NO_x$ was very well replicated at just 3% difference, also showing very low test-to-test variation. Though this difference equates to nearly 22% of the EU6 emissions limit, this is due to the generally very high $NO_x$ output of this Euro6b vehicle.

CO and PN showed the biggest variation between road and lab. The latter also showing large test-to-test variation. This is perhaps expected from a DPF fitted vehicle as PN emission is generally very low. It will be affected slightly by soot loading and as shown in Figure 14 bottom panel, the end result is mostly due to PN generated right at the start of the test. This could very likely include components that have condensed in the exhaust since last use. Regardless, the difference between road and lab is < 1% of the EU6 PN limit and therefore should not be too concerning. CO difference similarly

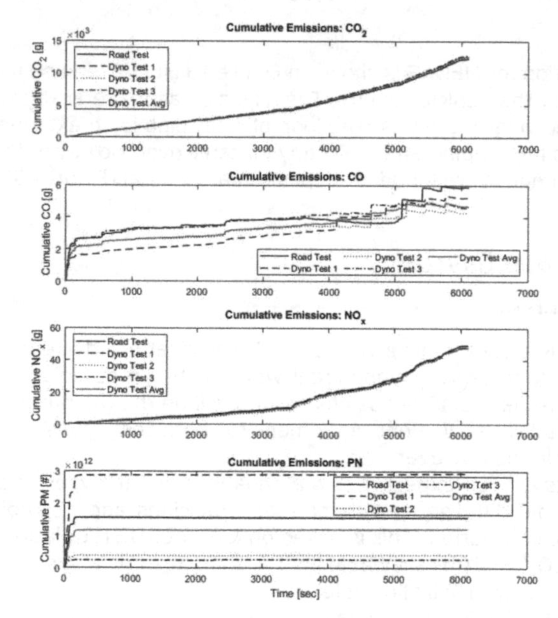


**Figure 14. Vehicle A cumulative emissions result for Nuneaton.**

makes up a small portion of the allowable emissions limit even though the difference between road and lab is quite large in percentage terms. Historically CO emissions measurement with PEMS has shown more variation compared to $CO_2$ for example.

lab results from HORIBA MEXA analyser are also included for comparison. Differences are expected between the lab analyser and PEMS. Historical correlation data for the PEMS unit used is given at the bottom of the table. Results are generally in agreement remembering the allowable differences between lab and PEMS given at the start of this section.

### 3.3 Altitude RDE replication (Vehicle A)

For the high altitude replication Vehicle A was modified as per earlier description for use with MEDAS (see Figures 13, 31, 32).

Presented are results of the HTM replication test using the Vehicle A road test over the Avila test route (Table 7). As before, the test was replicated three times and the results will compare the road test to the averaged lab test data. The test cell was set to the road test average ambient temperature of 32°C and 31% relative humidity. All three MEDAS units were used, allowing for full engine inlet air control. MEDAS was configured to follow the pressure, temperature and humidity profiles captured on the road

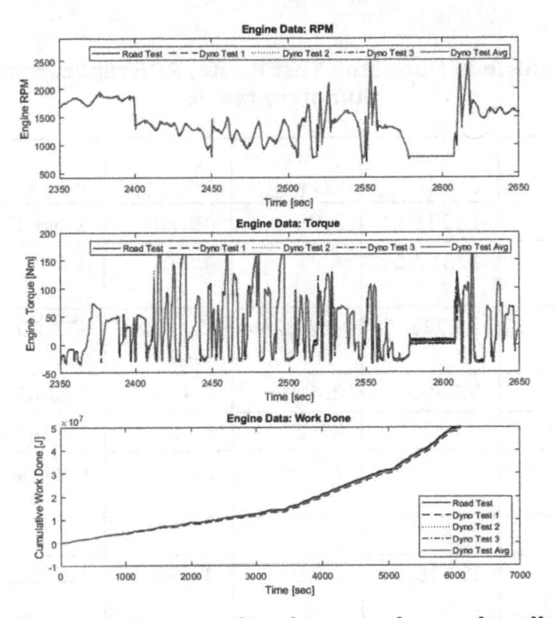


**Figure 15. Vehicle A engine speed and torque demand replication (detail view) with calculated whole-cycle work done by the engine.**

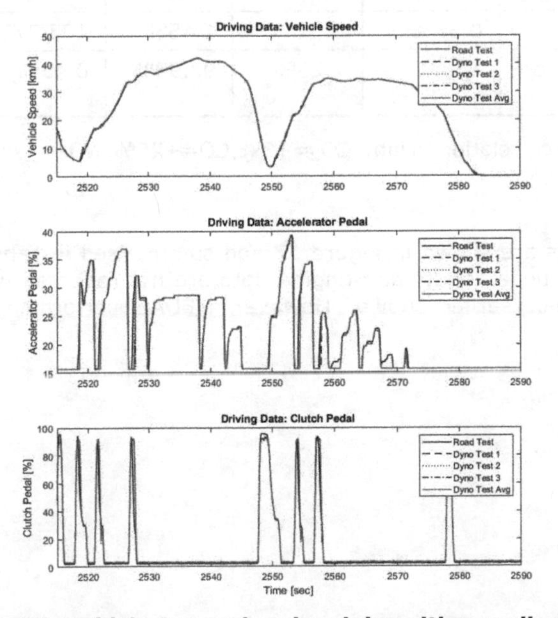


**Figure 16. Vehicle A speed and pedal position replication.**

test. PEMS unit for this test is, unfortunately, not the same unit as used on the road test. This is because the unit requires modification to work with MEDAS owing to the pressure differential between the room and sampling line. There is only one unit available that is modified for this and purchases some time after road testing. Performance is approximately similar for $CO_2$ and $NO_x$, but PN reads noticeably lower compared to the lab (approx -20%).

**Table 10. Vehicle A, Nuneaton Test Route, RDE replication with HTM. Summary result.**

| PEMS vs PEMS | $CO_2$ [g] | CO [g] | $NO_x$ [g] | PN [#] | Work Done [MJ] |
|---|---|---|---|---|---|
| Road Test | 12231.11 | 5.88 | 49.28 | 1.56e12 | 49.96 |
| Dyno Test (avg of 3) | 12557.67 | 4.83 | 47.79 | 1.20e12 | 49.09 |
| Road/Dyno Difference | 2.67% | -17.82% | -3.01% | -23.00% | -1.78% |
| CoV | 0.88% | 8.41% | 0.49% | 100.85% | 1.27% |
| EU 6 Limit for Test | - | 42.27 | 6.76 | 5.07e13 | - |
| Difference as % of EU6 Limit | - | 2.48% | 21.97% | 0.71% | - |
| PEMS vs Lab | | | | | |
| Dyno Test (avg of 3) | 12128.60 | 2.82 | 42.99 | 1.313e12 | - |
| Road/Dyno Difference | -0.84% | -52.10% | -12.75% | -16.05% | - |
| CoV | 0.55% | 8.51% | 1.45% | 103.31% | - |
| Difference as % of EU6 Limit | - | 7.25% | 92.93% | 0.49% | - |

Historical PEMS correlation to lab: $CO_2$=+3%,CO=+25%, $NOx$=+4.5%,PN=+5%

Emissions results are shown in Figure 17 and summarised in Table 11. For brevity, examples of pedal inputs and engine data are not repeated here as they are not different than earlier shown. However, MEDAS performance is shown in Figure 18.

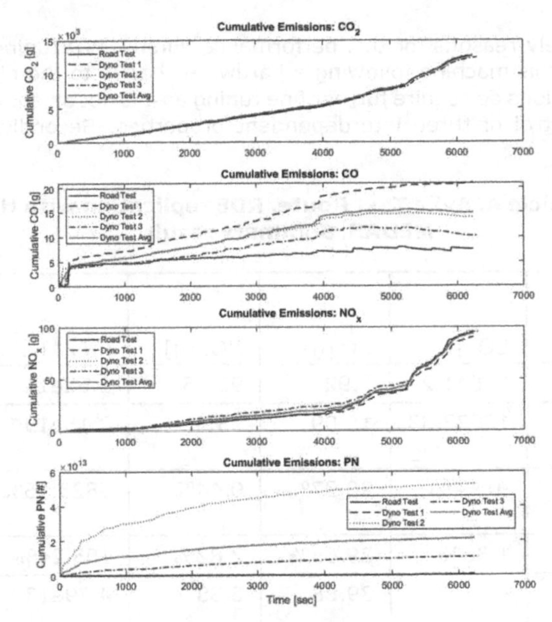


**Figure 17. Vehicle A cumulative emissions result for Avila.**

As with the Nuneaton replication, a very close match to the road test has been achieved. This time $CO_2$ is well within the target 2% range with similarly tight test-to-test variation. $NO_x$ is also extremely close being within 1% on average, though the test-to-test spread was a little wider than with the Nuneaton test. Notably, a one-off test where MEDAS was not used highlighted the need for replicating altitude; posting $CO_2$ and $NO_x$ results that were approximately 1 kg and 30 g less respectively, than the tests performed at altitude.

CO again shows a larger disparity, especially via the PEMS result which is significantly higher than on the road test, with wider test variation. However, the lab analyser shows a markedly different result that is a closer match and with less variation. This may indicate a problem with the PEMS CO analyser when used in conjunction with MEDAS even though it has been modified to work with the device. Alternatively, it could be due to the affect of MEDAS creating pressure differentials within the engine where none exist on the road test. The PN result shows an extremely large discrepancy and this will be commented on later in this section.

In general, MEDAS performed as expected. However, Figure 18 shows some less than ideal behaviour. Of note are the large pressure deviations shown occasionally during the test. There is a large one at around 1200 seconds and several more, smaller ones, closer to the end of the test. When the pressure swings like this, it causes the other two units controlling temperature and humidity to also swing (as all three conditions are inexorably linked). Though relatively short-lived, it is not ideal behaviour because it can take some time for the PID controller to re-balance the system. A second MEDAS unit on the HORIBA-MIRA site shows what the performance of the system is supposed to look like. Example Figures 33, 34 & 35 are located in the supplementary materials and show performance over a similar Avila altitude profile.

There are two likely reasons for this performance. Firstly, PID tuning improvements were carried out this machine following a hardware change to the pipe work. Potentially these alterations do require further fine tuning as it is not an easy task to balance the transient control of three interdependent properties. Secondly, control valves

**Table 11. Vehicle A, Avila Test Route, RDE replication with HTM utilising MEDAS. Summary result.**

| PEMS vs PEMS | $CO_2$ [g] | CO [g] | $NO_x$ [g] | PN [#] | Work Done [MJ] |
|---|---|---|---|---|---|
| Road Test | 12641.27 | 7.92 | 93.15 | 3.62e11 | 45.66 |
| Dyno Test (avg of 3) | 12623.33 | 15.09 | 93.56 | 2.15e13* | 45.42 |
| Road/Dyno Difference | -0.14% | 90.37% | 0.44% | 5823.15%* | -1.18% |
| CoV | 1.36% | 29.65% | 2.82% | 105.24% | 1.18% |
| EU 6 Limit for Test | - | 39.88 | 6.38 | 4.79e13 | - |
| Difference as % of EU6 Limit | - | 17.96% | 6.36% | 44.10*% | - |
| PEMS vs Lab* | | | | | |
| Dyno Test (avg of 3) | 12414.00 | 4.44 | 86.66 | - | - |
| Road/Dyno Difference | -1.80% | -44.02% | -6.97% | - | - |
| CoV | 0.70% | 19.14% | 5.13% | - | - |
| Difference as % of EU6 Limit | - | 8.75% | 101.72% | - | - |

* Suspected DPF Failure. See discussion text

** No lab exhaust flow or lab PN measurement with MEDAS

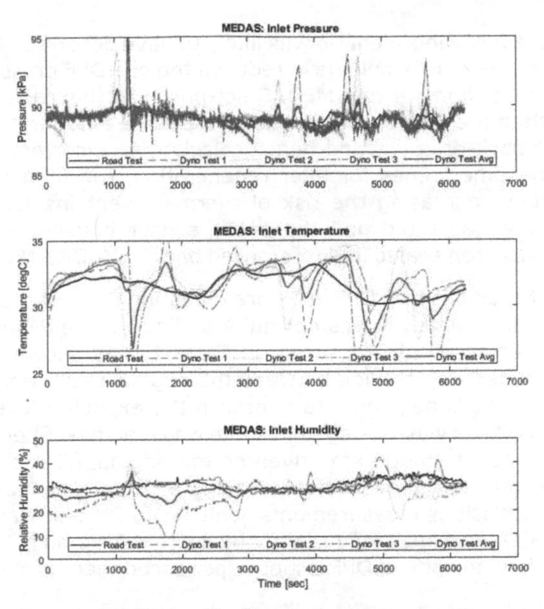


**Figure 18. Vehicle A MEDAS replication of atmospheric pressure, temperature and humidity. There was some spiking in the pressure control which caused the others to deviate.**

inside the machine are controlled by the building compressed air supply, which has been known to be unstable at times. Any drops in pressure (as it is only able to be supplied with the minimum pressure requirement) would lead to valve positions shifting without instruction, unexpectedly disturbing the system equilibrium. As the emissions replication result was still very good, it is not clear how much, if any, impact these spiking events events have had on the outcome. However, it is a clear area for improvement in the facility.

Regarding the PN emissions, they showed a sudden jump between tests 1 and 2. The vehicle had performed another test, in between these two tests, but it was not completed due to emergency shutdown. However, the vehicle had also initiated a DPF regeneration. When able to re-commence the test program, the PN emission then showed elevated levels. A number of tests were done off-line to confirm this, including further regeneration and re-testing, with PN levels remaining high. Examination by borescope revealed no visible damage on either DPF surfaces, but the face of the LNT did show a darkened patch suggesting there was potentially some PM escaping from the DPF.

Although not confirmed, the most logical theory amongst the team was that of localised melt damage inside the DPF. This could be put down to an omission made when modifying the vehicle for use with the MEDAS system. Vehicle A uses a DPF delta-pressure sensor that has only a single visible sampling point on the DPF canister, before the filter. So it was assumed that this was compared to the atmospheric reference that was inside the vehicle ECU. As per Figure 32 (supplemental material) the controller enclosure was modified to tie the ambient pressure sensor to the vehicle air- box and hence, MEDAS. However, it was later found that the ambient reference for the DPF sensor was situated on the backside of the DPF sensor package as per Figure 30.

Given the above, the following scenario was likely to have occurred. The application of MEDAS to the vehicle exhaust would have reduced the pre-DPF pressure by the difference between the test chamber and MEDAS set-point; In this case, approx. 10 kPa. Without also adjusting the ambient reference by the same amount, the ECU would see a reduced exhaust back-pressure and hence calculate an incorrect DPF soot loading estimate. When the time comes for filter regeneration, the filter would have been overloaded with soot, increasing the risk of thermal event inside. To remedy this, a new DPF assembly was fitted and the entire sensor housing was enclosed into a small box which was then sealed and referenced back to MEDAS (Figures 31).

Efforts were made to run the new DPF in before repeating the test program, but unfortunately time pressures meant it was not quite sufficient. The PN emissions results, pre and post DPF replacement are shown in Table 12. As shown, there is a large (2-orders of magnitude) step change between the original test 1 and test 2. The new DPF starts off lower, but is beginning to approach the expected level by test 3. It is clear that the original filter was damaged in some way. The new filter has rectified this somewhat, but with some 35,000 km driven on the original filter, the vehicle cannot be expected to perform exactly as before. Making a comparison to the original test program using lab emissions measurements (only PEMS PN unit was available at the time of re-testing) $CO_2$ was increased slightly by approx. 2% and $NO_x$ was reduced by approx. 7%. Therefore, in spite of DPF issues experienced there is still high confidence in the result.

**Table 12. Change in test to test PN for original DPF and new DPF. There is two order magnitude change between test 1 and test 2 for original filter.**

| Test No. | Original DPF | Replacement DPF |
|---|---|---|
| 1 | 8.00e11 | 6.43e10 |
| 2 | 5.29e13 | 5.87e10 |
| 3 | 1.07e13 | 1.47e11 |

### 3.4 Hybrid vehicle RDE replication (Vehicle B)

Experimentation with hybrid and electric vehicles is an important step for the HTM technique. With increasing numbers of PHEV and BEV becoming available and gaining popularity [20], HTM should be compatible if it is to become a useful development tool. As the dynamometer rollers are providing the speed control and thus braking effort, the crucial question is whether the regenerative aspect of vehicle braking can be replicated successfully by HTM.

For this type of vehicle, the expectation was that being able to replicate brake pedal position and timing would produce the desired outcome. The brake pedal was instrumented with a string potentiometer to achieve this (Figure 9).

Presented in this section are the results of HTM replication test using Vehicle B (Table 4), a hybrid vehicle, driving the Nuneaton test route (Table 6). The test was replicated three times and the results will compare the road test to the averaged lab test data. The test cell was set to the road test average ambient temperature of 23°C and 31% relative humidity.

Emissions result is shown in Figure 19 and summarised in Table 13. As shown the results are comparable to previous testing with vehicle A. $CO_2$ is within 2% and so too are $NO_x$ and PN which is an excellent outcome. $NO_x$ emissions show a high variation value, but this is likely a function of the extremely low $NO_x$ output of the vehicle (small absolute changes showing as large percentage changes). CO emission is once more, showing a larger difference during the lab testing and suggests that investigation as to why, is warranted. Like before, a large percentage difference road- to-lab, but a small proportion of EU6 limits.

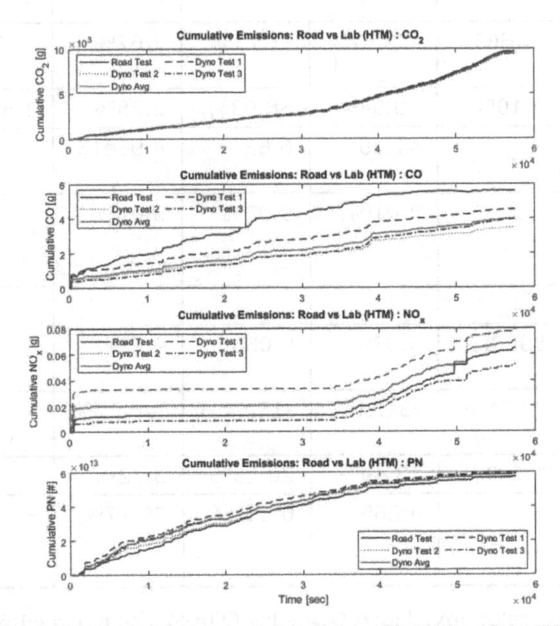


**Figure 19. Cumulative emissions result for vehicle B using HTM, Nuneaton.**

Equally as important for this test is replication of hybrid system performance. Concerning energy consumed and regenerated, Figure 20 and Table 14 show very close matching of energy consumed by the system as a whole (engine & hybrid system), energy consumed from the battery and energy regenerated back to the battery, with less than 2% difference road-to-lab. Figure 20, bottom panel, shows the SOC reported by the battery management system. It was difficult to get starting SOC to the right value pre-test because there is no way to charge the vehicle externally. However, as shown, SOC very quickly converges for all tests on to the road test values and then track very well for the remainder of the test.

Figures 21 to 24 serve to highlight the quality of the replication further. Figure 21 shows high-voltage battery voltage and current draw for a portion of the test. Both track the road test values very closely. Figures 22 and 23 show engine and MG speeds and loads respectively. MG2 speed is directly related to the speed of the vehicle, but this is not the case for the ICE and MG1 which have a looser relationship. MG1 is also responsible for sending energy to the battery whilst driving with the ICE active, so it is important to produce a good replication in this respect. As shown, speeds and loads are very well matched between road and lab for all three motors, including ICE firing

**Table 13. Vehicle B, Nuneaton Test Route, RDE replication with HTM. Summary result.**

| PEMS vs PEMS | $CO_2$ [g] | CO [g] | $NO_x$ [g] | PN [#] | Energy Consumption (total) [kWh] |
|---|---|---|---|---|---|
| Road Test | 9647.26 | 5.58 | 0.064 | 5.75e13 | 14.93 |
| Dyno Test (avg of 3) | 9487.33 | 3.93 | 0.063 | 5.90e13 | 14.75 |
| Road/Dyno Difference | -1.66% | -29.67% | -1.73% | 2.67% | -1.20% |
| CoV | 3.10% | 10.04% | 58.93% | 2.28% | 1.50% |
| EU 6 Limit for Test | - | 41.36 | 6.62 | 4.96e13 | - |
| Difference as % of EU6 Limit | - | 4.01% | 0.02% | 3.09% | - |
| PEMS vs Lab | | | | | |
| Dyno Test (avg of 3) | 8958.84 | 3.83 | 0.056 | 4.25e13 | - |
| Road/Dyno Difference | -7.14% | -31.55% | -13.64% | -26.04% | - |
| CoV | 1.59% | 13.49% | 20.29% | 3.32% | - |
| Difference as % of EU6 Limit | - | 4.26% | 0.13% | 30.17% | - |

Historical PEMS correlation to lab: $CO_2$=+3%,CO=+25%, $NO_x$=+4.5%,PN=+5%

**Table 14. Vehicle B energy consumed by powertrain (engine & hybrid system) with energy regenerated back to traction battery and estimated energy from battery.**

| Test No. | Total Energy (system) [kWh] | Energy Regenerated to Batt.[kWh] | Energy Consumed from Batt. [kWh] |
|---|---|---|---|
| Road Test | 14.93 | 2.78 | 2.20 |
| Dyno Test 1 | 15.06 | 2.78 | 1.91 |
| Dyno Test 2 | 14.55 | 2.73 | 1.84 |
| Dyno Test 3 | 14.65 | 2.74 | 1.86 |
| Dyno Test Avg | 14.75 | 2.75 | 1.87 |
| Difference | 1.50% | 0.79% | 1.34% |

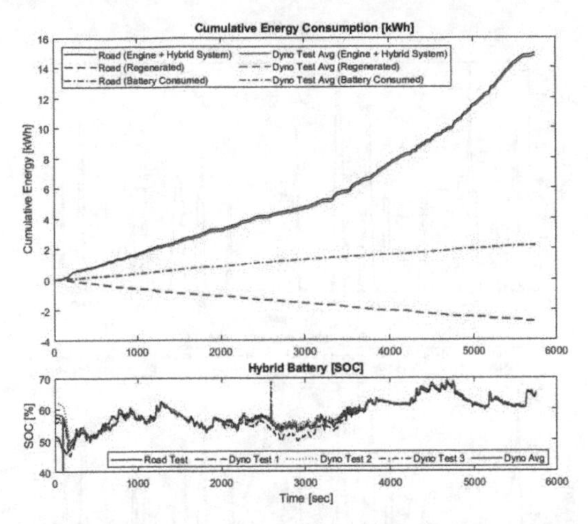


**Figure 20. Vehicle B energy consumption replication. Engine & Hybrid system combined, battery conumption and battery regeneration.**

and shut-off events. Finally, Figure 24 shows replication of hybrid system temperatures, including the battery. The latter is particularly critical given the sensitivity of battery performance in relation to temperature. Shown in each panel are temperatures for the ICE, MG unit, high-voltage inverter and high-voltage traction battery. All demonstrate very close matching with the road test.

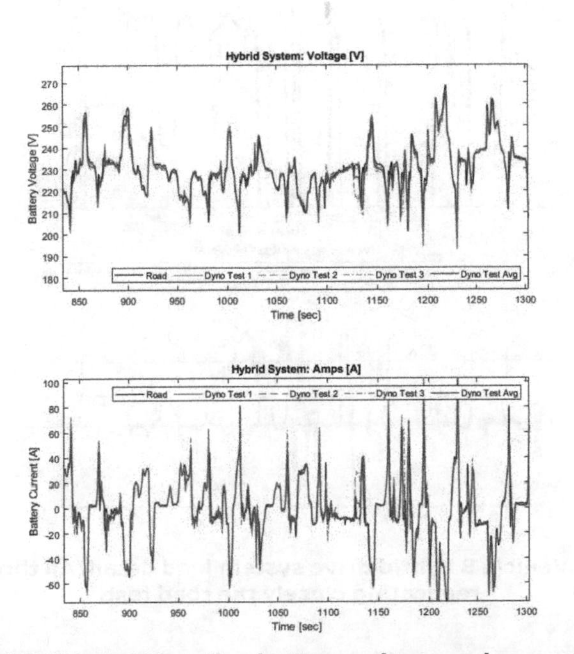


**Figure 21. Vehicle B hybrid traction battery voltage and current detail. Total pack voltage and current draw track closely to road test.**

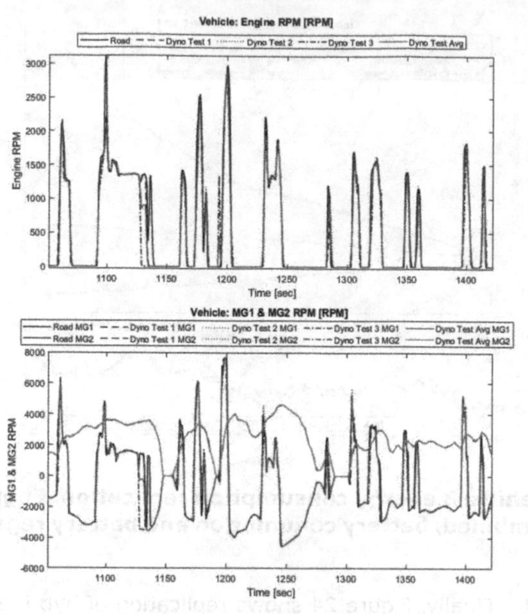


**Figure 22. Vehicle B hybrid drive system speed detail. ICE, MG1 & MG2 track closely to the road test. Including ICE firing and shut-off events.**

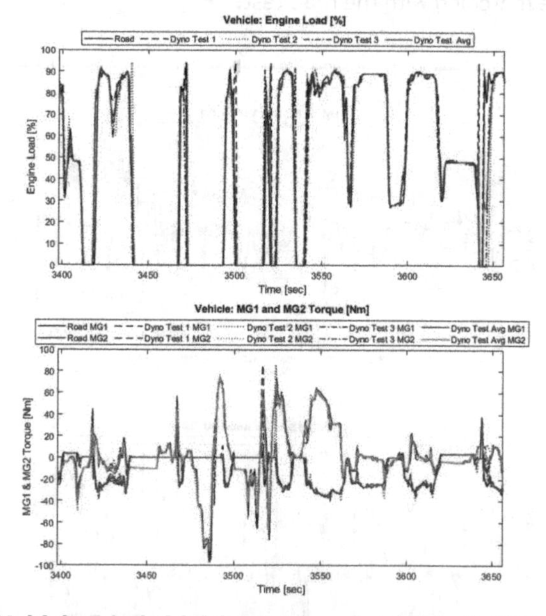


**Figure 23. Vehicle B hybrid drive system load detail. All three motors replicating closely the road test.**

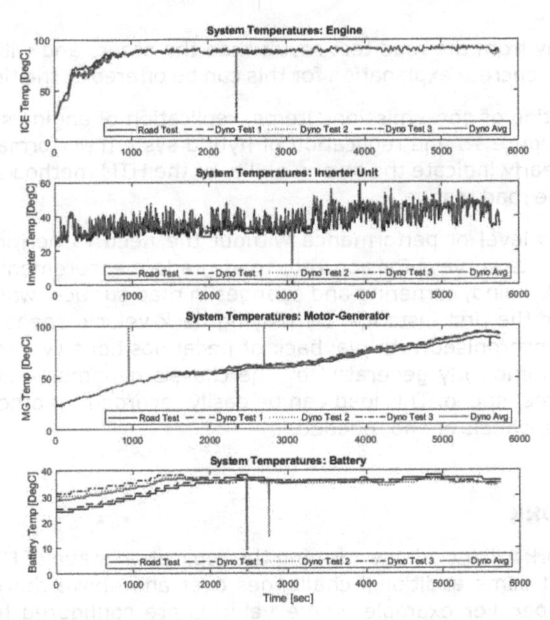


**Figure 24. Vehicle B powertrain temperatures. Motors and high-voltage electronics temperatures are well replicated.**

## 4 CONCLUSIONS

This paper has introduced, discussed and demonstrated the HORIBA Torque Matching (HTM) road to lab replication technique which is part of HORIBA's new suite of development tools under the RDE+ program banner. It has covered the successful application of HTM using a conventional diesel vehicle, to replicate both sea-level and high-altitude (in conjunction with HORIBA MEDAS) real- world road drives. Extending further, the effectiveness of HTM when applied to the hybrid vehicle case has also been successfully demonstrated.

$CO_2$ correlation was found to be in the range of 2% or less for all tests, meeting the project target. Very high correlation of $NO_x$ emissions at much less than 5% was also found for all tests. This shows HTM has the ability to be successfully applied to different vehicles types and test cell arrangements. Excluding the DPF failure on the high altitude replication test, PN emissions are shown to be more than acceptable. There was an approximately 20% difference between road and lab for the diesel sea-level test. However, the total PN value itself is significantly under the allowable emissions limits, with the difference representing < 1% of the limit. Additionally, the vast majority of PN was generated right at the start of the test. This PN could be from combustion products condensed in the exhaust pipe from last use. But it does highlight that there could be room for improvement in the way certain vehicles are prepared pre-test. This does not seem to have been an issue with the hybrid vehicle (gasoline, no exhaust filter) testing however, which shows very close match of PN emissions back to the road test and with low variation.

For both vehicles, $CO_2$ and $NO_x$ emissions generally have shown close matching with low test-to-test variation, along with calculated work done by the vehicle. This highlights repeat-ability as a strong characteristic of the HTM technique. However, CO emissions replication does raise some questions. For all tests, it is generally seen to be much further away from the road test result than the others and with higher test-to-test variation. No concrete explanation for this can be offered at the time of writing.

Regarding replication of non-emissions items, replication of engine speed and torque demand (as per Figure 15 and replication of hybrid system performance (as per Figures 21 to 24), clearly indicate the strong ability of the HTM method at the replication of all aspects of the road testing.

HTM achieves this level of performance without the need to perform vehicle coast down testing or route surveys for accurate road-grade measurement. It also handles additional loads like wind, cornering and changes in road surface, which are very difficult to measure in the first instance. By playing back vehicle speed (via the chassis dynamometer) synchronised with playback of pedal positions (via robot driver), the exact road load is implicitly generated by the chassis dynamometer controller as it controls to the target speed. This load can be easily recorded for other uses, meaning devices like torque wheels are not needed.

## 5 FUTURE WORK

Planned future work will include performing the replication stage of HTM on a full BEV. BEVs may present some additional challenges over and above the vehicles demonstrated in this paper. For example, some vehicles are configured to provide a one pedal driving dynamic where the driver is seldom, if at all, required to use the friction braking system. Others, will allow for free-wheeling or sailing and there are also differences between braking systems; some blending regenerative and friction braking and some not. However, based on the experience with the hybrid vehicle, there is high confidence that HTM will be capable of the same quality of replication shown.

Additionally, the implementation of a new driving robot (HORIBA ADS-Evo) will be undertaken alongside the development of the emulation stage of HTM. As a reminder, in the emulation stage, closed loop control of the pedal inputs will allow for matching a target road load, such as the one recorded during a replication test. This will allow for development teams to make whole sale changes to vehicle components and calibrations, then run those changes over the same load profile to gauge their effects on performance.

## ACKNOWLEDGEMENTS

The authors wish to thank the global HORIBA teams for their contributions to the work published in this paper.

## ACRONYMS

**AETC** Advanced Emissions Test Centre. 9, 10, 21

**AGVES** Advisory Group on Vehicle Emission Standards. 3

**BEV** Battery Electric Vehicles. 1, 3, 8, 14, 17

**CAN bus** Controller Area Network bus. 7

**CAZ** Clean Air Zone. 3

**$CH_4$** Methane. 2

**CO** Carbon Monoxide. 10–12, 14, 15, 17

**$CO_2$** Carbon Dioxide. 3, 5, 10–12, 14–17

**CPE** Cumulative Positive Elevation. 7

**CVS** Constant Volume Sampling. 9

**CVT** Continuously Variable Transmission. 3, 20

**DOC** Diesel Oxidation Catalyst. 6

**DPF** Diesel Particulate Filter. 6, 10, 13, 14, 16, 21

**ECU** Electronic Control Unit. 8, 13, 22

**EiL** Engine-in-the-Loop. 3

**GHG** Greenhouse Gas Emissions. 3

**HTM** HORIBA Torque Matching. 1, 3–6, 9–17, 20

**ICE** Internal Combustion Engine. 1, 3, 6, 14–16

**ISC** In Service Conformity. 2

**kWh** kilowatt-hour. 6

**LNT** Lean $NO_x$ Trap. 6, 13

**MEDAS** Multi-function Efficient Dynamic Altitude Simu- lator. 3, 9–13, 16, 21, 22

**MG** Motor-Generator. 6, 14–16

**$N_2O$** Nitrous Oxide. 2

**NA** Naturally Aspirated. 6

**NEDC** New European Drive Cycle. 1, 2, 6

**$NH_3$** Ammonia. 2

**NiMH** Nickle-Metal Hydride. 6

**$NO_2$** Nitrogen Oxide. 2

**$NO_x$** Nitrogen Oxides. 1, 2, 5, 6, 10–12, 14–17

**PEMS** Portable Emissions Measurement System. 3, 4, 7, 9–15

**PHEV** Plug-in Hybrid Electric Vehicles. 14

**PID** Proportional Integral and Derivative control. 4, 6, 12

**PM** Particulate Matter. 1, 13

**PN** Particulate Number. 2, 5, 9–17

**RDE** Real Driving Emissions. 2, 3, 6, 7, 9, 10, 12, 13, 15, 20

**RDE+** Real Driving Emissions Plus. 1, 3, 4, 16

**RLR** Road Load Replication. 3

**RLS** Road Load Simulation. 3, 5, 6, 20

**SOC** State of Charge. 14

**SUV** Sports Utility Vehicle. 2

**TWC** Three Way Catalyst. 6, 20

**WHO** World Health Organisation. 1

**WLTC** World-Harmonised Ligt-Vehicle Test Cycle. 2

**WLTP** World-Harmonised Light-Vehicle Test Procedure. 2

## REFERENCES


[1] European Environment Agency (EEA). *Air quality in Europe — 2019 report*. 10, European Environment Agency, 2019. ISBN 9789294800886.URL http://www.eea.europa.eu/publications/air-quality-in-europe-2013.

[2] Bernard Y. Current situation and perspectives on vehicles real-world emissions. Technical report, International Council on Clean Transportation, Brussels, 2019. URL https://environnement.brussels/sites/default/files/user_files/pres_20190423_colloquesortiethermique_icct.pdf.

[3] Grelier F. Cars with engines: Can they ever be clean? (Diesel-gate Report). Technical report, Transport & Enivronment, 2018. URL www.transportenvironment.org.

[4] Cozzi L and Petropoulos A. Growing preference for SUVs challenges emissions reductions in passenger car market, 2019. URL https://tinyurl.com/47dy66d7.

[5] van Mensch P, Cuelenaere RFA and Ligternik NE. Assessment of risks for elevated $NO_x$ emissions of diesel vehicles outside the boundaries of RDE. Technical report, TNO Earth, Life & Social Sciences, 2017. URL https://repository.tudelft.nl/view/tno/uuid:b0ff9bd6-41d0-4d88-89fe-012321e955be.

[6] ADAC. Clean diesels: Euro 6d Temp models tested — ADAC, 2019. URL https://www.adac.de/verkehr/abgas-diesel-fahrverbote/abgasnorm/test-euro-6d-temp/.

[7] Dimaratos A, Triantafyllopoulos G, Ntziachristos L et al. Real- world emissions testing on four vehicles. Technical report, 2017. URL http://www.emisia.com.

[8] Moody A and Tate J. In Service $CO_2$ and $NO_X$ Emissions of Euro 6/VI Cars, Light-and Heavy-dutygoods Vehicles in Real London driving: Taking the Road into the Laboratory. *Journal of Earth Sciences and Geotechnical Engineering* 2017; 7(1): 51–62. URL http://eprints.whiterose.ac.uk/111811/.

[9] Suarez-Bertoa R, Valverde V, Clairotte M et al. On-road emissions of passenger cars beyond the boundary conditions of the real-driving emissions test. *Environmental Research* 2019; 176: 108572. DOI:10.1016/j.envres.2019.108572. URL https://www.sciencedirect.com/science/ article/pii/S001393511930369Xhttps: //linkinghub.elsevier.com/retrieve/pii/ S001393511930369X.

[10] DieselNet. Emission Standards: Europe: Cars and Light Trucks, 2020. URL https://dieselnet.com/standards/eu/ld.php.

[11] Samaras Z, Hausberger S and Mellios G. Preliminary findings on possible Euro 7 emission limits for LD and HD vehicles. Technical Report October, Advisory Group on Vehicle Emission Standards, 2020. URL https://circabc.europa.eu/sd/a/fdd70a2d-b50a-4d0b-a92a-e64d41d0e947/CLOVEtestlimitsAGVES2020-10-27finalvs2.pdf.

[12] Continental Automotive GmbH. Worldwide Emission Standards and Related Regulations. Technical report, 2019. URL https://www.continental-

automotive.com/getattachment/8f2dedad-b510-4672-a005-3156f77d1f85/EMISSIONBOOKLET2019.pdf.
[13] Bernard Y, German J and Muncrief R. Worldwide Use of Remote Sensing to Measure Motor Vehicle Emissions. Technical Report April, The International Council on Clean Transport, 2019. URL www.theicct.org.
[14] Woopen H. Remote Sensing: Measuring Emissions From Cars As They Drive By. Technical Report February, OPUS RCE, 2018.
[15] Hull R. This is Money: As Birmingham plans to stop vehicles driving through the city, we reveal the areas set to ban cars or have a low emission zone within three years, 2020. URL https://tinyurl.com/357r794t.
[16] Roberts PJ, Mumby R, Mason A et al. RDE Plus - The Development of a Road, Rig and Engine-in-the-Loop Test Methodology for Real Driving Emissions Compliance. In *SAE Technical Papers*, volume 2019-April. DOI:10.4271/2019-01-0756. URL https://www.sae.org/content/2019-01-0756/.
[17] Roberts P, Mason A, Tabata K et al. RDE Plus - A Road to Rig Development Methodology for Whole Vehicle RDE Compliance: Engine-in-the-Loop and Virtual Tools. In *SAE Technical Papers*. 2020. DOI:10.4271/2020-01-2183. URL https://www.sae.org/content/2020-01-2183/.
[18] Roberts P, Mason A, Whelan S et al. RDE Plus - A Road to Rig Development Methodology for Whole Vehicle RDE Compliance: Overview. DOI:10.4271/2020-01-0376. URL https://www.sae.org/content/2020-01-0376/.
[19] Mason A, Roberts P, Whelan S et al. RDE Plus – A Road to Rig Development Methodology for Complete RDE Compliance: Road to Chassis Perspective. pp. 1–21. DOI: 10.4271/2020-01-0378. URL https://www.sae.org/content/2020-01-0378/.
[20] SMMT. January 2021 - SMMT UK New Car and LCV Registrations Outlook to 2022. Technical Report April 2021, SMMT, 2022. URL https://tinyurl.com/vx3kzdj4.

## SUPPLEMENTAL MATERIAL

**Table 15. HTM vs RLS comparison vehicle data.**

| Specification | Value |
|---|---|
| Vehicle Mass | 1300 kg |
| Engine Size | 1.6 Litre, 4 cyl |
| Engine Type | Turbo-Gasoline |
| Engine Output | 140 kW, 240 Nm |
| Transmission | CVT |
| Aftertreatment | TWC |
| Emissions Cert | Japan JC08 |

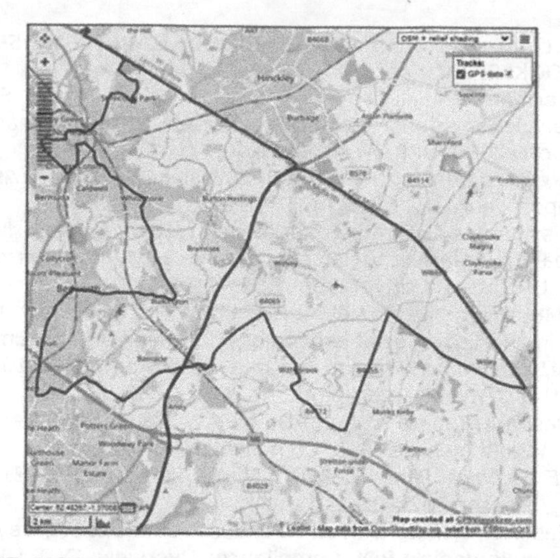

**Figure 25. HORIBA MIRA UK RDE route around Nuneaton area.**

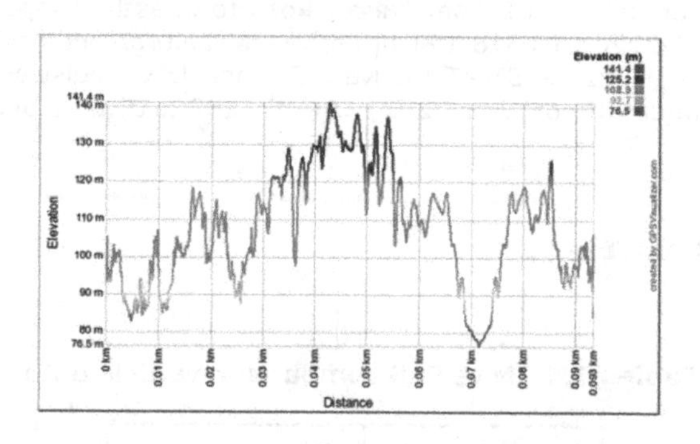


**Figure 26. HORIBA MIRA UK RDE route around Nuneaton area, elevation profile.**

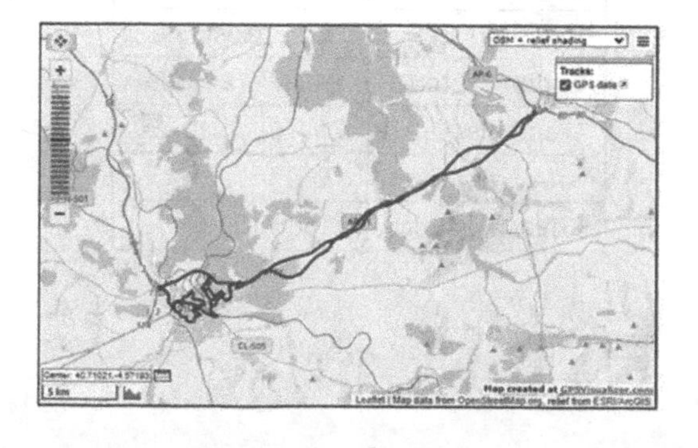

**Figure 27. HORIBA MIRA high altitude RDE route in Avila, Spain.**

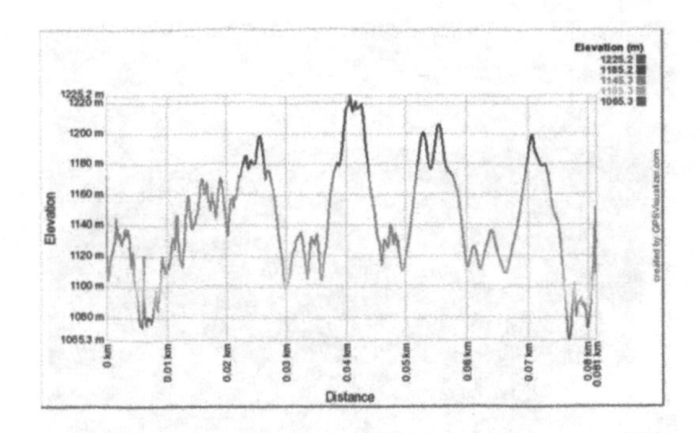


**Figure 28. HORIBA MIRA high altitude RDE route in Avila, Spain. Elevation profile.**

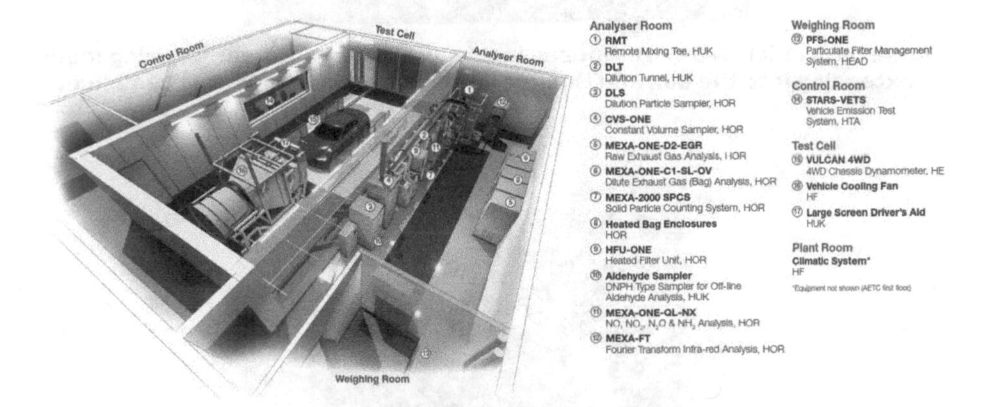


**Figure 29. HORIBA MIRA UK AETC facility.**

**Figure 30. Vehicle A DPF sensor. The backside of the sensor package houses a an ambient pressure reference that is used to determine a differential signal compared to pre-DPF pressure tapping location.**

**Figure 31. Vehicle A DPF re-housed inside an airtight enclosure. Wiring loom is extending into the box and the interior is referenced back to MEDAS inlet.**

**Figure 32. Vehicle A ECU is modified similar to the image above. Breather cap for detecting ambient pressure is removed and sealed with 6 mm nylon pipe inserted. Nylon pipe references to MEDAS inlet. ECU is not fully airtight but it is enough for MEDAS to maintain an pressure inside.**

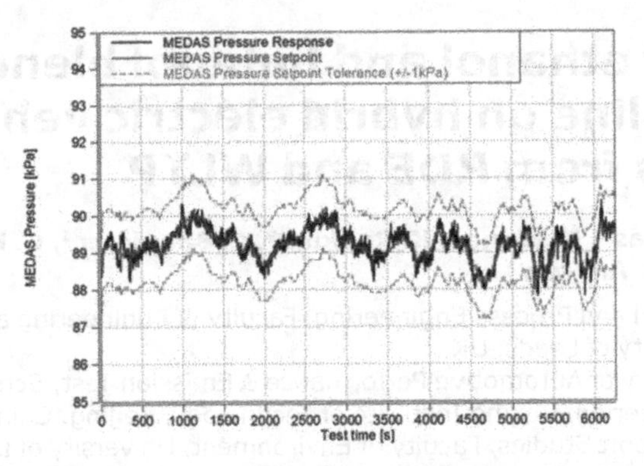


**Figure 33. Example MEDAS performance, from the second MEDAS device at HORIBA-MIRA. Ambient pressure replication of the same Avila test-route. Performance well within quoted performance envelope.**

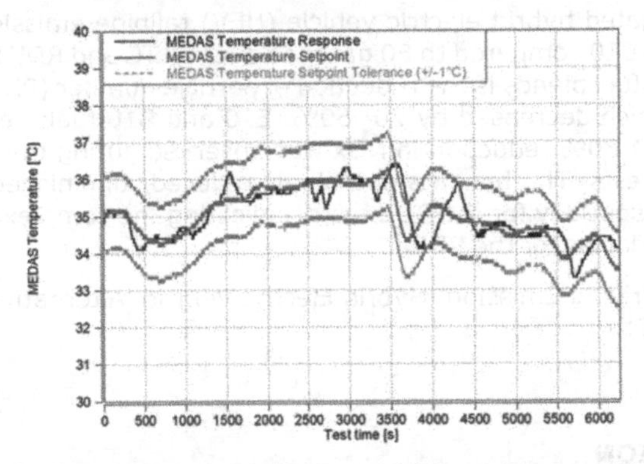


**Figure 34. Example MEDAS performance, from the second MEDAS device at HORIBA-MIRA. Ambient temperature replication of the same Avila test-route. Performance well within quoted performance envelope.**

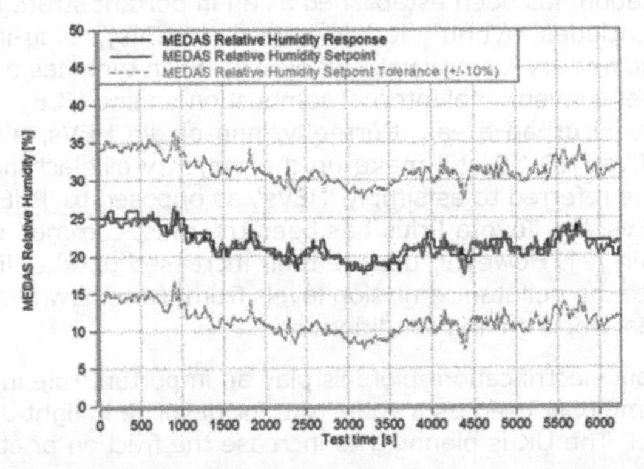


**Figure 35. Example MEDAS performance, from the second MEDAS device at HORIBA-MIRA. Ambient humidity replication of the same Avila test-route. Performance well within quoted performance envelope.**

# Impact of ethanol and butanol blends with gasoline on hybrid electric vehicle emissions from RDE and WLTP

**H. Li[1], D.B. Thomas[1], X. Wang[2], K. Ropkins[3], C. Bannister[4], G. Hawley[4], A.S. Tomlin[1], G.E. Andrews[1]**

[1]School of Chemical and Process Engineering, Faculty of Engineering and Physical Sciences, University of Leeds, UK
[2]National Laboratory of Automotive Performance & Emission Test, School of Mechanical Engineering, Beijing Institute of Technology, Beijing, China
[3]Institute for Transport Studies, Faculty of Environment, University of Leeds, Leeds, UK
[4]Automotive Engineering Powertrain & Vehicle Research Centre, Department of Mechanical Engineering, University of Bath, Bath, UK

## ABSTRACT

This paper investigated hybrid electric vehicle (HEV) tailpipe emissions from biofuel blends E5, E10 and B10 compared to E0 gasoline over WLTC and RDE tests. The greatest benefit from biofuel blends is the reduction of particle number (PN) emissions from the WLTC tests, which decreased by 20~50%. E10 and B10 fuels reduced CO emissions significantly. A 30% reduction in NOx was observed during the cold start phase of WLTC and RDE tests. The benefits of biofuels reduced/ diminished for the second and third phases, especially for the RDE tests, indicating the complexity and influence of traffic/engine variables for the RDE.

**Keywords:** Real Driving Emission, Hybrid Electric Vehicle, Alternative fuels, Ethanol, Butanol, WLTC, RDE.

## 1 INTRODUCTION

The transportation sector accounted for around 28% of total greenhouse gas (GHG) emissions in the UK in 2018, with 55% of this coming from passenger cars and taxis [1]. To reduce carbon emissions from the transport sector, especially for light duty vehicles, powertrain electrification has been established as an important strategy towards decarbonisation, which includes hybrid electric vehicles (HEVs), plug-in hybrid electric vehicles (PHEVs) and battery electric vehicles (BEVs). Each type has both efficiency and GHG advantages over conventional internal combustion engine (ICE) spark ignition (SI) vehicles, particularly in urban areas. Currently, non-plugin HEVs, also called Not Off Vehicle Charging HEVs (NOVC) still make up the majority of electrified vehicle fleets. These vehicles will be referred to as simply 'HEVs', as opposed to 'PHEV' for the plug-in variety. Of these HEVs, the Toyota Prius has been the most common generic model on roads in Great Britain [2]. However, despite their increased uptake, little research has been conducted as to the pollutant emission levels from these new hybrid technologies, particularly under real world driving conditions.

Apart from powertrain electrification, biofuels play an important role in decarbonization. The biofuel most commonly used as a substitute for gasoline in light duty SI passenger cars today is ethanol. The UK is planning to increase the fraction of ethanol in gasoline

DOI: 10.1201/9781003219217-15

from 5% (E5) to 10% (E10). The butanol isomers have some potential advantages over ethanol in that they have higher calorific values, are less hydrophilic, have a lower vapour pressures and can be blended to higher ratios without the risk of corrosion [3]. This makes both n- and iso-butanol possible alternatives to ethanol if production technologies can be improved. A great advantage of the use of these biofuels is that they can be used within current engine technologies, at low blending ratios without any modification, or in higher quantities given some engine modification. Combined with powertrain electrification, they have the potential to accelerate carbon reductions for road transport.

The aim of this work was to investigate how the deployment of E10 and the possible uptake of n-butanol (B10, n-butanol:gasoline = 10:90)) may affect the emissions of a HEV (Hybrid Electric Vehicle) (Toyota Prius). The regulated gaseous and particle number emissions over the WLTC (World-harmonised Light Duty Test Cycle) and RDE (Real Driving Emissions) tests including cold start are reported.

## 2 METHODOLOGY

### 2.1 Research vehicle and instrumentation

A Third Generation Toyota Prius HEV, of model year 2010 and Euro 5 emission compliance, was used for this research. The vehicle was instrumented with thermocouples for measuring temperatures across the TWC (Three-Way-Catalyst), coolant and lube oil.

An OBD (On-Board Diagnostics) data logger was used to collect ECU (Electronic Control Unit) information from the vehicle during testing. This was a HEM Data Corporation OBD Mini-logger, which was connected directly into the OBD port of the Toyota Prius. This unit logs data to a 1GB micro-SD card at 1 Hz. A very wide range of parameters are available for logging, including a database of Toyota Prius specific enhanced OBD parameter IDs purchased from HEM Data Corporation. The parameters include engine torque, vehicle and engine speed, lambda, fuel consumption, various temperatures and pressures, battery condition, EGR (Exhaust Gas Recirculation) and other parameters. More details can be found in reference [4].

### 2.2 Chassis dynamometer test facility and RDE test

The WLTC tests were carried out in the chassis dynamometer test cell at the Centre for Low Emissions Vehicle Research (CLEVeR) of University of Bath (UoB). The tests were performed on a 4WD AVL RoadSimTM 48" chassis dynamometer, within a test cell capable of temperature and humidity control. Table 1 presents main instruments for emission measurements.

**Table 1. Emission measurement instruments.**

| Apparatus | Measured Component | Measurement Principle | Range of Measurement |
|---|---|---|---|
| **Horiba MEXA 7400 and Horiba MEXA ONE GS** | NO | Chemiluminescence Detector (CLD) | 0-10 to 0-10000 ppm |
| | $NO_2$ | Chemiluminescence Detector (CLD) | 0-10 to 0-10000 ppm |
| | CO | Non-dispersive Infrared (NDIR) detector | 0-0.5 to 0-12 vol% |
| | $CO_2$ | Non-dispersive Infrared (NDIR) detector | 0-0.5 to 0-20 vol% |
| | $O_2$ | Magnetopneumatic detector | 0-1 to 0-25 vol% |
| | THC | Flame Ionisation Detector (FID) | 0-10 to 0-20000 ppmC |
| **Cambustion DMS-500** | PN | Electrical differential mobility spectrometry | 5nm-1μm |
| **Horiba OBS ONE PN** | PN | Condensation Particle Counter | 23nm-1μm |

For the RDE tests, a Horiba OBS-ONE PEMS system was installed into the research vehicle. This included the OBS-ONE GS (Gas) and particle number (PN) units for gaseous and particulate emission measurement along with a central control unit and B-size pitot tube with thermocouple and sample points that interfaced with the central control unit via a pitot flow meter module unit. Electricity was supplied to the equipment through a power supply (PS) unit. In addition to these units, there was an OBD connector, GPS tracker, and weather station. Fuel consumption was calculated based on fuel data input by the user.

A GPS tracking unit produced by Racelogic (VBOX Lite II module) was utilised to accurately log the movement of the vehicle at 20 Hz frequency. An RDE route around the city of Leeds that adhered to the requirements of the regulatory RDE test was designed and used.

A set of batteries were placed into the rear foot-wells of the Toyota Prius research vehicle to power the on-board equipment, allowing the vehicle's 12V power supply to remain unused during testing. The battery setup consisted of two 655HD Yuasa Cargo Heavy Duty 12V, 125Ah Batteries connected in series, together connected in parallel to two EXV140 Enduroline 12V, 142Ah Leisure Batteries also connected in series. This gave an overall capacity of 267Ah at 24V output, which was then inverted from direct current (DC) to alternating current (AC) by a Studer AJ 2400-24 inverter and used to power a 12-way multi-socket.

### 2.3 Fuels

Four testing fuels were used in the current research, including a pure fossil gasoline without ethanol as the base fuel (referred to as E0 with 95 RON), a 10% by volume blend of ethanol in the base fuel (referred to as E10), a 10% by volume blend of pure n-butanol in the base fuel (referred to as B10) and a 5% ethanol market gasoline from a petrol station. The E0 base fuel was made from typical gasoline components including paraffins, iso-paraffines, olefins, naphthene and aromatics with representative fractions stipulated by EN228.

Before starting the test of a new fuel, the fuel tank was drained as fully as possible and flushed three times. After this, the vehicle was driven a short distance (approximately 1 mile), either on the chassis dynamometer or on roads, to ensure that the new fuel had completely flushed out the old fuel, and to give the ECU some time to adjust to the new fuel type and its different combustion properties.

## 3 RESULTS AND DISCUSSION

### 3.1 Vehicle behaviour (Speed, SOC, coolant and TWC temperatures)

Figure 1 presents profiles of the vehicle velocity, engine speed, SOC (State of Charge), coolant and TWC temperatures of the WLTC tests for one of the E0, E10 and B10 tests each as examples. All the WLTC tests were commenced under identical cold start and initial hybrid battery SOC conditions and repeated twice for each fuel. In Figure 1c the temperatures given are the upstream TWC ("TWC 1"), downstream TWC ("TWC 2") and the engine-out exhaust gas temperature. The engine speed profiles show that the biofuels do not cause large differences in engine behaviour, except for a small extra burst of activity toward the end of the warm-up strategy of the B10 test at 100 s into the test, and an additional small duration ignition event at around 350 s for E10. The use of these fuels causes small changes to the temperatures visible in Figure 1c, but no significant changes to engine coolant temperature in Figure 1b. Figure 1b shows that the SOC across the three tests remains consistent, with the tests commencing and finishing with the same SOC (within 0.4% of one another). Over the course of the test, the motor in the B10 test appears to be doing more work, visible as a dip in SOC compared to the E0 and E10 tests. Generally, all values appear close to one another. From the first 100 s of the three tests, the E10 blend has the lowest exhaust gas temperature, followed by the E0 test, then the butanol at a marginally higher temperature. From this data, it can be inferred that the combustion temperature is the lowest for E10, in agreement with literature [5-7]. The finding that butanol has very similar or slightly higher exhaust gas temperature compared to gasoline is in agreement with [8,9] from PFI (Port Fuel Injection) research engine work. The lower exhaust temperatures from the E10 test could be attributed to more advanced combustion phasing due to a boost of RON by ethanol. As the boost effect of n-butanol to RON is moderate (lower than ethanol) and thus less advanced combustion phasing compared to ethanol, it is therefore the exhaust temperature from the B10 test was only slightly different from the E0 test. This will have consequent effect on engine out emissions, but the impact could be mitigated or diminished by the effect of the TWC for tailpipe emissions.

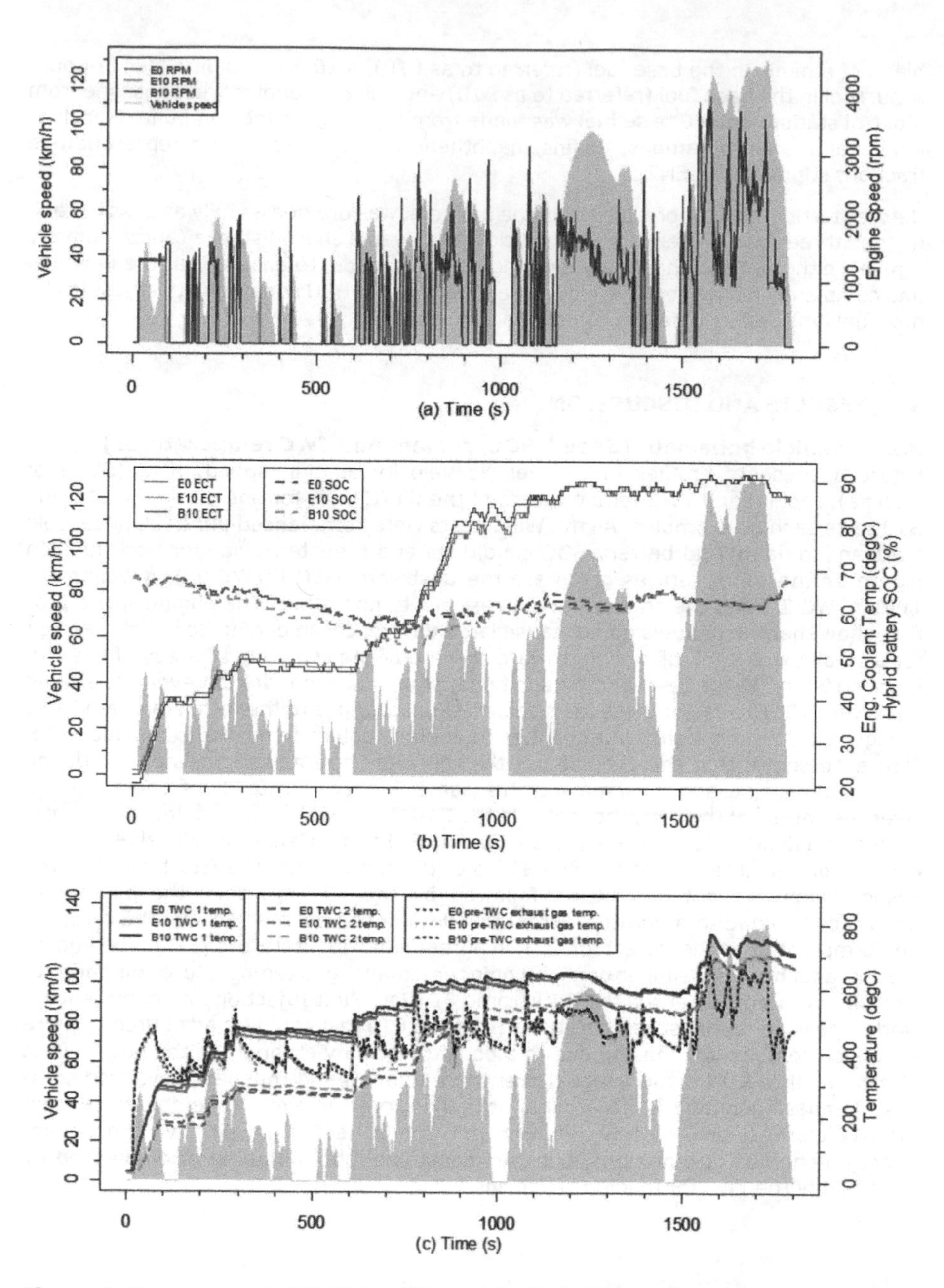


**Figure 1. One example WLTC test for each of E0, E10 and B10 fuels (the grey is the vehicle speed in each figure), outlining (a) Engine speed (b) SOC and engine coolant temperature profiles and (c) Temperatures around the TWC.**

### 3.2 Fuel consumption

Figure 2 shows the average fuel consumption from repeated RDE tests for four fuels. For the RDE tests, 3-5 repeats for each fuel were done. The energy densities of ethanol and n-butanol are lower than that of gasoline (approximately 27, 34 and 43 MJ/kg or 21, 28 and 32 MJ/l respectively), so some fuel consumption penalty is expected. Little research has been done to assess this penalty for HEVs specifically, but the results presented in Figure 2 show that the E10 blend gave the highest fuel consumption (~15%), followed by the B10 (~10%), compared to the E0 fuel. The gap in fuel consumption between E5 and E0 is very small but the E5 fuel was purchased at a commercial petrol station while the E10 and B10 fuels were splash blended using E0. The hydrocarbon composition of the E5 could be different from the E10 and B10 fuels.

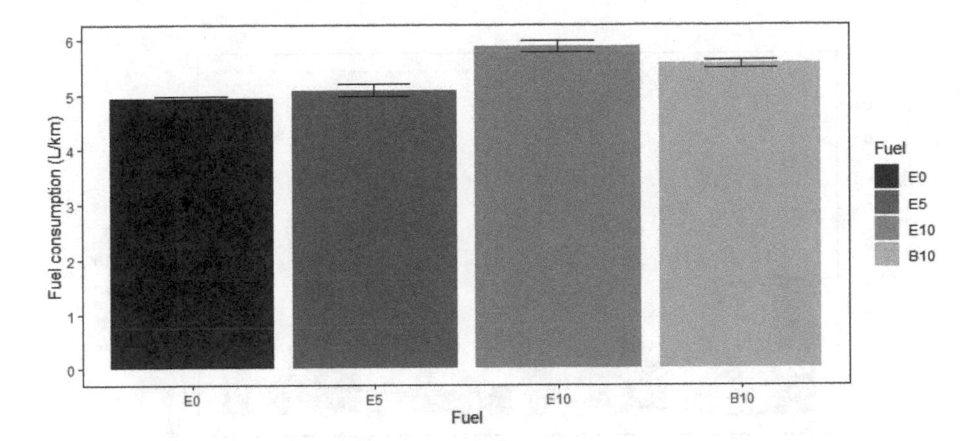


**Figure 2. Fuel consumption across repeated HEV RDE tests fuelled by E0, E5, E10 and B10. Uncertainties are the standard errors of the distributions.**

### 3.3 Gaseous emissions

#### 3.3.1 *CO emissions*

CO emissions from the WLTC and RDE tests are presented in Figures 3 and 4 respectively. The WLTC results in Figure 3 indicate that both E10 and B10 decreased CO emissions compared to E0 (by 51% and 33% in phase 1, and 28% and 22% for the total WLTC, respectively), in agreement with literature [5,10,11]. The results in this research shows the same trend as reported by Suarez-Bertoa [12] for the PHEV, i.e. decreasing CO with ethanol blend while [12] also reported higher CO for their MHEV. The total WLTC emission factor of 110 ± 10 mg/km for E10 currently presented bridges the gap between their PHEV (62 mg/km) and their MHEV (at 238 mg/km) for E10.

Over the total test, both E10 and B10 gave very similar CO emissions, while in the urban section, B10 resulted in a 36% increase in CO emissions compared to E10 but it was within the range of the error bars. All three fuels gave emissions below the Euro 6 limit of 1g/km.

The RDE test results in Figure 4 show that E0, E10 and B10 follow the same trend as the WLTC for the urban section (21% and 5.5% decreases for E10 and B10 compared to E0), but then E10 - and to a lesser extent, E5 - have greater CO emissions for phases 2 and 3, which are defined as vehicle speed ranges of 60-90km/h and above 90km/h respectively. Canakci et al [5] found that at higher vehicle speeds on the chassis dynamometer, SI vehicle CO emissions increased with ethanol blending, so there is some precedent for this observation as the RDE maintains higher average speeds in the rural driving section, and maintains a high speed for much longer in motorway section, than the WLTC.

B10 also had higher emissions over the rural and motorway (and hence whole RDE) compared to E0, but to a lesser degree than E10. Elfasakhany [13-14] presented a trend of decreasing CO emissions with n-butanol blends (and ethanol blends) compared to E0 from their single cylinder research engine for low engine speeds, but butanol tests had comparable and higher CO emissions at higher engine speeds. This tendency was also presented by Liu et al [15] from their NEDC testing where at low speeds E0 was greater than E20 and B20, while at high speeds E20, then B20 were greater than E0.

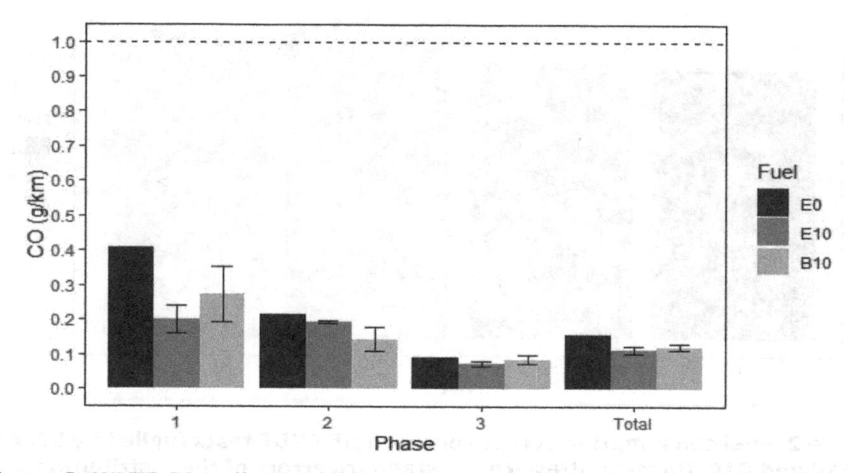


**Figure 3. WLTC CO pollutant emissions per km from phases 1, 2, 3 and total, for E0, E10 and B10 fuels. The Euro 6 emission limit (1g/km) is indicated by a dashed line on the graph.**

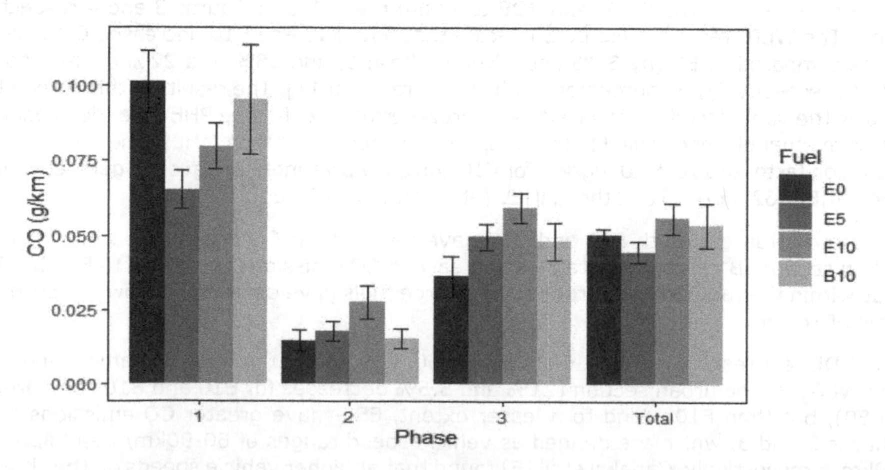


**Figure 4. RDE CO emissions (g/km) from phases 1, 2, 3 and total, for E0, E5, E10 and B10 fuels.**

Figure 5 shows representative transient and cumulative CO emissions from E0, E10 and B10 fuelled tests over the WLTC cycle. The magnitude of E0-fuelled CO spikes appear to be generally larger than those from E10 or B10. Li et al [16] found that during rich combustion, E10 gave lower levels of CO emissions than E0. HEV engine re-ignition events are often accompanied by fuel enrichment [4], so these lower emissions with E10 support this finding. Additionally, the cold start emissions appear to be larger for E0 than B10, and E10 has the smallest emissions. These results agree with [15], who reported improved cold start CO performance resulting from butanol blends to gasoline from a GDI SI engine.

Figure 5b shows a range of cumulative CO mass emissions across a selection of nine representative RDE tests (three each of E0, E10 and B10). Little common trend can be seen across repeats with the same fuel, indicating that real-world variables are having a greater impact on the emissions than the fuel type. So the chassis dynamometer tests are more suitable for the study of fuel types.

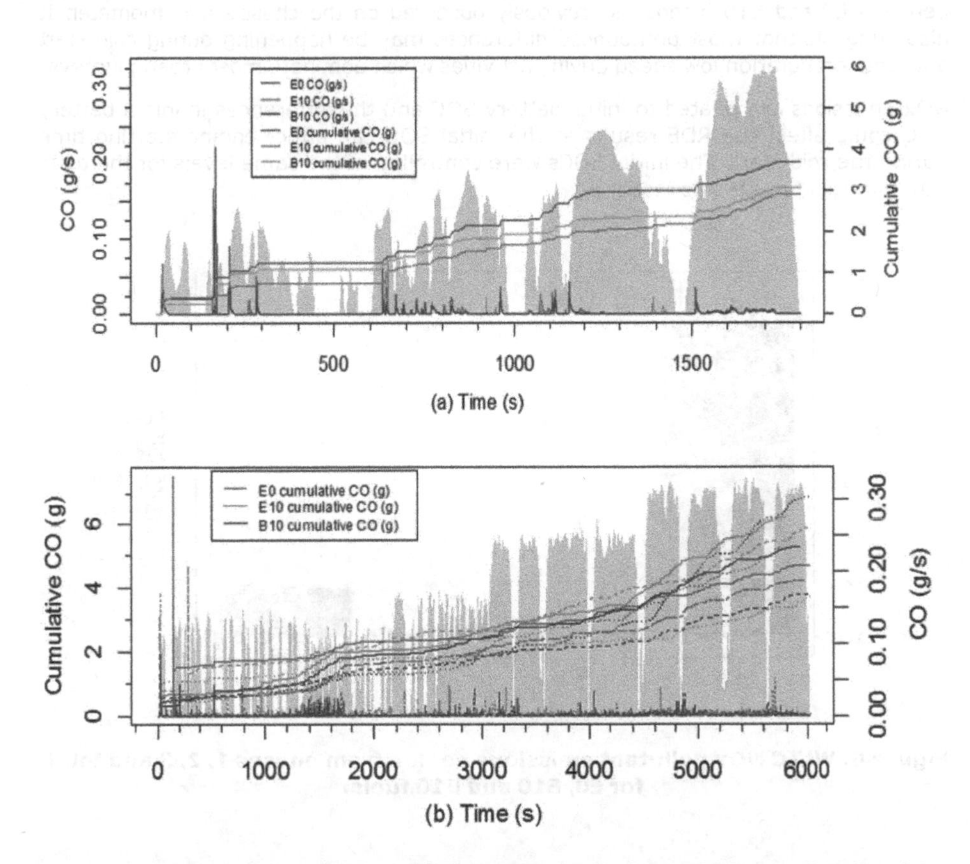


**Figure 5. (a) CO emissions from WLTC tests fuelled by E0, E10 and B10, with speed indicated in grey. (b) CO emissions across 9 replicate RDE tests (3 each of E0, E10 and B10), with example RDE speed in grey.**

### 3.3.2 *NOx emissions*

Figure 6 shows that the emissions of NOx over the WLTC test are very similar across all three fuel types, All fuels are far below the Euro 6 emission limit of 0.06g/km.

These results agree with the majority of previous SI work on the topic for E10 and B10 [12]. As with CO, the NOx results of the HEV in the current study bridge the gap between the MHEV and PHEV noted in [12] for both fuels.

Figure 7 shows that during RDE testing, the difference in NOx emissions between fuel types was greater than from the WLTC testing, with most of this difference originating within the urban section. A 36% decrease was witnessed for both E10 and B10 compared to E0 in the urban RDE section, with 24% and 18% lower emissions for E10 and B10 respectively over full RDE. All results are far below the Euro 6 emission limit of 0.06g/km. These findings agree with refs. [5, 17], who found that at lower vehicle speeds, NOx emissions decreased with inreasing ethanol within the blend, but as vehicle speed increased more similar emissions levels were seen between fuels. More recently, Liu et al [15] found the same for both E20 and B20 compared to E0. The current work is the first to display from on-road testing that NOx emissions between fuels converge as vehicle speed increases and there are only slight overall differences between E0 and E10 blends as previously observed on the chassis dynamometer. It also suggests that most pronounced differences may be happening during cold-start and/or short duration low speed driving activities which dominate most urban journeys.

NOx emissions are related to initial battery SOC and the differences in initial battery SOC could affect the RDE results as the initial SOC can affect engine starting time during the cold start. The initial SOCs were controlled at the same levels for the tests with different fuels in this research.

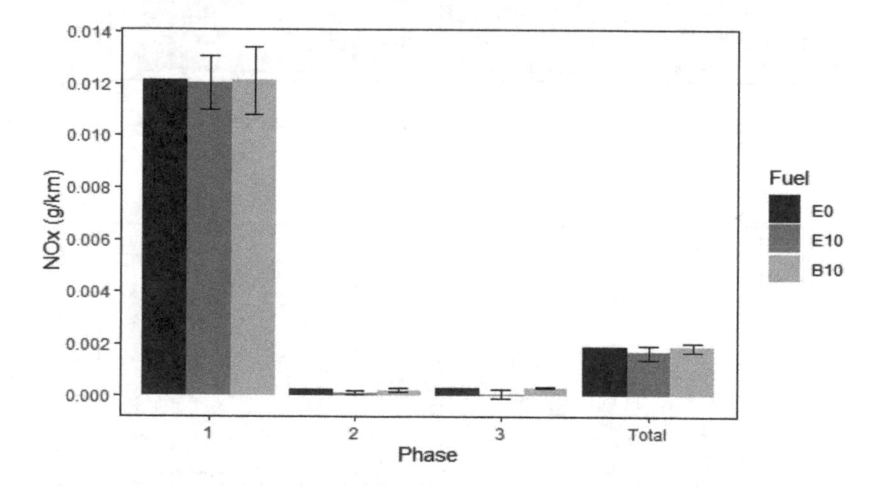


**Figure 6. WLTC NOx pollutant emissions per km from phases 1, 2, 3 and total, for E0, E10 and B10 fuels.**

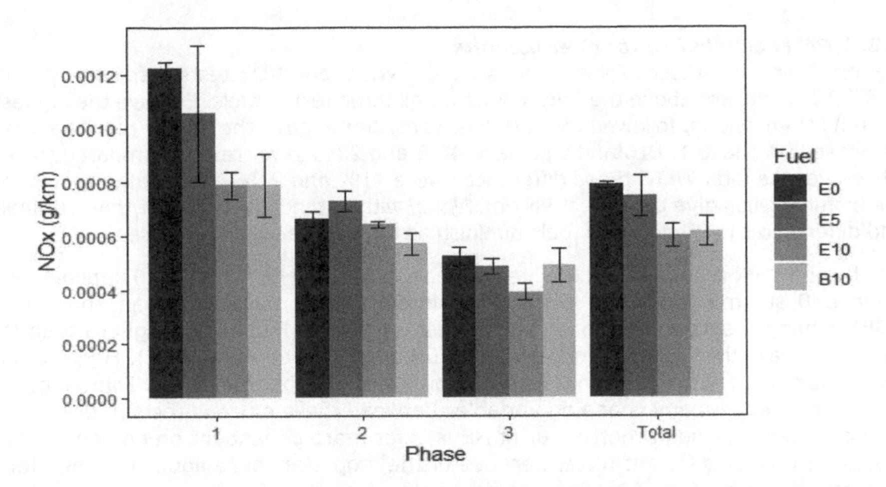


**Figure 7. RDE NOx pollutant emissions per km from phases 1, 2, 3 and total, for E0, E10 and B10 fuels.**

### 3.3.3 *Total hydrocarbon (THC) emissions*

The emissions of THC over the WLTC tests as shown in Figure 8 are very similar across all three fuel types, and well below the EU 6 emissions limit of 0.1g/km. THC emissions are slightly lower under E10 fuelling (6% in phase 1, 3% in total WLTC) and comparable under B10 fuelling (1.6% lower under phase 1, but 4% higher under total WLTC), compared to E0. THC emissions were not measured for the RDE tests as the PEMS did not include a hydrocarbon analyser.

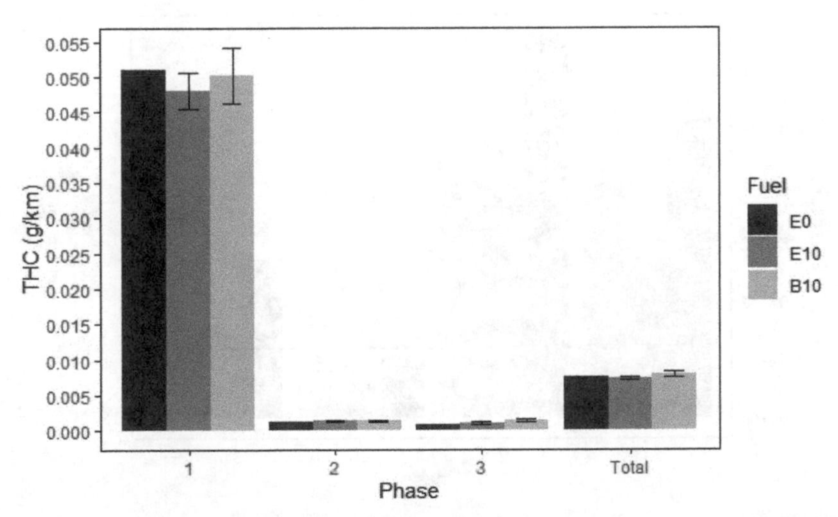


**Figure 8. WLTC THC pollutant emissions per km from phases 1, 2, 3 and total, for E0, E10 and B10 fuels.**

### 3.3.4 *PN (Particle Number) emissions*

Figures 9 and 10 present PN emissions for the WLTC and RDE tests respectively. The WLTC PN results are above the Euro 6 limit for all three tested fuels. E0 gave the highest overall PN emissions, followed by B10. E10 consistently gave the lowest results across all phases. In phase 1, E10 and B10 gave 40% and 23% lower results compared to E0, while over the total WLTC these differences were 41% and 21%. The results show that the biofuel blends give the lowest PN emissions, with butanol being higher than ethanol, and differrences in PN between fuels diminish as the vehicle speed increases.

The trend for PN results of the RDE tests measured by PEMS in Figure 10 appears less clear. E10 still maintains the lowest emissions (4% decrease for urban and whole RDE), while E5, surprisingly, has the greatest emissions. B10 shows higher PN emissions generally than E0 (4% increase over urban, 1% over whole RDE). However all PN emissions are within the range of the error bars for all four fuel types. This indicates that there are too many changing variables between these on-road tests to draw reliable conclusions. The PN behaviour of HEVs is far more dependent on engine behaviour than it is for different fuels, because of the stop-start behaviour. It is therefore unsurprising that on-road testing has not shown such clear correlations between biofuel use and PN emissions for HEVs than previous on-road studies have shown for conventional SI vehicles [10].

It should be pointed out that DMS500 used in the WLTC tests is an electrical mobility spectrometer with a measurement size range of 5-1000 nm while Horiba OBS ONE PN unit used in the RDE tests is a Condensation Particle Counter (CPC) particle instrument which measures the solid particle number concentration with a particle size range of 23-1000 nm. The later is recommended by the Particle Measurement Programme (PMP) for the regulated WLTC test.

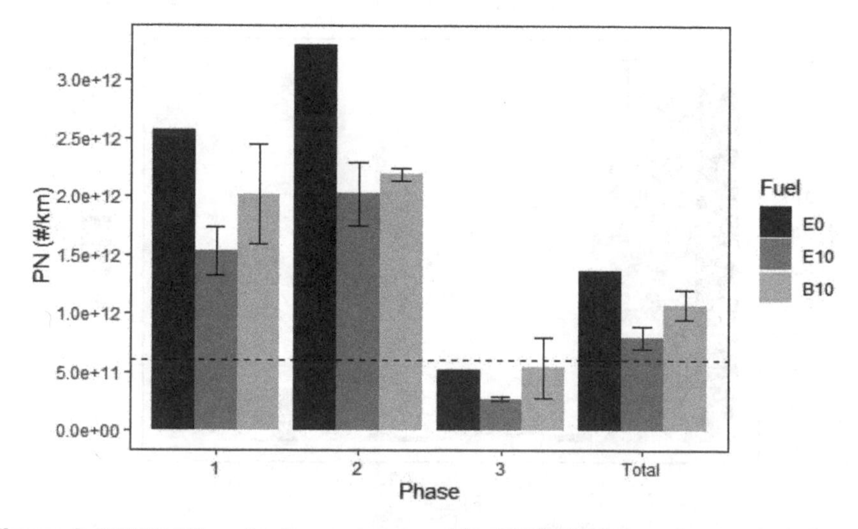


**Figure 9. WLTC PN emissions measured by DMS500 from phases 1, 2, 3 and total, for E0, E10 and B10 fuels. The Euro 6 PN emissions limit of $6x10^{11}$ #/km is indicated by a dashed line.**

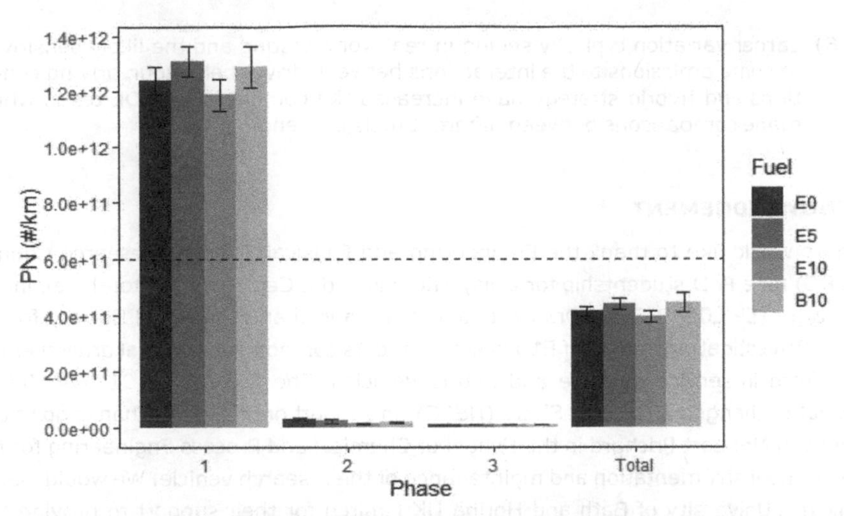


**Figure 10. RDE PN emissions measured by PEMS from phases 1, 2, 3 and total, for E0, E10 and B10 fuels. The Euro 6 PN emissions limit of $6x10^{11}$ #/km is indicated by a dashed line.**

## 4 CONCLUSION

Tailpipe emissions and fuel consumption were investigated over WLTC and RDE tests using a HEV vehicle (Toyota Prius) in a chassis dynamometer lab and on an RDE compli-ent route. The tests were dividied into the first phase (cold start and urban driving), the second (rural driving) and third (motorway driving) phases. Four fuels were tested: pure fossil gasoline (E0), E5 (5% ethanol, RDE only), E10 (10% ethanol) and B10 (10% n-Butanol). The following conclusions can be drawn from the current research:

1) The greatest benefit from biofuel blends (E10 and B10) is the reduction of PN emissions over all phases of the WLTC tests, which were decreased by 20-50% compared to E0. However, the HEV has not shown the same and clear PN trends under RDE testing because of a large variation of traffic conditions and engine activities in the real world driving. The PN results are very susceptible to differences in any other variables that change the engine speed trace of the vehicle (e.g. initial SOC which was investigated by the authors but not reported due to the page limit). This fact is important to note when assessing future experimental HEV PN results.
2) Fuel consumption from the E10 and B10 tests was 10 to 15% higher than the E0 tests. Mininmal difference between the E0 and E5 was observed.
3) The E10 and B10 fuels reduced CO emissions signficantly in the first phase (cold start) of the WLTC and RDE tests. The benefits of biofuels reduced or diminished for the second and third phases, especially for the RDE tests, which could be attributed to an artefact caused by the increased distance of the RDE test minim-ising the impact of the cold start section, where most of the difference lies.
4) There is no discernible difference in NOx emissions between fuels from the WLTC tests. However, the E10 and B10 fuels did show an approximately 30% reduction in NOx emissions in the cold sart or the first phase of the RDE tests, and reductions for the trip totals as well.
5) THC emissions were only measured in the WLTC tests and there were no noticeable differences observed between fuels.

6) Larger variation typically seeing in real world testing and the likely sensitivity of some emissions to the interactions between driver behaviour, driving conditions and hybrid strategy have increased the complexity of RDE tests, which made comparisons between different fuels challenging.


## ACKNOWLEDGEMENT

Authors would like to thank the Engineering and Physical Sciences Research Council (EPSRC) for a PhD studentship for Daisy Thomas in the Centre for Doctoral Training in Bioenergy (EP/L014912/1). Drs Hu Li and Xin Wang thank the Royal Society for an award "Investigating impact of E10 gasoline and its compositions on real driving emissions from in-service gasoline and hybrid vehicles, The IEC\NSFC\191747 - International Exchanges 2019 Cost Share (NSFC)" in support of this work. Thanks go to the technician Mr Scott Prichard in the School of Chemical and Process Engineering for his help on the instrumentation and maintenance of the research vehicle. We would like to thank the University of Bath and Horiba UK Limited for their support to provide the chassis dynamometer test facility and PEMS.


## ABBREVIATIONS

| | |
|---|---|
| BEV | Battery Electric Vehicle |
| CPC | Condensation Particle Counter |
| ECU | Electronic Control Unit |
| ECT | Engine Coolant Temperature |
| EGR | Exhaust Gas Recirculation |
| GHG | Greenhouse Gas |
| HEV | Hybrid Electric Vehicle |
| ICE | Internal Combustion Engine |
| NOVC | Not Off Vehicle Charging HEVs |
| OBD | On-Board Diagnostics |
| PEMS | Portable Emission Measurement System |
| PFI | Port Fuel Injection |
| PHEV | Plug-in Hybrid Electric Vehicle |
| PMP | Particle Measurement Programme |
| PN | Particle Number |
| RDE | Real Driving Emissions |
| SI | Spark Ignition |
| SOC | State Of Charge |
| THC | Total Hydrocarbon |
| TWC | Three-Way-Catalyst |
| WLTC | World-harmonised Light Duty Test Cycle |

## REFERENCES


[1] Department for Transport, Decarbonising Transport - Setting the Challenge. 2020.

[2] Parry, T. 2019.Vehicle Licensing Statistics: Annual 2019 [Online]. Available from: https://assets.publishing.service.gov.uk/government/uploads/system/uploads/attachment_data/file/882196/vehicle-licensing-statistics-2019.pdf.

[3] Lapuerta, M.; García-Contreras, R.; Campos-Fernández, J.; Dorado, M. P. Stability, Lubricity, Viscosity, and Cold-Flow Properties of Alcohol–Diesel Blends. Energy Fuels 2010, 24 (8), 4497–4502.

[4] Thomas.B.D. Real World Driving Emissions from Hybrid Electric Vehicles and the Impact of Biofuels. PhD Thesis, Unviersity of Leeds. 2020.

[5] Canakci, M., Ozsezen, A.N., Alptekin, E. and Eyidogan, M. 2013. Impact of alcohol–gasoline fuel blends on the exhaust emission of an SI engine. Renewable Energy. 52, pp.111–117.

[6] Eyidogan, M., Ozsezen, A.N., Canakci, M. and Turkcan, A. 2010. Impact of alcohol-gasoline fuel blends on the performance and combustion characteristics of an SI engine. Fuel. 89(10),pp.2713–2720.

[7] Masum, B.M., Kalam, M.A., Masjuki, H.H., Palash, S.M. and Fattah, I.M.R. 2014. Performance and emission analysis of a multi cylinder gasoline engine operating at different alcohol–gasoline blends. RSC Advances. 4, pp.27898–27904.

[8] Singh, S.B., Dhar, A. and Agarwal, A.K. 2015. Technical feasibility study of butanol-gasoline blends for powering medium-duty transportation spark ignition engine. Renewable Energy. 76, pp.706–716.

[9] Varol, Y., Oner, C., Oztop, H.F. and Altun, S. 2014. Comparison of Methanol, Ethanol, or n-Butanol Blending with Unleaded Gasoline on Exhaust Emissions of an SI Engine. Energy Sources Part A: Recovery Utilization and Environmental Effects. 36(9),pp.938–948.

[10] Vojtisek-Lom, M., Beranek, V., Klir, V., Stolcpartova, J. and Pechout, M. 2015. Effects of n-Butanol and Isobutanol on Particulate Matter Emissions from a Euro 6 Direct-injection Spark Ignition Engine During Laboratory and on-Road Tests. SAE International Journal of Engines. 8(5),pp.2338–2350.

[11] Hernandez, M., Menchaca, L. and Mendoza, A. 2014. Fuel economy and emissions of light-duty vehicles fueled with ethanol–gasoline blends in a Mexican City. Renewable Energy. 72, pp.236–242.

[12] Suarez-Bertoa, R. and Astorga, C. 2016. Unregulated emissions from light-duty hybrid electric vehicles. Atmospheric Environment. 136, pp.134–143.

[13] Elfasakhany, A. 2014. Experimental study on emissions and performance of an internal combustion engine fueled with gasoline and gasoline/n-butanol blends. Energy Conversion and Management. 88, pp.277–283.

[14] Elfasakhany, A. 2017. Investigations on performance and pollutant emissions of spark-ignition engines fueled with n-butanol–, isobutanol–, ethanol–, methanol–, and acetone–gasoline blends: A comparative study. Renewable and Sustainable Energy Reviews. 71, pp.404–413.

[15] Liu, H., Wang, X., Zhang, D., Dong, F., Liu, X., Yang, Y., Huang, H., Wang, Y., Wang, Q. and Zheng, Z. 2019. Investigation on blending effects of gasoline fuel with n-butanol, DMF, and ethanol on the fuel consumption and harmful emissions in a GDI vehicle. Energies. 12(1845),pp.1–21.

[16] Li, Y., Gong, J., Deng, Y., Yuan, W., Fu, J. and Zhang, B. 2017. Experimental comparative study on combustion, performance and emissions characteristics of methanol, ethanol and butanol in a spark ignition engine. Applied Thermal Engineering. 115, pp.53–63.

[17] Kalita, M., Muralidharan, M., Subramanian, M., Sithananthan, M., Yadav, A., Kagdiyal, V., Sehgal, A.K. and Suresh, R. 2016. Experimental Studies on n-Butanol/Gasoline Fuel Blends in Passenger Car for Performance and Emission. SAE Technical Papers. 2016-Octob.

# Ultra-low NOx and PN with integrated emission control systems for light-duty gasoline and heavy-duty diesel vehicles

**J. Demuynck, P. Mendoza Villafuerte, D. Bosteels**
Association for Emissions Control by Catalyst, AECC AISBL, Belgium

## ABSTRACT

Public data shows Euro 6d and Euro VI-D vehicles overall have low on-road emissions. But there remain areas where higher emissions occur for both light- and heavy-duty. Two ultra-low emission demonstrator vehicles were built to investigate further emission reductions through an integrated approach of emission control technologies.

On a gasoline car, a close-coupled three-way catalyst (TWC) is put in combination with an underfloor catalysed gasoline particulate filter (cGPF), second TWC and ammonia slip catalyst (ASC). The work focuses on further reducing the initial cold-start emissions by using a ccTWC substrate with maximised surface area in combination with early closed-loop lambda control activation. The emission control system is integrated within a 48V mild-hybrid powertrain.

On a diesel truck, the system integrates a close-coupled DOC, a catalysed DPF, dual-SCR system (one in a close-coupled position before the DPF), with twin AdBlue® dosing. Both SCR catalysts contain an ammonia slip catalyst. The innovative system layout allows ultra-low NOx emissions in even the most challenging low-load and urban driving conditions including cold-start.

Pollutant emissions were evaluated for both vehicles over a broad range of operating conditions, with tests in the lab and on the road. Results show ultra-low pollutant emissions are achieved for all regulated pollutant emissions, while controlling the currently non-regulated emissions (PN10, $NH_3$ and $N_2O$ are measured with a prototype PEMS).

The demonstrator vehicles show efforts to further reduce the impact of internal combustion engines on air quality. The impact on climate change is reduced in the case the vehicle operates on a sustainable renewable fuel.

## 1 INTRODUCTION

European emission standards for light and heavy-duty have promoted innovation in catalyst and filter technology design, emissions control system layout and system control within an integrated approach of powertrain development. This continuous emission reduction has had a significant impact on European air quality. State-of-the-art systems of light-duty Euro 6d vehicles on the market, of which two examples are shown in Figure 1, evolved to integrated systems with combination of technologies for both diesel and gasoline passenger cars.

DOI: 10.1201/9781003219217-16

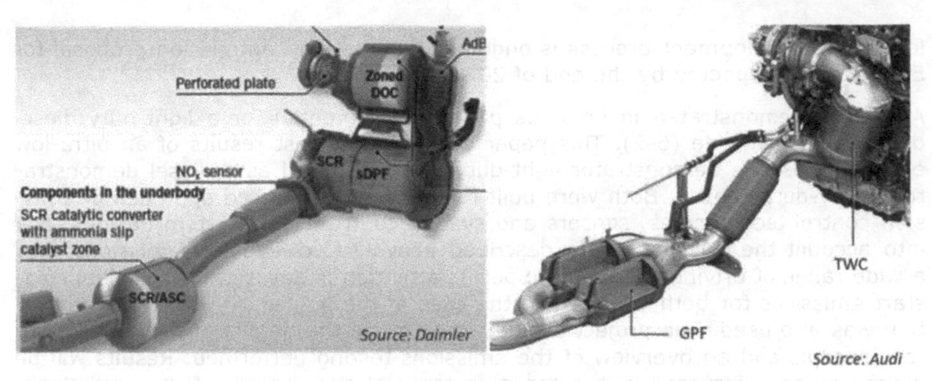


**Figure 1. Emission control system layouts of Euro 6d diesel (left) and gasoline (right) passenger cars.**

Same is depicted for heavy-duty vehicles, with the example of an announced system in Figure 2, adding close-coupled catalysts to allow a faster heat-up of the emission control system.

**Figure 2. Announced closed-coupled emission control system layout for heavy-duty diesel trucks.**

The exemplary systems shown above illustrate common design criteria to handle real-world operation emissions. No matter whether it is a car, van, truck or bus, it is important to have following elements: (a) close-coupled catalysts for cold-start and low speed/load driving in the city, (b) underfloor catalysts for high speed/load driving on the motorway and (c) total catalyst and filter volume to cope with peak engine pollutant flow.

On-road NOx and PN emissions have reduced significantly as a consequence of the evolution in emission control systems and this is confirmed by OEM data at Type Approval (1-2) and independent third-party testing (3-5). Further evolution for light- and heavy-duty is expected towards Euro 7/VII for which the legislative development process is ongoing. A European Commission proposal for Euro 7/VII is expected by the end of 2021.

AECC has demonstrated in previous papers the potential for a light-duty diesel demonstrator vehicle (6-7). This paper describes the test results of an ultra-low emission gasoline demonstrator light-duty vehicle as well as a diesel demonstrator heavy-duty vehicle. Both were built following an integrated approach of emission control technologies, sensors and system controls. The system designs took into account the design criteria described above to achieve low emissions over a wide range of driving conditions. Specific attention is devoted to the initial cold-start emissions for both vehicles. In the case of the light-duty vehicle, hybridisation was also used. The project set-up will describe the details of the demonstrator concepts and an overview of the emissions testing performed. Results will be presented and discussed with a focus on some of the aspects of the preliminary data obtained. Further work is intended, the next steps will be described as part of the outlook.

## 2 DEMONSTRATOR VEHICLES SET-UP

### 2.1 Light-duty gasoline

#### 2.1.1 *Vehicle and powertrain characteristics*

The demonstrator vehicle was built starting from a Euro 6d C-segment base car. The vehicle powertrain consists of a 4-cylinder, 110kW gasoline engine with direct injection. The engine is equipped with variable valve timing and cylinder deactivation. The powertrain includes a 48V mild-hybrid system in P0 configuration (belt starter- generator). The electric component can deliver up to 9kW as a motor and captures a maximum of 12kW as a generator. An open engine control unit was available to implement the control measures described below.

#### 2.1.2 *Emission control system layout and controls*

The original emission control system of the vehicle was removed and replaced by the layout shown in Figure 3. It consists of a close-coupled three-way catalyst (TWC) in combination with an underfloor catalysed gasoline particulate filter (cGPF), second TWC and ammonia slip catalyst (ASC). TWC and GPF are technologies already used in vehicles on the market. Focus in this project has been on implementing a close-coupled TWC substrate with maximised surface area for enhanced cold-start performance. The catalysed GPF is combining high filtration efficiency for particulates in combination with minimised impact on back-pressure and $CO_2$ emissions. The total TWC volume is spread over a total of three components and targets to control peak engine-out emissions flow. The work investigated the ASC operation strategy for gasoline vehicles in addition to improved lambda control, which is one of the most novel aspect of the project as this technology is not yet implemented on gasoline vehicles on the market. The ASC implemented relies on a combination of storage and oxidation functionality to reduce $NH_3$ emissions. The emission control components used in the project were bench aged, targeting 160k km.

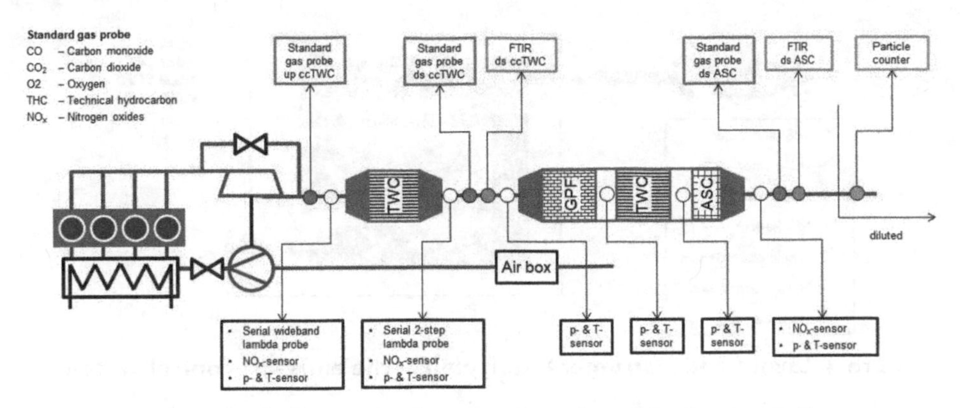


**Figure 3. Emission control system layout and instrumentation.**

The system control relies on as early as possible activation of closed-loop lambda control to match the early light-off of the close-coupled TWC to support minimising gaseous emissions during the cold-start phase. A combination of wideband lambda sensor (before TWC) and 2-step lambda sensor (after TWC) is used. Engine bench calibration work was conducted to fine-tune the lambda control strategy to the catalysts implemented. The control strategy aims for zero NOx emissions in steady state conditions after the close-couple TWC, with minimised emissions of CO and $NH_3$. The demonstrator work relies on passive regeneration of the particulate filter.

## 2.2 Heavy-duty diesel

### 2.2.1 *Vehicle and powertrain characteristics*

The heavy-duty vehicle was built on a N3 4x2 tractor equipped with a 12.8l engine with high pressure EGR and homologated to Euro VI-C. The rated power of the engine is 450hp at 1600rpm.

### 2.2.2 *Emission Control Technology layout*

The design considered the OEM original system layout and packaging of the truck. Much of the new emission control system was installed within a box equivalent to that of the baseline configuration.

As shown in Figure 4, a close-coupled Diesel Oxidation Catalyst (ccDOC) was fitted directly behind the turbine for fast CO and HC control and to allow optimal heat transfer into the emissions control system. The outlet cone of the DOC was modified to integrate a urea injector, allowing the use of the downpipe and compensator for optimal mixing of the injected urea before entering the close-coupled Selective Catalytic Reduction (ccSCR) catalyst. The ccSCR has a zone coated Ammonia Slip Catalyst (ASC) to ensure minimized secondary emissions creation. This close-coupled system was integrated with the purpose of improving the cold-start, low temperature and city driving emission performance.

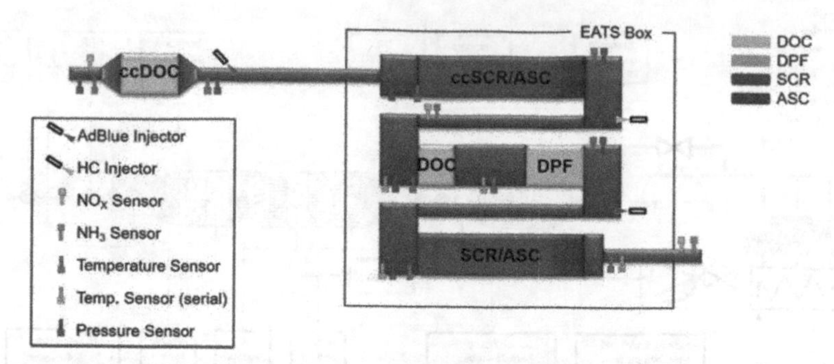


**Figure 4. Layout and instrumentation plan of the emission control system.**

Downstream of the close-coupled SCR system, the system layout resembles that of a conventional truck emission control system design, containing a DOC and Diesel Particulate Filter (DPF) with integrated Hydrocarbon (HC) dozer for DPF regeneration support. Downstream of the DPF, there is a second urea injector and mixing pipe before the second SCR with an integrated ASC to minimize ammonia slip.

## 3 EMISSIONS TESTS

Pollutant emissions were evaluated for both vehicles over a broad range of operating conditions, with tests in the lab and on the road.

### 3.1 PEMS equipment integration

The on-road results include tests performed with and without Portable Emissions Measurement System (PEMS). For the tests where the PEMS was not available, measurements from on-board sensors installed in the tailpipe were used. But most of the tests relied on the results of an enhanced PEMS that was fitted to the demonstrator vehicles as shown in Figure 5 (LD gasoline at the top, HD diesel at the bottom).

The PEMS kit (AIP) contained NO (Chemiluminescence Detector) and $NO_2$ (Photoacoustic spectroscopy) analyzers to determine tailpipe NOx speciation, as well as CO and $CO_2$ measurement devices (Non-Dispersive Infrared Sensor). In addition to the gaseous measurements, PN10 measurement equipment was also fitted. NO is measured using chemiluminescence spectroscopy. $NO_2$ is measured using photo acoustic sensor technology. CO and $CO_2$ were measured using nondispersive infrared spectroscopy (NDIR) and the PN10 is measured using a condensing particle counter (CPC).

**Figure 5. PEMS installed on the LD gasoline (top) and HD diesel (bottom) vehicle.**

### 3.2 Light duty emissions tests

A range of on-road and chassis dyno tests were conducted to investigate the emissions performance of the vehicle. The ambient and driving conditions covered include conditions within and beyond the Euro 6d RDE boundary conditions. The driving style was varied and covered normal, smooth and aggressive (beyond Euro 6d RDE). An RDE derived trace is also used for tests on the chassis dyno, this test is at the boundary of the Euro 6d RDE dynamic conditions and is labelled to be aggressive (a).

The chassis dyno RDE test is conducted at 23°C and -10°C. WLTC is conducted at 23°C as well. A period of wintertime during the calibration phase of the project was used to collect on-road test data when ambient temperatures varied between -5°C and 2°C in addition to tests with ambient temperature around 20°C. Most of this data was collected over a shorter on-road test of around 20 km, compared to the RDE road test of around 80 km that was conducted at -5°C. This test was used to collect calibration input data, covering shorter periods of urban, rural and motorway driving. The results of this type of tests will be shown as well and will be labeled as 'CaliTest'. Tests are always conducted with a cold-start after the vehicle has been soaked at the ambient temperature of the test.

Most of the data in this paper will cover the lambda control as it is implemented on the base Euro 6d vehicle, but finetuning the calibration parameters to the catalysts implemented. This will be labelled as 'serial lambda control'. A further finetuning of the calibration will be looked at as well for NOx, which will be labelled as 'calibration modification'. The second calibration status is with following two modifications to decrease initial engine-out emissions and decrease emission control heat-up time: 1) earlier closed-loop lambda control activation with variation of start lambda value (from 9 to 2.5 seconds after engine start); 2) retarded spark timing (between 3 and 20 degrees crank angle).

### 3.3 Heavy-duty emissions tests

For this vehicle, this paper covers on-road measurements only. The tests were defined to include a broad range of driving conditions which covered the largest possible range from the operating map of the engine. Initial results show the system performing robustly during in-service conformity routes as well as during urban delivery routes, these initial results were previously reported (8). The results reported in this paper extend the range of tests covered including results obtained from further on-road measurements with PEMS.

Urban and regional delivery conditions as well as regular in-service conformity routes were conducted. These trips were conducted with different payloads and on different days to have some ambient temperature variation. The urban delivery trips were used to understand the impact of driving conditions which include a percentage of stop and go operation in keeping the system at optimal operating temperature.

## 4 RESULTS AND DISCUSSION

### 4.1 Light-duty gasoline

#### 4.1.1 *NOx emissions*

Figure 6 gives an overview of the cumulative NOx emissions with the serial lambda control. RDE data is plotted in blue for an on-road test at -5°C with normal driving (n) and the aggressive RDE on chassis dyno (RDE a) at -10°C and 23°C. On-road data is included from the shorter calibration test (CaliTest) in orange as well, with ambient temperatures between -5°C and 23°C depending on the weather conditions. The 3 different driving styles described above have been tested: normal (n), smooth (s) and aggressive (a).

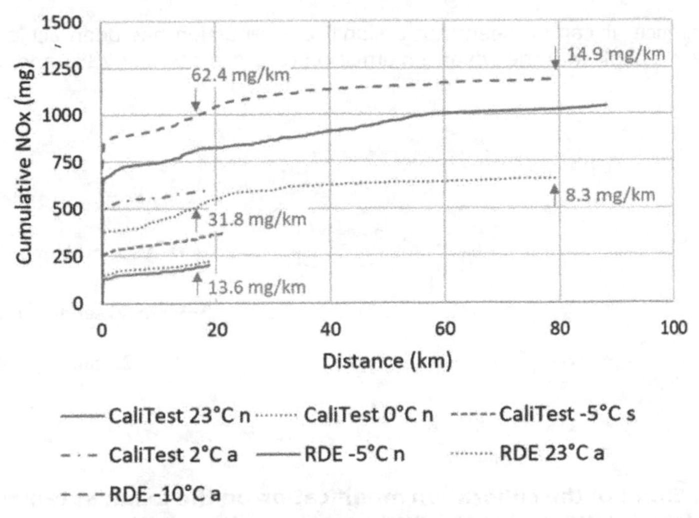


**Figure 6. Cumulative NOx emissions.**

The majority of the emissions are emitted in the first kilometer of the test during the initial cold-start phase. The amount of cold-start NOx emissions is 140 mg at 23°C with normal driving. This increases when the vehicle is tested at a lower ambient temperature and with a more aggressive driving style. A maximum of 846 mg is measured for the aggressive RDE test on chassis dyno at -10°C within the range of test conditions covered. Emissions during the rest of the test, once the close-coupled TWC reached its light-off temperature, are independent from the ambient temperature.

Figure 6 also includes the range of mg/km values obtained from an evaluation at the distance indicated by the arrow. Urban NOx emissions are evaluated at the Euro 6d RDE minimum trip distance of 16 km including the cold-start. These vary from 13.6 mg/km (CaliTest 23°C n) over 31.8 mg/km (RDE 23°C a) up to 62.4 mg/km (RDE -10°C a) for the range of test conditions covered. Total RDE NOx emissions are evaluated at the end of the trip. These vary between 8.3 mg/km and 14.9 mg/km.

As described above, a calibration modification was applied to investigate further emission reduction potential on the RDE aggressive test on the chassis dyno at 23°C and -10°C. The effect of the modification on the exhaust temperature before the ccTWC can be seen in Figure 7, by comparing the green and blue line for each test. The ccTWC heats up faster, about 1 second is gained in both tests. The resulting decrease in the initial cold-start NOx emissions at 23°C is shown in Figure 8. The reduction is not observed to the same extent at -10°C. The $COR_2R$ impact of this measure is evaluated to be 0.2 % over the entire RDE test and 1 % on WLTC.

For reference, also the measurement of the base Euro 6d vehicle and its serial emission control system is added in Figure 8, plotted in a black colour. These tests cannot be compared directly as the base vehicle contained a fresh emission control system whereas an aged status was tested for the advanced emission control system. But despite this ageing difference, it can be seen that a significant reduction has been achieved in the initial cold-start peak with the advanced emission control system at 23°C and -10°C.

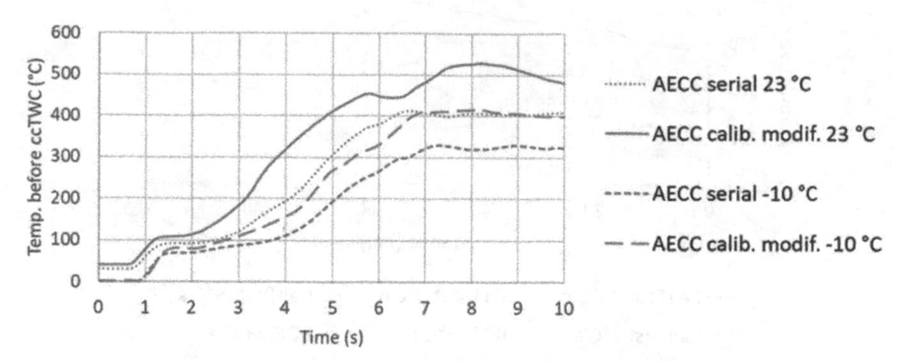


**Figure 7. Effect of the calibration modification on the exhaust temperature before ccTWC during the RDE aggressive test on chassis dyno.**

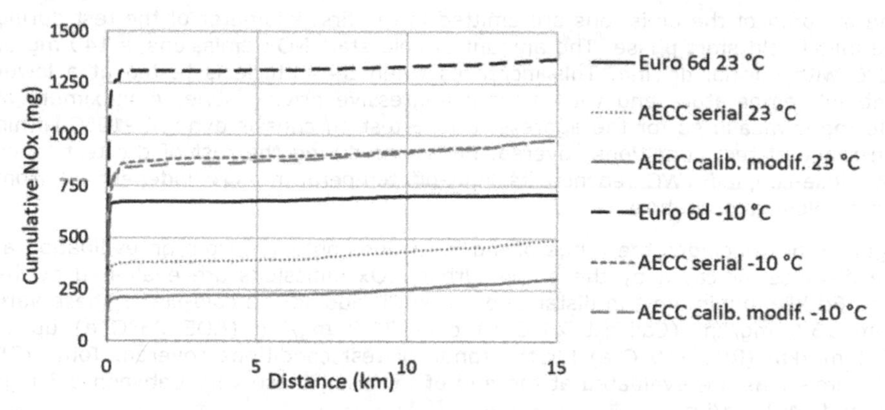


**Figure 8. Effect of the calibration modification on the NOx emissions during the RDE aggressive test on chassis dyno.**

### 4.1.2 *PN emissions*

It is to be noted that the soot and ash accumulation during the ageing of the emission control system supports the filtration efficiency. The PN emissions of the fresh filter are not examined in this paper. Figure 9 shows all PN10 results measured with the serial lambda control, which corresponds to the NOx results of Figure 6. A similar effect described for NOx can be observed for the particulate emissions. The initial cold-start effect is even more pronounced as the rest of the test produces near-zero emissions. The higher amount of particulate emissions during the initial cold-start is due to a combination of increase in engine-out emissions and lower filtration efficiency of the filter.

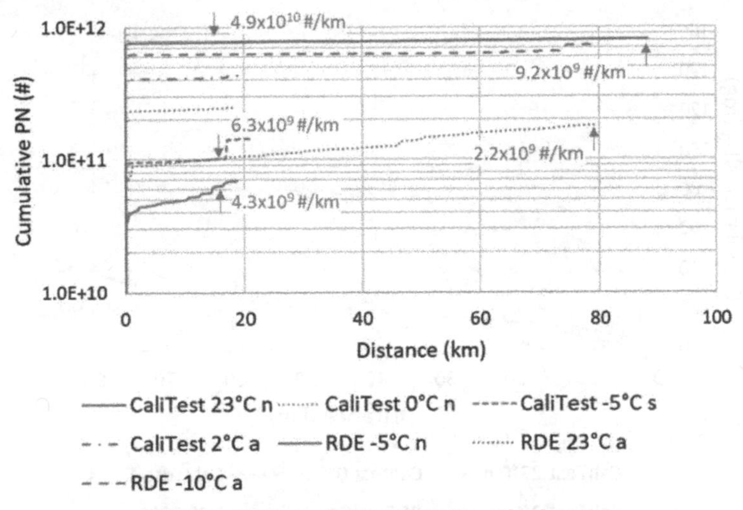


**Figure 9. Cumulative PN10 emissions.**

The initial accumulated cold-start PN emissions are $4x10^{10}$ particles for the CaliTest with normal driving at 23°C. This increases for lower ambient temperatures and with more aggressive driving up to $7.6x10^{11}$ particles. The effect is not as clear as for NOx. The impact of the filter status at the beginning of the test results in a higher test-to-test variability and this is mixed with the effects tested. As a consequence, it is not the RDE aggressive chassis dyno test at -10°C that results in the highest PN10 emissions, but the on-road RDE test at -5°C with normal driving. Minor emission slip is observed during the remaining part of some of the emission tests.

Evaluation of the urban PN10 emissions at 16 km including the initial cold-start shows emissions vary from $4.3x10^9$ #/km (CaliTest 23°C n) over $6.3x10^9$ #/km (RDE 23°C a) up to $4.9x10^{10}$ #/km (RDE -5°C n). Total RDE PN10 emissions vary between $2.2x10^9$ #/km and $9.2x10^9$ #/km.

### 4.1.3 *$NH_3$ emissions*

Figure 10 shows the collected $NH_3$ data over the range of tests with serial lambda control. In contrast to the other species explained above, there is no cold-start peak under moderate conditions at around 20°C and with normal driving. The storage functionality of the ASC is capable of capturing the cold-start effect that is visible with TWC only emission control system. The $NH_3$ emissions increase mainly due to the aggressive driving style, because the lambda value is drifting to richer conditions due to the harsh accelerations. Emissions start to occur earlier during the cold-start phase, as the storage function is not capable to capture all emissions coming. Additionally, a higher slip of $NH_3$ emissions occur during the warm phase.

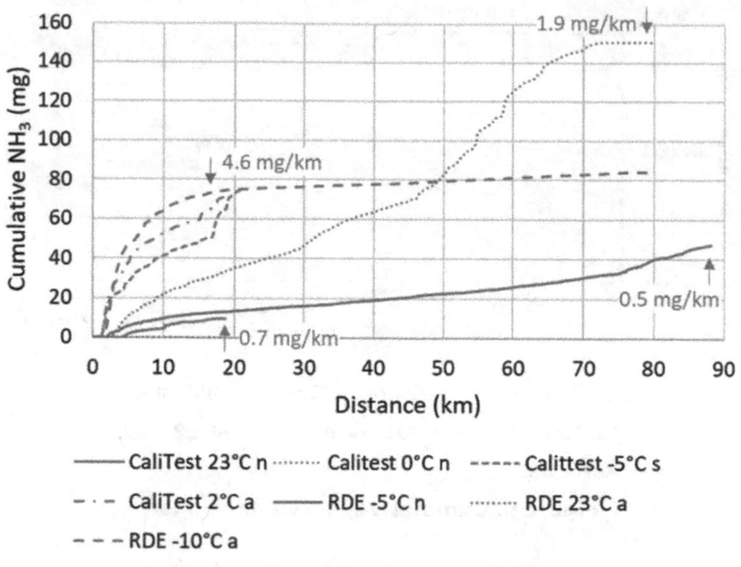


**Figure 10. Cumulative $NH_3$ emissions.**

The overall emissions achieved at the tailpipe are however significantly lower compared to the Euro 6 reference data which is reported to be between 10 and 40 mg/km for normal driving (7). The data in Figure 10 results in values between 0.5 mg/km (total value of RDE -5°C n) and 4.6 mg/km (urban value of RDE -10°C a).

#### 4.1.4 *Other emissions*

CO emissions are dominated by an initial cold-start phase. Emissions increase towards low ambient temperature and aggressive driving style. Evaluation of urban CO emissions at 16 km including the initial cold-start shows it varies from 111 mg/km (CaliTest 23 °C n) up to 368 mg/km (RDE -10 °C a).

The range of $NR_2RO$ emissions measured vary between 0.7 and 5.4 mg/km. These are also dominated by an initial cold-start effect.

### 4.2 Heavy-duty diesel

#### 4.2.1 *NOx emissions*

An overview of the NOx emissions obtained during the trips is shown in Figure 11. The results show the integrated emissions over the urban, rural or motorway shares of operation vs. the average speed driven during each operation share. Similarly, in the case of the urban delivery trips, the emissions have been integrated vs. the average speed of the trip. As it can be seen, the results confirm the initial findings reported in (8). The NOx emissions during the urban operation of the in-service trips range from 48 to 188 mg/kWh. The urban driving trips emissions show higher emissions compared to the urban share of operation of the in-service trips, these range from 228 to 363 mg/kWh. The urban delivery trips contain a significant amount of stop and go operation, including varying the duration of the stops. All stops were held at idle to ensure worst case temperature conditions.

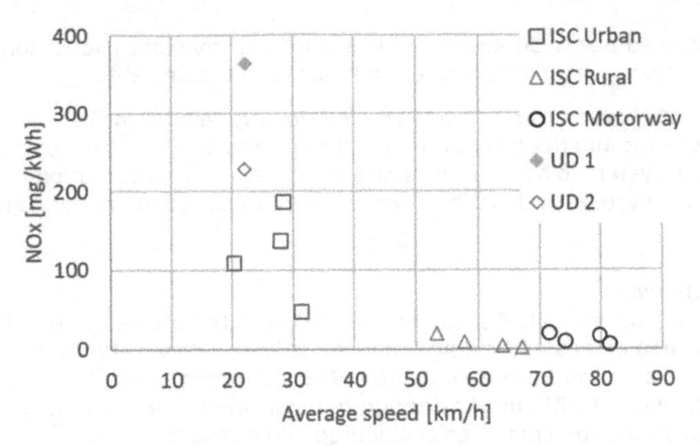


**Figure 11. Speed binned NOx emissions for ISC and urban delivery trips.**

Figure 12 shows the cumulative NOx emissions of the reported trips. It is important to highlight that most of the trips show one single emission event at the beginning of the trip. Depending on the testing conditions, the magnitude of the NOx emissions can vary up to three times. In every ISC (test 1-4), the total tailpipe cumulative NOx are below 5 g at the end of the trip. To put these resulting emissions in context, the engine is producing on average 5.5 g/kWh of engine out emissions. The cumulative engine out emissions in an in-service conformity test for this application vary between 600 to 800 g of total produced NOx. The system is showing a high efficiency on reducing NOx emissions, mostly above 98 %. These conversion rates are in good agreement with those reported in recent studies (9, 10).

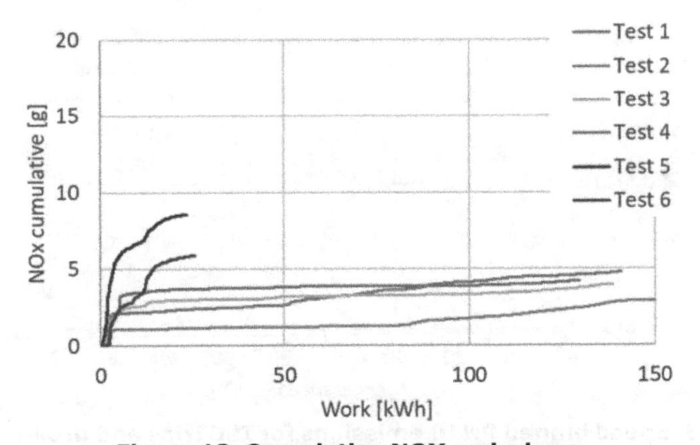


**Figure 12. Cumulative NOX emissions.**

There are two urban delivery trips represented by tests 5 and 6. These tests show higher cumulative tailpipe emissions. The highest concentrations are produced during the first part of the trip (i.e. cold start). The system quickly heats up reaching 200 °C around 200 seconds into the trip. From that moment, the urea starts to be injected and the system begins to convert the NOx. The nature of the operation where the speed is controlled below 50 km/h and the stops are constant allows some NOx slip and thus, both tests show some further emissions are accumulated.

As discussed, the reason for these high emissions is related in first instance to the cold-start, but later into the trip to the number of stops and the effect of these stops in challenging the system to hold its operating temperature. This operation remains challenging for the system and further work needs to be conducted to enhance this behaviour.

### 4.2.2 *PN emissions*

Figure 13 shows the particulate number (PN10) results obtained during these tests. The data has been binned in the same manner as the other pollutants. The particulate emissions have been measured with the PEMS instrument described above and are reported as measured without considering the instrument measurement uncertainty. Crankcase emissions have not been considered in the results.

As it can be seen, results show the bulk of the PN emissions are produced during the beginning of the trip, and particularly during the cold-start phase. This can be attributed to the state of the filter at the beginning of the tests.

An item that needs to be considered to interpret these results is that the day before the start of each test, 1 hour of motorway driving is done without urea dosing to reduce the $NH_3$ loading in the SCR to minimum and to allow passive regeneration of the DPF. The intention of this preconditioning is to have similar starting conditions of the system for every test. Following this preconditioning, the initial status of the filter at the start of a test is assumed to be almost empty. The particulates generated can be attributed to the blow off from the cold pipes and combustion chambers at the initial start of the engine. After this initial emission peak the PN behaviour is consistent with very low particulate emissions produced.

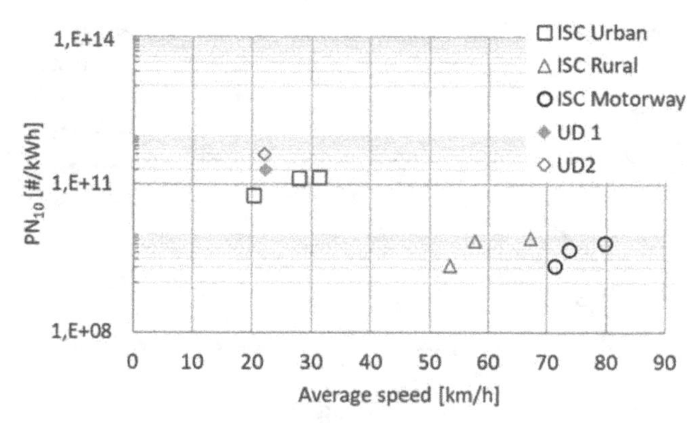


**Figure 13. Speed binned PN10 emissions for ISC trips and urban delivery trips (logarithmic scale).**

In the case of particulates, it is worth to look at the cumulative emission traces to understand where the particulates are generated and to confirm what Figure 13 is presenting. For this purpose, Figure 14 (Test 3) and Figure 15 (Test 6) have been included below.

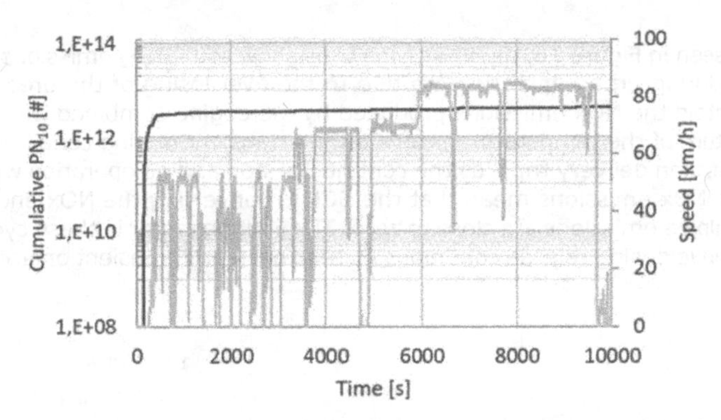


**Figure 14. Test 3, ISC PN10 cumulative emissions (logarithmic scale).**

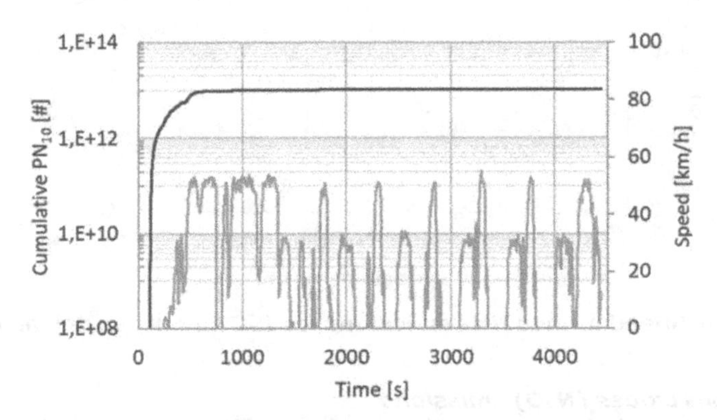


**Figure 15. Test 6, urban delivery PN10 cumulative emissions (logarithmic scale).**

In both Figure 14 and Figure 15, PN10 emissions are generated within the initial 300-500 seconds from start. This is within the heat up of the engine and emission control systems. As it can be seen, once the initial (cold-start) operation is finished, the observed PN emissions during the rest of the trip are very low.

### 4.2.3 *$NH_3$ emissions*

Figure 16 shows the ammonia emission results for the trips. Ammonia emissions can be generated by an overdose of urea solution, commonly known in the European market as AdBlue®. Current Euro VI emission standards include a 10 ppm average limit for $NH_3$ emissions. During the tests conducted, the ammonia results have shown a robust slip control from the emissions control system. When there were no PEMS data available, the data reported have been taken from the $NH_3$ sensors installed in the vehicle at an equivalent position for consistency.

As already described, this system includes ammonia slip catalysts after each SCR, this allows to control any significant ammonia slippage. Thanks to the close coupling of the catalysts these reach operating temperature much faster than a conventional under-floor system. Thus, the dosing of AdBlue®in this system starts very early in the trip, once the system is under regime temperature (over 200 °C), the ccSCR will start converting NOx. It is to be noted that the preconditioning described in section 4.2.2 also reduces the $NH_3$ loading in the SCR to minimum and this has also an impact on the ammonia measured at tailpipe.

As it can be seen in Figure 16, the in-service trips show some $NH_3$ emissions. The emissions found during urban operation are due to the overdosing of the urea during this period to contain the NOx emissions produced by the engine combined with the rapidly changing nature of the temperature profile during transient driving conditions. It is different in the urban delivery trips, as the continuous stop and go operation with the consequent high NOx emissions mean that the SCR is converting the NOx and hence the measured tailpipe emissions are close to zero. The limiting factor in these cycles are the ability to achieve dosing release conditions instead of steady transient operation.

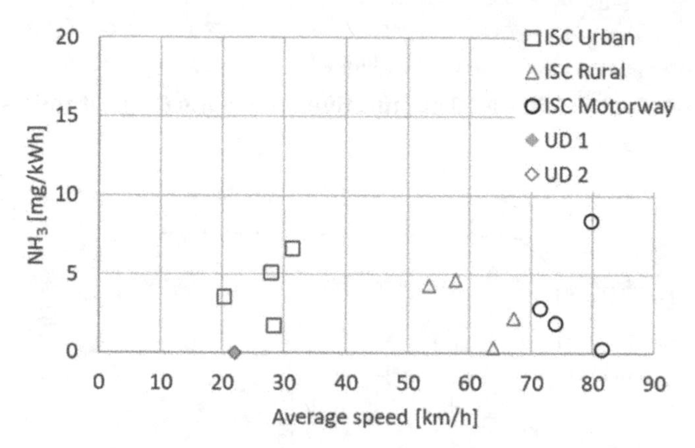


**Figure 16. Speed binned $NH_3$ emissions for ISC and urban delivery trips.**

### 4.2.4 *Nitrous oxides ($N_2O$) emissions*

Figure 17 shows the $N_2O$ emissions. The complexity of $N_2O$ emissions relies in the various mechanisms from which these are produced. The $N_2O$ can be formed over a DOC during light off conditions. Additionally, $N_2O$ emissions can be generated as a by-product of the chemical reactions within the SCR. It is also produced via unselective oxidation of unreacted $NH_3$ within the ASCs. As such, each of these mechanisms needs to be optimized to achieve the lowest tailpipe $N_2O$ values.

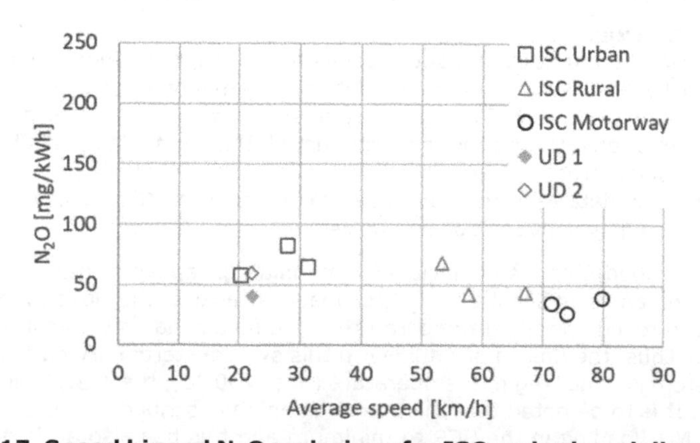


**Figure 17. Speed binned $N_2O$ emissions for ISC and urban delivery trips.**

There is an emissions trade-off between the $N_2O$, $NH_3$ and NOx emissions from emission control systems. In the demonstrator vehicle, the goal was to keep $N_2$ O emissions as low as possible but with the firm objective to achieve lowest possible NOx emissions. Figure 17 shows how the $N_2O$ emissions are well controlled by the system, it is important to note that nitrous oxides are produced during the whole trip.

## 5 SUMMARY

This paper described emission test results conducted with a light-duty gasoline car and heavy-duty diesel truck. It showed further efforts are undertaken to reduce the impact on air quality. The work focused on reducing the initial cold-start emissions and emission control under urban driving conditions. On both vehicles, an emission control system was investigated within an integrated powertrain approach. It consisted of close-coupled and underfloor components.

A multitude of emission tests were conducted to characterise the emission performance of the two systems over a wide range of driving conditions. For light-duty gasoline, this included conditions beyond current Euro 6d RDE boundary conditions. For heavy-duty diesel, this included low-load and urban driving routes. It was shown that the NOx and PN emissions are controlled at an ultra-low level. The remaining emissions are from the initial cold-start phase, which has been reduced as much as possible. The initial cold-start peak increases as a consequence of lower ambient soaking temperature (investigated on the light-duty gasoline vehicle only for the moment) and driving style/route. This applies to both NOx and PN. The levels of regulated pollutants are achieved while controlling the currently un-regulated pollutants as well. PN10, $NH_3$ and $N_2O$ have been measured.

## 6 OUTLOOK

Investigation of low ambient temperature impact will also be done for the heavy-duty diesel vehicle. The investigations will then continue for both vehicles to further reduce the pollutant emissions based on the observations described in this paper. The current system performance is determined by the time needed to achieve a certain exhaust temperature during the cold-start phase. Either to reach light-off of the TWC on the light-duty gasoline vehicle and DOC on the heavy-duty diesel vehicle or to be able to start AdBlue® dosing on the heavy-duty diesel vehicle. To further reduce the initial cold-start emissions, an electrically heated catalyst will be investigated as an example of active thermal management that was possible to be implemented on the demonstrator vehicles. For light-duty gasoline, this will be combined with further calibration change measures, mainly exploring the possible support from the 48V hybrid system.

In addition, the work will also look into the reduction of the impact on climate change. This will be done by repeating tests on a renewable e-fuel for both vehicles. These fuels are derived from renewable electricity and captured $CO_2$.

With both low pollutant and $CO_2$ emissions, internal combustion engines are part of the solution to achieve the long-term near-zero emission goals.

## ACKNOWLEDGEMENTS

The authors would kindly like to thank members of AECC (Association for Emissions Control by Catalyst) and IPA (International Platinum Group Metals Association) for the financial support, the supply of catalyst parts and for their highly valuable contributions to this study. The authors would also like to thank IAV (light-duty) and FEV (heavy-duty) for the engineering and testing work. The authors appreciate the respective OEMs, VW and Daimler, allowed to use the base vehicle. AIP is acknowledged for the supply of the prototype PEMS equipment.

## REFERENCES

[1] ACEA RDE monitoring database, https://www.acea.be/publications/article/access-to-euro-6-rde-monitoring-data

[2] JAMA RDE monitoring database, http://www.jama-english.jp/europe/publications/rde.html

[3] Handbook of Emissions Factors (HBEFA), release 4.1, https://www.hbefa.net/e/index.html

[4] Green NCAP, https://www.greenncap.com/

[5] Emissions Analytics, EQUA real-world test database, https://www.emissionsanalytics.com/equa-databases

[6] J. Demuynck, C. Favre, D. Bosteels, F. Bunar, J. Spitta, A. Kuhrt, Diesel vehicle with ultra-low NOx emissions on the road, SAE 14th International Conference on Engines & Vehicles paper 2019-24-0145, 2019.

[7] J. Demuynck, C. Favre, D. Bosteels, G. Randlshofer, F. Bunar, J. Spitta, O. Friedrichs, A. Kuhrt, M. Brauer, Integrated diesel system achieving ultra-low urban and motorway NOx emissions on the road, 40th International Vienna Motor Symposium, 2019.

[8] Mendoza Villafuerte, P., Demuynck, J., Bosteels, D., Wilkes, T., Robb, L., Schönen, M.; Demonstration of Extremely Low NOx Emissions with Partly Close-Coupled Emission Control on a Heavy-Duty Truck Application, 42nd International Vienna Motor Symposium, 2021

[9] Nilsson, M., Birgersson, H., Müller, W., Gabrielsson, P., Senar Serra, E.; Next Generation Global Emission Solution Platform with Dual Urea Dosing – Meeting Future Emission and Efficiency Requirements, 42nd International Vienna Motor Symposium, 2021.

[10] Selleri, T., Melas, A., Joshi, A., Manara, D., Perujo, A., Suarez-Bertoa, R., An Overview of Lean Exhaust deNOx Aftertreatment Technologies and NOx Emission Regulations in the European Union. Catalysts 2021, 11, 404, 2021.

## *Session 7: Powertrain development systems and analysis*

# Research on experimental design method of engine system based on D-optimal design

**B.C. Yang[1], T. Chen[1], Y.F. Feng[1], H. Shi[1], H. Zhao[1,2]**
[1]State Key Laboratory of engine combustion, Tianjin University, China
[2]Brunel University, London, UK

**ABSTRACT**
In this paper, a few experimental points were used to calibrate the throttle model, fuel injection model and semi-predictive combustion model of the engine system based on the D-optimal experimental design method. The results show that the number of calibration test points can be reduced by more than 90% on the premise of ensuring the accuracy of each sub-model. At the same time, a set of general calibration experiment design methods for each sub-model in the engine system is established by comparing the distribution of experimental points selected from each sub-model.

**Keywords:** engine modelling, D-optimal, DOE, model calibration

## 1 INTRODUCTION

The increasing energy crisis and the continuous improvement of vehicle emission requirements put forward higher requirements on engine thermal efficiency, fuel consumption and pollutant emission and other performance indicators. Engine calibration technology is one of the key technologies to improve engine performance, but the process of traditional calibration completely depends on the bench test, and the demand for test points is usually thousands or more than tens of thousands, which requires a lot of experimental resources, and makes the calibration work cost is high and the cycle is long[1-2].

The workload of bench calibration can be effectively reduced by establishing a virtual engine model to replace the actual engine for model calibration. However, it is necessary to ensure that the model has a sufficiently high prediction accuracy before using a virtual engine to replace the actual engine for simulation calibration. The improvement of model accuracy depends on a large number of experimental data, which contradicts the purpose of reducing calibration workload by virtual engines. Therefore, how to establish a high-precision engine model based on a small amount of experimental data is an important basis for realizing virtual calibration, reducing cost and shortening cycles.

The objective of optimal experimental design is to obtain a combination of experimental points that can maximize the information matrix. And there are many corresponding evaluation criteria, including Latin Hypercube Sampling, D-optimal, V- optimal and A-optimal, etc. To reduce the experimental points needed for model calibration, Anthony Gullitti et al. [3] used the intelligent experimental design system developed by IAV Company, based on the D-optimal experimental design method, and supplemented by the V-optimal experimental design method and space-filling method, and finally completed the model building with a small amount of experimental data (269 experimental points). The ROOT MEAN SQUARE ERROR of the model is less than 2.5%, which has high accuracy. Tianhong Pan et al. [4] used the Latin Hypercube

DOI: 10.1201/9781003219217-17

Sampling algorithm to select only 650 experimental points from 4688381250 experimental points to build the model and the $R^2$ (coefficient of determination: indicating the degree of overlap between the predicted value and the actual value) of the model was greater than 0.92, which could meet the research requirements.

The above methods can not only ensure the accuracy of the model but also greatly reduce the experimental points required for the model establishment, which can reduce the cost and period of calibration. But these methods are only applicable to the specified engines, and the test points need to be re-selected for other engines, which makes their expansion ability poor. The D-optimal design method has advantages in the application of online calibration and is suitable for a wide range because of its simple calculation and a relatively low requirement on hardware computing capability[5]. In order to improve the universality of the experimental design scheme among different engines, a study on experimental design based on the D-optimal design method was carried out in this paper. A small number of experimental points were selected to calibrate the throttle model, fuel injection model and semi-predictive combustion model respectively. And a general experimental design method is established by analyzing the distribution of selected experimental points, which greatly reduces the requirement of the number of experimental points for model calibration, and is of great significance for the virtual calibration of engines.

## 2 METHOD OF MODELLING

### 2.1 Basic engine parameters

This study is based on a 1.3L GDI (Gasoline Direct Injection) engine with turbocharging and dual VVT. The specific structural parameters are shown in Table 1.

Through the sweep-point experiment, the experimental data of 2600 working points were obtained. The distribution is shown in Figure 1.

**Table 1. Basic parameters of the engine.**

| Parameter | Value |
|---|---|
| Number of cylinders | 4 |
| Number of stroke | 4 |
| Bore/mm | 76 |
| Stroke/mm 74 | 74 |
| Displacement/L | 1.342 |
| Maximum compression ratio | 10 |
| Injection method | Direct injection |
| Number of valves | 4 |
| Intake mode | VVT |

### 2.2 Throttle model

The throttle can be regarded as a standard throttle valve, and the airflow through the throttle can be treated as a one-dimensional isentropic stable flow from the perspective of fluid mechanics [6]. The calculation formula is as follows:

$$m_{ub} = A_{eff}\rho_{is}U_{is} = C_D A_R \rho_{is} U_{is} \tag{1}$$

$$\rho_{\text{is}} = \begin{cases} \rho_0 (P_r)^{\frac{1}{\gamma}}, & P_r > \left(\frac{2}{\gamma+1}\right)^{\frac{\gamma}{\gamma-1}} \\ \rho_0 \left(\frac{2}{\gamma+1}\right)^{\frac{1}{\gamma-1}}, P_r > \left(\frac{2}{\gamma+1}\right)^{\frac{\gamma}{\gamma-1}} \end{cases} \tag{2}$$

$$U_{is} = \begin{cases} \sqrt{RT_0}\left\{\frac{2\gamma}{\gamma-1}\left[1 - P_r^{\frac{\gamma-1}{\gamma}}\right]\right\}, & p_r > \left(\frac{2}{\gamma+1}\right)^{\frac{\gamma}{\gamma-1}} \\ \sqrt{\gamma RT_0}\left\{\frac{2}{\gamma+1}\right\}^{\frac{1}{2}}, & P_r \leq \left(\frac{2}{\gamma+1}\right)^{\frac{\gamma}{\gamma-1}} \end{cases} \tag{3}$$

Where, $A_{eff}$ represents the effective circulation area; $\rho_{\text{is}}$ represents the upstream stagnation density; $U_{is}$ represents the isentropic density of the throat outlet; $C_D$ represents the flow coefficient; $A_R$ represents the reference flow area; $P_r$ represents the absolute pressure ratio (static pressure at the exit/total pressure at the entrance); R is the gas constant; $T_0$ represents upstream stagnation temperature;$\gamma$ is the adiabatic index.

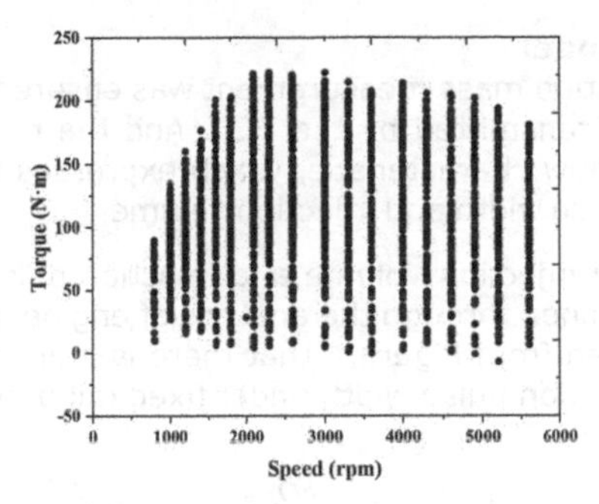


**Figure 1. Distribution of 2600 experimental points.**

The flow coefficients of each throttle opening were calculated by a one-dimensional isentropic stable flow equation according to the experimental data obtained. Wang Zhuwei et al from the University of Shanghai for Science and Technology have studied the throttle flow coefficient and found that there is a certain variation rule between flow coefficient, throttle angle and pressure ratio[7]. Therefore, the relationship between them is also studied in this paper, as shown in Figure 2:

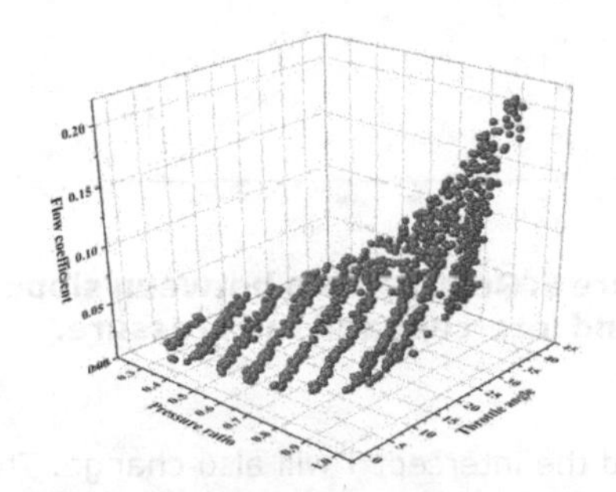


**Figure 2. Relationship between throttle flow coefficient and throttle opening and pressure ratio.**

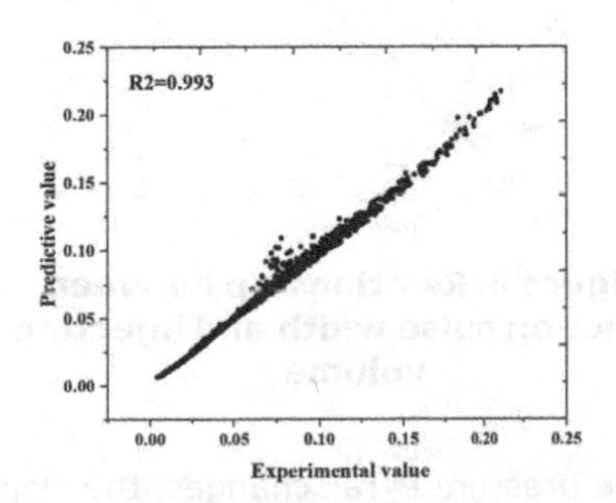


**Figure 3. Comparison of simulation results of throttle flow coefficient model.**

It can be seen from Figure 2 that the throttle flow coefficient is in smooth plane distribution with the throttle opening and the pressure ratio, and it mainly increases with the increase of the throttle opening. Therefore, a throttle flow coefficient model based on pressure ratio correction is built in this paper, as shown in Equation (4):

$$\begin{cases} C_D = p_0 + p_1\theta + p_2 p_r + p_3\theta^2 + p_4\theta p_r + p_5\theta^3 + p_6\theta^2 p_r \\ p_r = \frac{\text{pim}}{p\_up\_thr} \end{cases} \tag{4}$$

Where, $C_D$ is the throttle flow coefficient;$\theta$ is the throttle opening; $p_r$ is the ratio of the throttle rear-end pressure to the front end pressure;$p_{im}$ is the intake manifold pressure;$p\_up\_thr$ is the throttle front pressure. The accuracy of the model is shown in Figure 3 after the experimental data is brought into the model.

As can be seen from Figure 3, $R^2$ of the flow coefficient calculated by the model is 0.993. This indicates that the established throttle flow coefficient model has high accuracy.

## 2.3 Fuel injection model

The accuracy of fuel injection mass measurement was ensured by receiving the signal of injection pulse width transmitted by the ECU. And the main characteristic of the injector is the injection flow characteristic, which expresses the relationship among rail pressure, injection pulse width and injection volume.

The relationship between injection volume and injection pulse width under different rail pressures can be obtained through the analysis of engine injection data, as shown in Figure 4. It can be seen from Figure 4 that there is a linear relationship between injection volume and injection pulse width under fixed rail pressure. And the injection model is:

$$T_{inj} = kQ_f + t \tag{5}$$

Where, $T_{inj}$ represents the injection pulse width, $Q_f$ represents the injection volume of a single cycle, k represents the slope, and t represents the intercept.

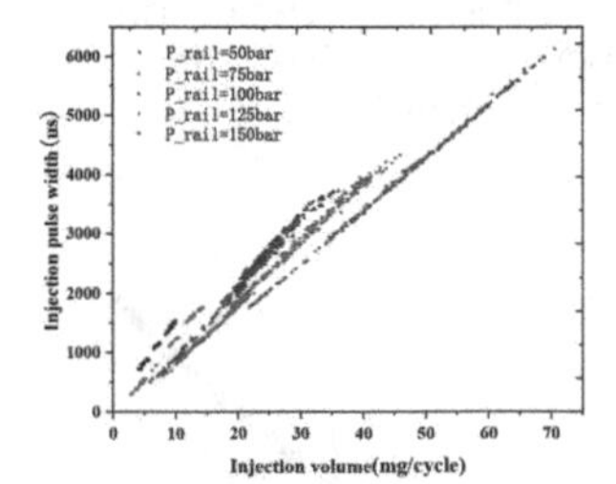


**Figure 4. Relationship between injection pulse width and injection volume.**

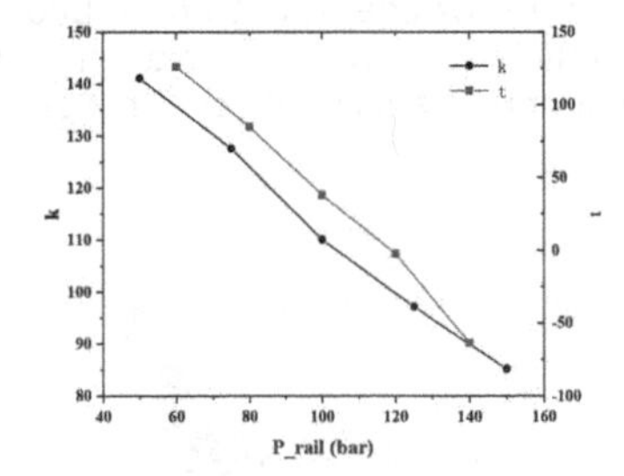


**Figure 5. Relationship between slope and intercept and rail pressure.**

As the pressure P_rail changes, the slope K and the intercept T will also change. The changing relationship between slope k and intercept t with rail pressure is shown in Figure 5, and a good linear relationship between them is shown in Figure 5. And the regression model of slope k and intercept t with rail pressure (injection flow coefficient model) is established:

$$k = m_0 \times P\_rail + m_1 \tag{6}$$

$$t = n_0 \times P\_rail + n_1 \tag{7}$$

The model accuracy is shown in Figure 6 after the experimental data is brought into the model:

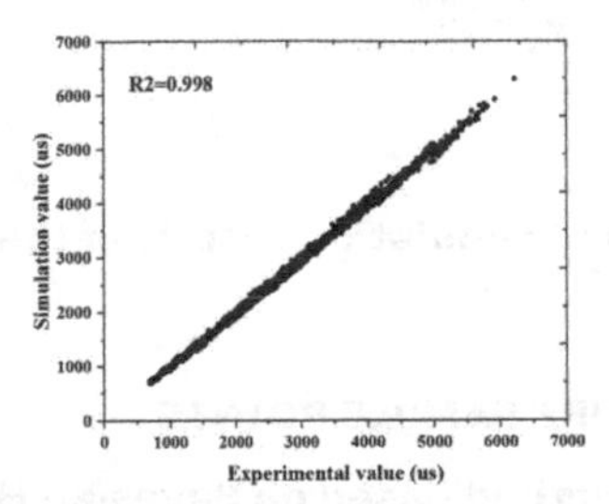


**Figure 6. Comparison of simulation results of injection pulse width.**

It can be seen from Figure 6 that the $R^2$ of injection pulse width calculated by the model is 0.998, which indicates that the established injection model has high accuracy.

## 2.4 Semi-predictive combustion model

The Wiebe combustion model is usually chosen for combustion calculation when the virtual engine is used instead of the real engine for simulation calculation [8]. However, $CA_{50}$ and $CA_{10-90}$ are unknown before the experiment and cannot be directly used as the input of the combustion model. Therefore, the Wiebe combustion model alone cannot be used to predict the heat release rate of combustion under variable working conditions. In this research, a semi-predictive combustion model is established, which enables the virtual engine model to predict the values of $CA_{50}$ and $CA_{10-90}$ under variable operating conditions, and it can predict the values of $CA_{50}$ and $CA_{10-90}$ according to the operating conditions of the engine. The formula is shown as follows:

$$\theta_{50} = a_0 + a_1\lambda^2 + a_2\lambda + a_3p_{im}^2 + a_4p_{im} + \frac{a_5}{T^2} + a_6\theta_{spark}^2 + \frac{a_7}{RPM^2} + a_8IVO + a_9RGF^2 \tag{8}$$

$$CA_{10-90} = b_0 + b_1\theta_{50} + b_2\theta_{50}^2 + b_3\theta_{spark}^2 + b_4\theta_{spark} + b_5p_{im}^2 + b_6\lambda^2 + \frac{b_7}{RPM^2} + b_8RGF^2 \tag{9}$$

$$CA_{50} = \theta_{50} + \theta_{spark} \tag{10}$$

Where, $CA_{10-90}$ is the duration of the Wiebe combustion curve, $CA_{50}$ is the Angle "anchoring" the Wiebe curve to TDC, λ is the excess air coefficient; $p_{im}$ is the intake manifold pressure (bar); T is the inlet air temperature (K); $\theta_{spark}$ is the ignition advance Angle; RPM is the engine speed (r/min); IVO is the opening phase of the intake valve; RGF is the residual exhaust gas rate in the cylinder.

The model accuracy is shown in Figure 7 after the experimental data is brought into the model. It can be seen from Figure 7 that the $R^2$ values of $CA_{50}$ and $CA_{10-90}$ calculated by the combustion model are 0.96 and 0.87, respectively. Due to the complexity of the combustion model, it is difficult to make an accurate prediction of $CA_{10-90}$. However, this accuracy can meet the basic requirements of the combustion model, and there is still a large space to further improve the prediction accuracy of $CA_{10-90}$.

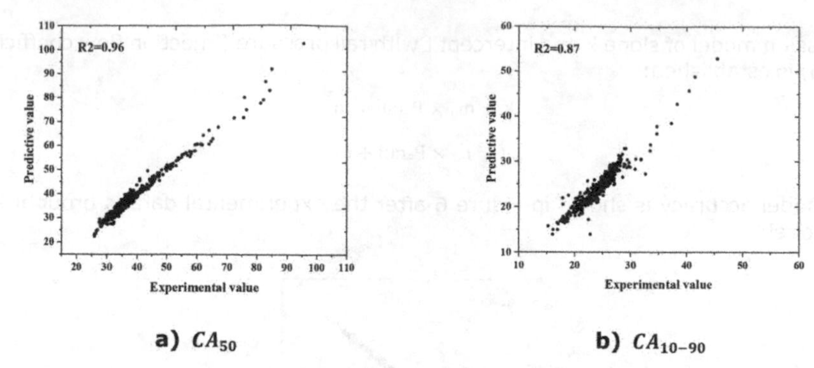


**a)** $CA_{50}$ **b)** $CA_{10-90}$

**Figure 7. Comparison of simulation results of the combustion model.**

## 3 SELECTION OF FEATURE SAMPLE POINTS

### 3.1 D-optimal design method based on Bayesian Modification

Experimental design refers to the rational arrangement of experiments and the use of fewer experimental points to obtain relatively ideal model calibration results, so as to reduce the calibration cycle and cost[9]. The experimental design methods include classical design method, space-filling design method and optimal design method. Among them, the optimal design method can properly establish the design criteria that can reflect the purpose of the experiment, and obtain the optimal design scheme to arrange the experiment, so as to save the experiment cost. The optimal design method is usually adopted if you have a deep understanding of the engine and the optimal model has been selected[10].

Regression models can generally be expressed as the following matrix [11]:

$$y = X\beta + \in \tag{11}$$

Where y is the output; β is the parameter to be estimated; X is the coefficient matrix. The least-square estimation of this model is $\hat{\beta} = (X^T X)^{-1} X^T y$, and $X^T X$ is required to be non-degenerate. Model of the covariance matrix of the $cov(\hat{\beta}) = \sigma^2 (X^T X)^{-1}$, and sigma error factors by experiment. The matrix $M = X^T X$ is the information matrix that contains the information of the model and the experimental points.

D-optimal design refers to maximizing the determinant value of the information matrix $(X^T X)$, which is as follows:

$$|X^{*T} X^*| = \max(|X^T X|) \tag{12}$$

In some cases, the sample points obtained according to the D-optimal design may have duplicate values, which not only wastes computational resources but also has no help to the fitting of model coefficients. Therefore, the Bayesian Modification (BM) method proposed by Dumouchel and Jones is adopted to add coefficient terms of higher-order at the end of the established regression model, which can effectively solve the problem of repeated points [12].

The process of D-optimal design based on Bayesian modification includes data standardization, determination of candidate point matrix, selection of regression model, determination of experimental points, the establishment of information matrix and optimization design. In this study, it is programmed to quickly obtain the experimental design scheme according to the Fedorov algorithm, as shown in Figure 8.

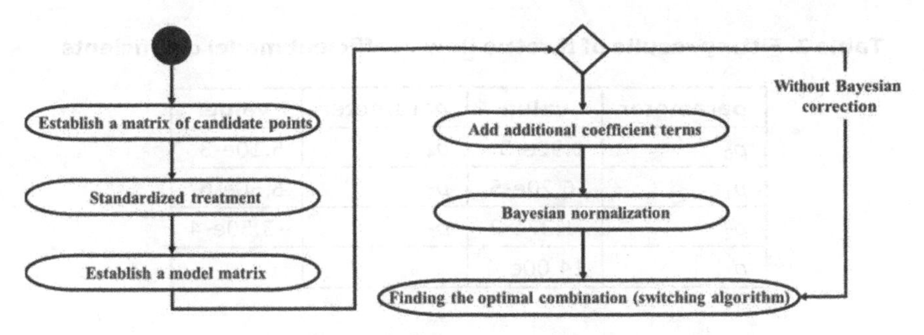


**Figure 8. Implementation flow of D-optimal design.**

### 3.2 Application of D-optimal in the throttle model

The throttle flow coefficient model based on pressure ratio modification is experimentally designed according to the principle of D-optimal design. The input of the model is throttle opening θ and the ratio of throttle rear-end pressure to front end pressure $p_r$, and the number of target samples selected is set as 20. The distribution of the final selected sample points is shown in Figure 9.

It can be seen from Figure 9 that the sample points are mainly distributed on the edge of the interval composed of throttle opening and pressure ratio, and only one point is about the centre of the interval. According to the geometric feature analysis of the selected points, the points at the edge cover the whole interval and the points with sharp shape changes on the edge are called feature points. The overall shape of the surface is mainly affected by these feature points because the throttle flow coefficient, θ and pressure ratio $p_r$ are quadratic functions and the shape of the surface is smooth.

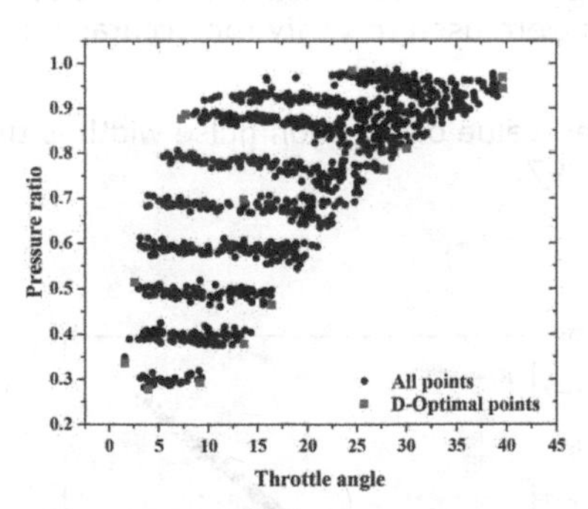


**Figure 9. Distribution of sample points in the throttle flow coefficient model.**

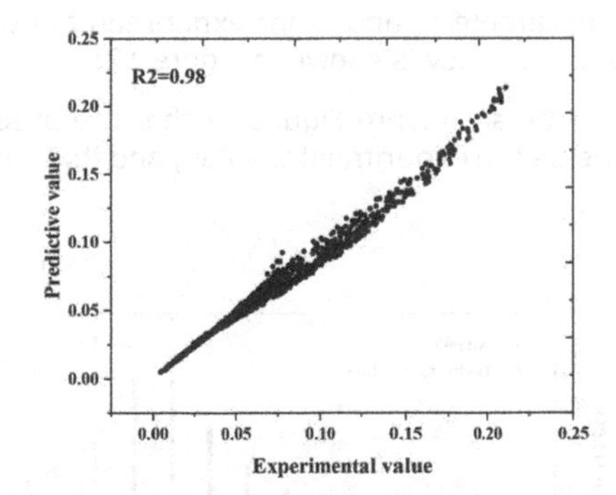


**Figure 10. Comparison of experimental and predicted throttle flow coefficient results.**

The 20 selected sample points were substituted into the throttle flow coefficient model for solving, and the values of each coefficient in the model could be obtained, as shown in Table 2:

**Table 2. Fitting results of throttle flow coefficient model coefficients.**

| parameter | value | parameter | value |
|---|---|---|---|
| $p_0$ | 6.90e-3 | $p_4$ | 5.10e-3 |
| $p_1$ | -6.20e-5 | $p_5$ | 5.50e-6 |
| $p_2$ | -1.37e-2 | $p_6$ | -3.50e-4 |
| $p_3$ | 14.00e-4 | | |

Next, the prediction accuracy of the model is verified. The remaining non-pressurized conditions (throttle opening less than 90 degrees, about 1300 experimental points) were substituted into the model for testing, and the results were shown in Figure 10. It can be seen from Figure 10 that the predicted value of the throttle flow coefficient is basically consistent with the experimental value, and its $R^2$ is 0.98.

It can be seen that the throttle flow coefficient model based on the calibration of 20 sample points can accurately predict the throttle flow coefficient under different throttle opening and pressure ratios under all non-pressurized conditions, which can achieve the purpose of using a small number of experimental points to calibrate the model.

### 3.3 Application of D-optimal in the fuel injection model

Similarly, the injection model is designed experimentally according to the principle of D-optimal design. The inputs of the model are injection pressure and injection volume, and the number of target samples selected is set as 16. The distribution of the final selected sample points is shown in Figure 11.

The 16 selected experimental points were brought into the injection model to solve the parameters, and other experimental points were used to verify the accuracy. The model accuracy is shown in Figure 12:

It can be seen from Figure 12 that the predicted value of injection pulse width is the same as the experimental value, and its $R^2$ is 0.997.

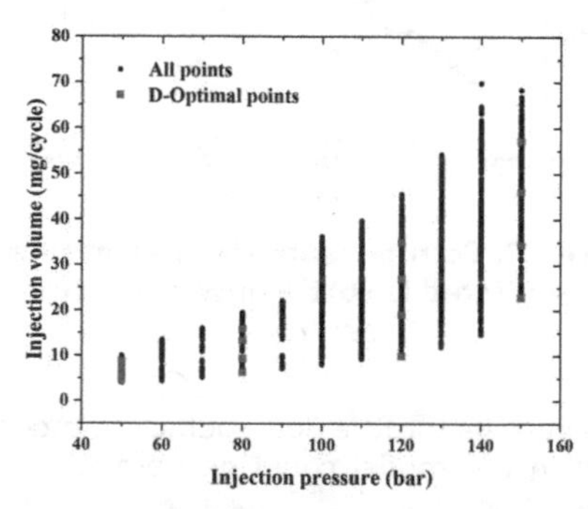


**Figure 11. Distribution of sample points in the injection model.**

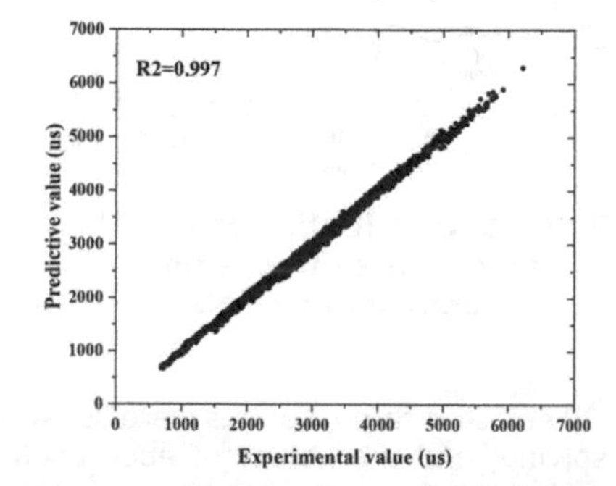


**Figure 12. Comparison of experimental and predicted injection pulse width results.**

### 3.4 Application of D-optimal in the semi-predictive combustion model

Similarly, the semi-predictive combustion model is designed experimentally according to the principle of D-optimal design. And the number of target samples selected is set to 54. The selected experimental points are put into the model to solve the model parameters, and the model accuracy is shown in Figure 13:

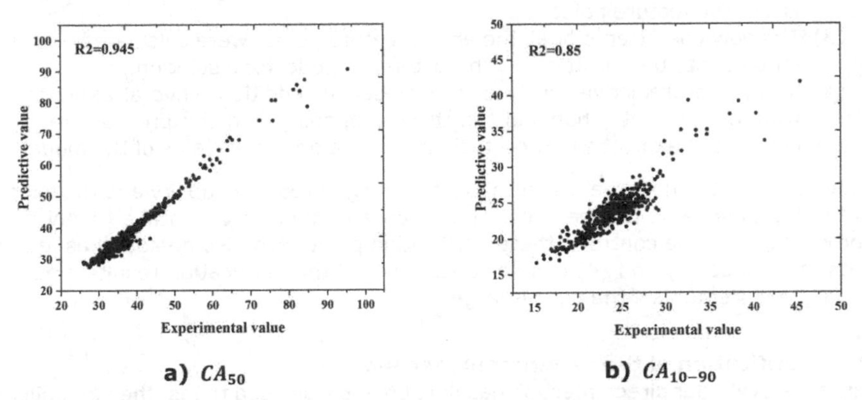


**a)** $CA_{50}$ **b)** $CA_{10-90}$

**Figure 13. Comparison of the experimental and predicted values for $CA_{50}$ and $CA_{10-90}$ in the combustion model.**

It can be seen from Figure 13 that the predicted values of $CA_{50}$ and $CA_{10-90}$ in the combustion, model are basically consistent with the experimental values, and their $R^2$ are 0.94 and 0.85, respectively.

## 4 GENERAL EXPERIMENTAL SCHEME

### 4.1 Design of experimental scheme

The above sample points are targeted at the engine in this study, which may not apply to engines of different types. In order to improve the expansion abillty of the experimental design scheme, this paper summarizes a general experimental design scheme by analyzing the selection rules of experimental points under different models.

This paper takes the throttle model as an example to study the general experimental scheme. In the throttle model, the experimental points selected based on the D-optimal method are mainly distributed on the edge of the interval composed of throttle opening θ and pressure ratio $p_r$. This range is not the same for different engines. In this study, a universal experimental design scheme applicable to different engines is summarized to solve this problem by analyzing the distribution characteristics of selected experimental points and combining the feasibility of real bench experiment operation, as follows:

(1) Turn off the turbocharger and set the throttle opening to the minimum allowable opening $\theta_{\min}$ (such as 1 degree). When the engine speed is gradually increased from idle speed to the maximum allowable speed, the minimum and maximum values of the throttle gate rear pressure ratio are recorded, and these two values are recorded as feature points under the $\theta_{\min}$ opening degree.

(2) A certain number of throttle opening points are selected evenly within the throttle opening range. And the selection points can be denser when the throttle opening is small, and sparser when the throttle opening is largely

based on the characteristics of the throttle. In this study, the throttle opening is adjusted to 3 degrees, 5 degrees, 7 degrees, 10 degrees, 15 degrees, 20 degrees, 25 degrees, 30 degrees, 35 degrees and 40 degrees respectively, and the rotating speed is adjusted according to Step (1) to obtain the corresponding feature points under each opening. Besides, when the throttle opening is adjusted to 15 degrees and the ratio of the rear end of the throttle to the front end pressure is 0.7 by adjusting the engine speed, the working point is also the feature point.

(3) The flow coefficients of all the above feature points were calculated and substituted into the throttle flow characteristic model for a solution.

(4) Verify the predictive ability of the model. In addition, several experiments were carried out (changing the throttle opening and the pressure ratio) to compare the throttle flow coefficient with the predicted value of the model.

It indicates that the flow coefficient model has a high predictive ability and the experimental design scheme can be used for virtual calibration of the intake model if the error is small. On the contrary, these verification points can also be added as feature points, and steps (3) and (4) can be repeated until the verification results meet the accuracy requirements if the error is large.

## 4.2 Verification of the experimental scheme

Next, a 1.4L cylinder direct injection gasoline engine was used to test the versatility of the above experimental design. The engine previously obtained 600 test points through sweeping points. According to the above experimental design scheme, 20 sample points were selected and substituted into the throttle flow characteristic model for the solution, and then the remaining 580 test points were used for verification.

The validation results are shown in Figure 14. It can be seen that the shape of the theoretical value and the predicted value of the flow coefficient is basically the same, and its $R^2$ is 0.972.

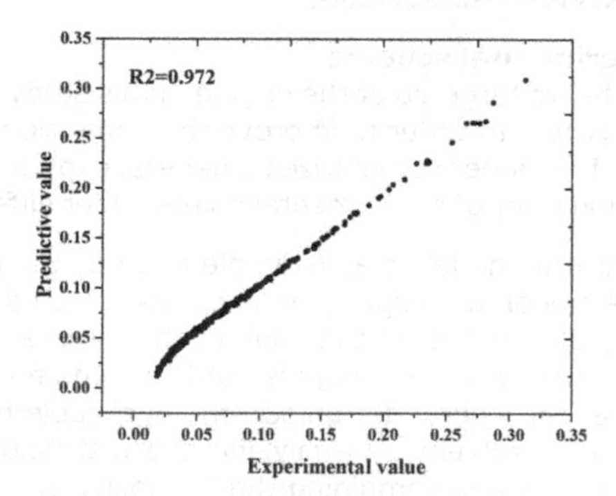


**Figure 14. Comparison of experimental and predicted throttle flow coefficient results.**

The above research shows that the experimental design of throttle flow characteristics is suitable for different gasoline engines, and accurate throttle flow characteristics can be obtained through a small number of test points calibration.

Similarly, according to the point selection rules of the fuel injection model and semi-predictive combustion model, the accuracy of the 1.4L engine model was verified by a similar experimental design scheme, and their $R^2$ are 0.99, 0.93 and 0.84, respectively.

## 5 CONCLUSION

In this study, the throttle model, fuel injection model and semi-predictive combustion model were established respectively through the correlation analysis among variables, and the key experimental points were selected for these three models through the experimental design method based on D- optimal. According to the distribution law of selected experimental points, the respective experimental schemes were designed, and their versatility was verified. Finally, the following conclusions were drawn:

(1) The established throttle flow coefficient model, injection pulse width model and semi-predictive combustion model have high prediction accuracy and the $R^2$ values of the model prediction results and the experimental results are 0.993, 0.998, 0.96 and 0.87, respectively.
(2) Based on the D-optimal experimental design method, 20, 16 and 54 experimental points were selected to identify the model parameters of the throttle model, fuel injection model and semi-predictive combustion model, and the $R^2$ values of the models were 0.98, 0.997, 0.945 and 0.85 respectively, which can reduce the number of test points needed for calibration by more than 90% while ensuring the accuracy of the model.
(3) By analyzing the distribution of selected experimental points of each model, a general experimental scheme was designed and verified in another engine. The verification results show that the $R^2$ values of the model can reach 0.972, 0.99, 0.93 and 0.84 respectively when the number of experimental points is reduced by more than 90%, which proves the feasibility of the general experimental scheme.

## REFERENCES


[1] R. Diewald, T. Cartus, M. Schüler, & H. Bachler. (2009) Model based calibration methodology. SAE Tech. doi: 10.4271/2009-01-2837.
[2] A. Dhand & M. O'Mahony. (2011) Optimisation of an existing Automatic Transmission Calibration for maximizing Fuel Economy using AVL CRUISE. Researchgate.
[3] S. Jiang, D. Nutter, & A. Gullitti. (2012) Implementation of model-based calibration for a gasoline engine. SAE Tech. doi: 10.4271/2012-01-0722.
[4] T. Y. K. & S. H. LEE. (2012) Development of an Engine Calibration Model using Gaussian Process Regression. International Journal of Automotive Technology. doi: 10.1007/s12239.
[5] K. Röpke & C. Von Essen. (2008) DoE in engine development. Quality and Reliability Engineering International. doi: 10.1002/qre.941.
[6] Z. Anwei, L. Jujiang, L. Sicong, & W. Jian. (2012) Research on Engine Charge Model based on Throttle Plate Position. Small internal combustion engine and Motorcycle. doi:CNKI:SUN:XXNR.0.2012-05-012.
[7] Z. W. Wang, B. N. Wang, & Z. D. Zhang. (1999) Experimental study on flow characteristics of throttle. doi:10.16630/j.cnki.1002-5855.1999.04.005.
[8] Y. Wang, G. Conway, J. McDonald, & A. Birckett. (2018) Predictive GT-power simulation for VNT matching to EIVC strategy on a 1.6 L turbocharged GDI engine. SAE Tech. doi: 10.4271/2018-01-0161.

[9] D. Wang, S. Zhang, R. L. Liu, & Y. Shi. (2016) Research on Key Technologies of Electronic Control Engine Based on Model Calibration. FRONTIER DISCUSSION. doi:CNKI:SUN:SDQE.0.2017-14-014.

[10] Y. Cai. (2020) Research on Engine Calibration Model using Gaussian Process Regression. https://kns.cnki.net/KCMS/detail/detail.aspx?dbname=CMFD202101&filename=1020974471.nh.

[11] F. Triefenbach. (2008) Design of Experiments: The D-Optimal Approach and Its Implementation As a Computer Algorithm. doi:10.1.1.159.6913.

[12] Yamamoto, M., Yoneya, S., Matsuguchi, T., & Kumagai, Y. (2002). Optimization of Heavy Duty Diesel Engine Parameters for Low Exhaust Emissions Using The Design of Experiments. SAE Transactions, 111, 2009–2014. http://www.jstor.org/stable/44743216.

# Life cycle analysis of power electronics and electric machines for future electrified passenger cars

**C. Antoniou, A. Cairns, C. Gerada**

Fluids & Thermal Engineering (FLUTE) and Power Electronics, Machines & Control (PEMC) Research Groups, The University of Nottingham, UK

**S. Worall, R. Townend, P. Dahele, G. Day**

GKN Automotive Ltd, UK

**S. Simplay**

Belcan International Ltd, UK

## ABSTRACT

The reported work was concerned with Life Cycle Analysis of automotive electric machines and power electronics. The aim was to improve the modelling of these parts and examine the prospects of onshoring manufacturing within the UK. The motor and inverter unit were accountable for 12-17% of the embedded carbon emissions in the electric powertrain. The benefit in onshoring the manufacturing of these parts varied between 7-33% reduction in equivalent $CO_2$ relative to key regions of interest, primarily associated with the clean UK electricity grid. The work has demonstrated the potential of battery "rightsizing" and the environmental benefits of range extension via renewable fuels.

## 1 INTRODUCTION

Modern society is highly dependent on road transport, with substantial international growth predicted over the coming decades. According to a study commissioned by the World Business Council for Sustainable Development (1), worldwide personal transport is projected to grow at an annual rate of 1.7% between 2000 and 2050, resulting in ownership of 2 billion Light-Duty Vehicles (LDVs)[1] and more than 70 trillion passenger-kilometres travelled per year. LDVs, mostly passenger cars currently run primarily on petroleum-derived fuels, account for over 40% of the global transport energy demand and are expected to do so until 2050 (2). Specifically in the European Union (EU), transport is the biggest source of carbon emissions, contributing 27% to total anthropogenic Greenhouse Gas (GHG) emissions - with passenger cars representing 44% of these (3). In the pursuit to achieve the global Paris climate agreement goals of limiting the global temperature rise to 1.5°C, transport emissions must be reduced to zero by 2050 at the very latest. In response to this, global leaders are pushing to eliminate fossil fuel-powered vehicles, which are major contributors to climate change and air pollution. The United Kingdom (UK) has the ambition to become the fastest Group of Seven (G7) country to decarbonise cars and vans, with plans to accelerate

DOI: 10.1201/9781003219217-18

1. Light-Duty Vehicles include cars, Sport Utility Vehicles (SUVs), mini vans and personal-use light trucks.

a greener transport future through a two-step phase-out of petrol and diesel vehicles announced in November 2020 (4). The first step will see the termination of new petrol and diesel car sales brought forward to 2030, while step two mandates all new cars and vans having zero tailpipe emissions from 2035 onwards.

The aforementioned measures have resulted in most major manufacturers in the automotive industry moving away from traditional internal combustion engine vehicles (ICEVs) and towards electrified propulsion methods – including hybrid electric vehicles (HEVs), plug-in hybrids (PHEVs), fuel cell vehicles (FCVs) and battery electric vehicles (BEVs) (5). Apart from low to zero tailpipe emissions, EVs offer additional advantages over conventional ICEVs. EVs have a superior tank-to-wheel energy efficiency as they can convert 70-90% of the chemical energy stored in the battery to movement (6), not including potential charging losses. EVs feature highly efficient individual powertrain components (battery, motor, power electronics and transmission) (7) and the ability to apply regenerative braking, which can supply roughly 10-20% of the total energy used depending on driving style and conditions (8). For plug-in vehicles, some of the in-use efficiency benefits are offset by conversion losses during fossil fuel-based electricity generation and losses in the transmission, distribution and charging (9). Charging losses, predominately occurring in the power electronics used for AC-DC conversion, can be as high as 20%, depending on charging rates and State of Charge (SoC) (10, 11). Additional advantages of EVs include lower maintenance requirements (12) and noise pollution levels (13). However, for the full adoption of electrified vehicles significant challenges remain around the required infrastructure and the recycling of key drivetrain components.

The currently reported work was concerned with Life Cycle Analysis (LCA) of the electric powertrain components, focusing on the GHG impacts in the production stages of the electric traction motor and power inverter unit. The material extraction and manufacturing processes for these components involve energy intensive processes, the environmental impact of which is often underestimated by commonly adopted LCA tools that usually assess whole vehicles. The aim of the study was to improve the modelling of these parts in the popular open access transport life cycle model "GREET" (The Greenhouse Gases, Regulated Emissions and Energy Use in Technologies Model) developed by the Argonne National Laboratory (14) and use this revised tool to understand the potential benefits of onshoring related manufacturing within the UK, where the electricity mix has a high percentage of renewables and is less carbon intensive relative to key European and international regions. The upgrade was achieved by adopting recently published, scalable inventories for an Interior Permanent Magnet (IPM) assisted synchronous reluctance machine and an Insulated Gate Bipolar Transistor (IGBT) inverter unit, both of which are representative of modern automotive applications.

## 2 RELATED PRIOR LCA

In earlier stages of this research, previous LCA studies of EVs have been reviewed. There are LCA studies that focus only on specific components of EVs, such as the high-voltage battery and power electronics (15-18), mostly based on confidential Life Cycle Inventories (LCIs). Several others assess the environmental impacts of BEVs and HEVs by examining the whole vehicle (19-23).

In their 2008 paper, Samaras et al. (19) presented a case study comparing the life cycle GHG emissions of traditional ICEVs and HEVs/PHEVs in the US. As with most LCA studies, the Global Warming Potential (GWP) impact was evaluated as carbon dioxide equivalent ($CO_2$-eq) GHG emissions with a time horizon of 100 years (GHG-100). This study employed the assumption that automotive manufacturing for all the vehicles considered was identical, except for the addition of batteries for the hybrid models. These included a HEV featuring a 16 kg Lithium-ion (Li-ion) battery and three PHEVs with an all-electric range (AER) of 30, 60 and 90 kilometres (km), featuring Li-ion batteries weighing 75-250 kg. For the use stage, three electricity mix scenarios with varying carbon intensities were tested – a low-carbon mix, the U.S. average and a carbon-intensive grid. These had carbon intensities of 200, 670 and 950 grams $CO_2$-eq per kilowatt-hour (g $CO_2$-eq/kWh) respectively. The functional unit was set to 1 km and the vehicle lifetime was assumed to be 240,000 km.

In all cases the use phase was responsible for the majority of GHG emissions, either through the gasoline fuel cycle or electricity generation. Also, all electrified vehicles had a lower overall GWP impact than their combustion equivalent. Under the U.S. average electricity mix, HEVs and PHEVs reduced life cycle GHG emissions by 29% and 32% compared to ICEVs respectively. It was found that PHEVs with larger battery capacities are more sensitive than HEVs to the GHG intensity of the electricity mix, overtaking HEVs in terms of total GWP impact at around 725 g $CO_2$-eq/kWh. In addition to the conventional ICEVs, the authors investigated the life cycle impacts of using an 85% by volume cellulosic ethanol blend (E85) instead of pure gasoline. Under the average and carbon-intensive mixes, it was found that fuel efficient vehicles that do not use grid electricity (ICEVs and HEVs) minimised GHG emissions. In contrast, with a low-carbon electricity portfolio PHEVs utilising electricity for propulsion had lower life cycle GHG emissions than ICEVs and up to 5 g $CO_2$-eq/km more than HEVs.

In their 2013 study, Hawkins et al. (24) have conducted a comparative LCA between a conventional ICEV and a first-generation BEV representative of a typical small European car, including vehicle production, use and end of life together with all relevant supply chains. The foreground LCI was compiled using secondary data sources. To ensure comparability, a common generic vehicle glider was established, and the functional unit was set to 1 km driven under European average conditions. For the EV, a use phase efficiency of 623 kilojoules/km (kJ/km) was assumed – representative of a Nissan Leaf – while for the diesel and petrol ICEVs fuel economies of 5.35 and 6.85 litres/100 km (L/100 km) were used respectively, based on comparable Daimler production models at the time. Road efficiencies of EVs in other LCA studies have ranged from 400 kJ/km (25, 26) to 800 kJ/km (27, 28).

For both vehicles, the use phase was found responsible for the majority of the GWP impact, occurring either directly through fuel combustion or indirectly during electricity production. When powered by average European electricity, EVs were found to reduce GWP by 10-14% and 20-24% compared to diesel and petrol ICEVs respectively – under the assumption of a 150,000 km lifetime. Assuming an efficiency of 900 kJ/km, the studied EVs would have a GWP footprint between that of the base case diesel and gasoline ICEVs. Conversely, a fuel consumption between 4 and 5 L/100 km would allow the ICEV to break even with the base case EVs in terms of GWP.

In contrast to ICEVs, almost half of an EV's life cycle GWP was associated with its production. The GWP from EV production was estimated to be 87-95 g $CO_2$-eq/km, roughly twice the 43 g $CO_2$-eq/km from an ICEV. Battery production was found responsible for 35-41% of the total EV production phase GWP, while the traction motor for 7-8%. Other powertrain components, notably the inverter and passive battery cooling system, contributed to 16-18% of the embedded GWP of EVs. Excluding the glider, these translate to roughly 60%, 12% and 28% for the battery, motor and inverter-battery cooling system respectively as a fraction of the total GWP for the e-powertrain. Due to the relatively high production impacts of EVs, assuming higher vehicle lifetimes exaggerates the benefits of EVs over conventional ICEVs while lower lifetimes have the opposite effect, with 100,000 km being the tipping point where an EV has a GWP impact indistinguishable from that of a diesel ICEV.

In their 2018 case study, Del Pero et al. (29) have implemented a comparative LCA between an ICEV and an EV, following a "cradle-to-grave" approach and capturing the whole vehicle life cycle subdivided into production, use and end-of-life stages. The LCI was mainly based on primary data and the assessment examined a variety of impact categories to both human and eco-system health. The functional unit was set to 150,000 km driven under conditions reproduced by an analytical vehicle dynamics simulation model. The use phase GWP through fuel combustion for the ICEV was determined using Euro 5 emission levels while the European average mix was assumed for the generation of the electricity consumed by the BEV.

For the ICEV, the use phase was found to be responsible for more than 80% of the GWP impact. Exhaust gas emissions during operation represented the main contributor (about 70%) while the rest was attributable to the fuel supply chain. On the other hand, the environmental burden of the EV regarding GHG emissions was equally attributable to both production and use phases, with the later having a slightly higher footprint. The manufacturing impact of the EV was 80% higher than the one of the ICEV, with the drivetrain representing around 55% of the total EV production GWP. Nevertheless, the high production impact of EVs is compensated by the lower use phase impact, resulting in an overall 36% reduction in total GHG emissions. For the use phase break-even analysis, two additional electricity mixes with opposite environmental profiles were used. The hydro-dominated Norwegian grid and coal-based Polish grid allowed a comprehensive overview of the effects of grid carbon intensity on EV use phase impacts. The break-even point was found at about 45,000 km using the European average mix. Assuming the Norwegian mix, the break-even distance reduced to 30,000 while using the Polish grid resulted to no break-even point occurring up to 250,000 km.

## 3 METHODOLOGY AND SYSTEM DETAILS

LCA involves compiling an inventory of the environmentally relevant flows associated with all processes involved in the extraction and processing of raw materials, manufacturing, distribution, use, recycling and end-of-life disposal of a product and translating this inventory into impacts of interest (30, 31). The following sections describe the methodology employed to construct this framework and provide a summary of the data and assumptions adopted for the LCA study.

### 3.1 Goal and scope definition

The goal of this study was to provide a comparative LCA of an electric and conventional ICE powertrain for medium sized cars, focusing on the effect of different electricity mixes utilised during the production and use phases in key regions of interest. An additional objective of the study was to examine the potential benefits of D+ class vehicle battery "rightsizing", employing a significantly smaller battery and an E10 Range-Extender (RE) ICE also able to operate on pure bioethanol (E100) with minor modifications.

The modelling of electricity generation, extraction and processing of raw materials, powertrain manufacturing and fuel cycles was realised using GREET 2020, the latest version of a popular open access transport life cycle model developed by the Argonne National Laboratory. Adaptations upon the GREET inventories regarding the traction motor and inverter unit were made using LCI models recently published by the Swedish Life Cycle Centre (32-35). These provide a scalable mass estimation and manufacturing inventory for IPM assisted synchronous reluctance machines and IGBT inverter units. The motor LCI model is tailored for neodymium-iron-boron magnets with added dysprosium, referred to as Nd(Dy)FeB. The capabilities of the traction motor model lie within a rated maximum power of 20-200 kW and a rated maximum torque of 48-477 Nm, while for the inverter model a rated nominal power of 20-200 kW and a nominal battery voltage of 250-700 V. The traction motor and inverter unit are also referred to as the Power Electronics, Machines and Drives (PEMD) unit.

The system boundaries comprehend the entire life cycle of the powertrains examined apart from the end-of-life disposal, the impact of which is relatively low and will be investigated in the future. All other non-propulsion vehicle subsystems common to both EVs and ICEVs are assumed to have comparable environmental footprint and their life cycle impact was deemed beyond the scope of this study. The functional unit was set to 1 km travelled by the respective powertrains and the lifetimes for all cases were set to 150,000 km. The key environmental impact category of interest at the time of reporting this work was that of GWP, calculated as $CO_2$-eq emissions with a time horizon of 100 years.

### 3.1.1 *Electricity generation*

The focus of this study was to investigate the GWP benefits of onshoring the manufacturing of e-powertrain components to the UK, where the electricity mix has a favourable carbon footprint relative to many European and international countries due to the high percentage of renewables. Comparisons were made to grids on either end of the carbon intensity scale – specifically to the coal-based China (CN) and nuclear-dominated France (FR). The study was extended to cover the average European mix and additional regions of interest to the automotive industry – namely the United States (US), Germany (DE), Italy (IT), Poland (PL), Romania (RO) and the Czech Republic (CZ).

Power generation data were acquired by the International Energy Agency (IEA) for 2019 (36). Although GREET features US-based data for processes associated with electricity generation, these were assumed to be sufficiently representative on a worldwide scale. Transmission and distribution efficiency was set to 0.95.

### 3.1.2 *Baseline powertrains*

The e-powertrain includes the high-voltage battery, traction motor and power inverter unit – with the exclusion of the On-Board Charger (OBC) and DC-DC converter which powers the low-voltage systems of a BEV. Specifications were set to model a medium sized EV, such as the 2020 Tesla Model 3 Standard Range Plus (SR+). The adopted model vehicle utilises a single 200 kW permanent magnet synchronous motor with a maximum torque of 450 Nm, a 200 kW inverter unit and a 50 kWh battery operating at a nominal voltage of 360V. The estimates of the motor and inverter mass were 83.3 and 18.6 kg respectively. The selected battery chemistry was Li-ion with a cathode combination of nickel-manganese-cobalt in equal parts (NMC111), a popular choice among automotive manufacturers. Although the default energy density in GREET is 215 Wh/kg (37), a more realistic value of 160 Wh/kg was assumed for the high-voltage battery pack after attaining real world data (38).

For the conventional ICE, a comparable 2020 BMW 3 Series 2.0 L was selected as the combustion equivalent – specifically the 330i for petrol and the 320d for diesel models. The vehicles use inline-four (I4) turbocharged four-stroke engines with estimated masses of 150 and 180 kg for the Spark Ignition (SI) and Compression Ignition (CI) versions respectively. Both engines were typical of the state of the art within the European market. The SI engine was rated at 190 kW and was based around central direct fuel injection with a twin-scroll turbocharger and variable valvetrain. The diesel unit was rated at 140 kW and based around a common rail design with twin turbo boosting arrangement.

For the Range-Extender e-powertrain, the PEMD specifications were kept identical to the full electric version to ensure comparability. The battery capacity was halved to 25 kWh after related prior work at the University of Nottingham which estimated that a medium sized RE EV with a battery capacity of 25 kWh can operate in fully electric mode for ~80% of its lifetime with appropriate driver behaviour. The study was based on curve fitted data for electric mode share utilised in GREET and vehicle travel statistics acquired by the UK Department for Transport (39-41).

For on-board electricity generation, a MAHLE 0.9 L twin cylinder, four-stroke SI engine was emulated (42). The reported engine mass is 50 kg, with an additional 20 kg permanent magnet generator modelled using the LCI of the main traction motor. The key characteristics of the baseline powertrains are summarised in Table 1.

**Table 1. Key characteristics of the powertrains considered.**

| Characteristics | ICEV | | BEV | RE EV |
|---|---|---|---|---|
| | SI | CI | | |
| Power (kW) | 190 | 140 | 200 | 200 |
| Battery capacity (kWh) | - | - | 50 | 25 |
| Engine weight (kg) | 150 | 180 | - | 50 |
| Motor weight (kg) | - | - | 83.3 | 83.3 + 20 |
| Inverter weight (kg) | - | - | 18.6 | 18.6 |
| Battery weight (kg) | - | - | 233 | 116 |

### 3.1.3 *Use phase*

The first substage of the use phase is electricity production for the e-powertrain and fuel cycle for the conventional ICE, which consists of fossil fuel extraction or feedstock production, refining and distribution. For the e-powertrain, a road efficiency of 470 kJ/km was adopted, representative of the EPA range of a TESLA Model 3 SR+ (43). As with electricity generation, the fuel cycle of diesel and 10% ethanol petrol (E10) which is assumed to be used by both the conventional ICE and Range-Extender SI engines, was modelled using GREET.

The second substage of the use phase is tailpipe emissions during operation. While effectively zero for the e-powertrain, the use phase GWP through fuel combustion for the conventional ICE was determined using 2020-2021 EU $CO_2$ emission standards for fleet average targets which amount to 95 g $CO_2$/km (44). This translates to a fuel consumption of approximately 4.1 L/100km and 3.5 L/100 km for the SI and CI versions respectively.

Regarding the Range-Extender use phase emissions, a combination of Charge Depleting (CD) and Charge Sustaining (CS) modes was implemented at an 80:20 ratio. CD mode utilised the same electricity pathway and road efficiency as the BEV powertrain – the battery weight savings were assumed to offset the additional ICE mass, while the CS emissions were estimated using peak efficiency figures published by the RE ICE manufacturer (42) and fuel thermodynamic properties. The reported system efficiency (combined efficiency of the engine, generator and inverter) was 0.31, resulting in specific $CO_2$ emissions for CS mode of 102 g $CO_2$/km.

#### 3.1.4 *Logistics*

To assess the impact of the transportation and distribution of manufactured powertrain components on the life cycle GHG emissions, two distinct scenarios were examined. Specifically, Scenario 1 featured a UK manufactured PEMD while Scenario 2 featured a UK manufactured inverter unit and a traction motor manufactured in Italy. For each scenario, the logistics of exporting the PEMD units to Germany, China and the eastern US were mapped and the life cycle GHG emissions were compared to both local manufacturing and importing from China instead (only for Germany and US).

Deep sea container shipping was used to model exports to the US, while exports to China involved both marine and diesel rail freight. Journeys from the UK and within continental Europe utilised road freight, electric rail freight and RORO cargo shipping depending on the scenario.

The freight distances were estimated using digital mapping platforms and estimated specific $CO_2$ emissions from different modes of transport were acquired from secondary sources (45). These are summarised in Table 2.

**Table 2. Specific emissions for the modes of transport considered.**

| Mode of transport | Specific emissions |
|---|---|
| | g $CO_2$/tonne km |
| Road freight – Diesel HGV | 200 |
| RORO cargo ferry | 100 |
| Rail freight – diesel | 50 |
| Rail freight – electric | 25 |
| Container ship – ocean | 25 |

### 3.2 Model improvements

The uncertainty in calculating the life cycle GHG emissions of PEMD units in GREET is closely related with the quality of the inputs and data used by the LCA practitioner. In this case scalable inventories mentioned in previous sections were adopted in the default modelling of GREET to integrate an improved LCI with refined Bill-of-Materials (BoM) and intricate manufacturing processes involved in the fabrication of modern automotive e-machines.

#### 3.2.1 *Mass estimation and inventory update*

For medium sized electric passenger cars, GREET offers the default model "EV300" which is targeted for vehicles with up to 300 miles range. The software includes base versions of a traction motor and electronic controller, each with their respective inventories. The default model served as the starting point for further iterations of the inventory towards the modelling of a Tesla Model 3 equivalent e-powertrain. The initial motor and inverter masses were 122.5 and 77.6 kg respectively, whereas for a 200 kW powertrain they were estimated using the scalable LCI models as 83.3 and 18.6 kg.

Tables 3 and 4 show the difference in mass compositions for the PEMD, before and after the adoption of the scalable inventories. The absence of rare earth elements in the default motor model justifies the relatively high total mass, as modern permanent magnet machines have superior specific power than induction machines. The high default inverter mass was attributed to older generation power electronics with less power density than modern machines.

**Table 3. Traction motor mass break-down.**

| Material | EV300 | Scalable LCI |
|---|---|---|
| | % | % |
| Aluminium | 36 | 33 |
| Steel | 36 | 7 |
| Copper | 28 | 9 |
| Electrical steel | - | 46 |
| Nd(Dy)FeB magnets | - | 3 |
| Other | - | 2 |

**Table 4. Inverter mass break-down.**

| Material | EV300 | Scalable LCI |
|---|---|---|
| | % | % |
| Aluminium | 47 | 44 |
| Steel | 5 | 2 |
| Copper | 8 | 31 |
| Plastic products | 24 | 18 |
| Styrene-butadiene rubber | 4 | 0 |
| Electronic components | - | 1 |
| Other | 12 | 4 |

### 3.2.2 *Manufacturing processes*

This study was focused on the production impact of e-powertrain components, so there was an interest to identify the most energy intensive processes involved and assess the relative impact of different PEMD components. Table 5 presents the electricity intensity of several processes in the manufacturing and assembly pathway of the PEMD, as described in the scalable LCIs adopted for the purposes of this research.

**Table 5. Electricity usage of modelled processes in PEMD manufacturing.**

| Process | Electricity consumption | Specific electricity consumption |
|---|---|---|
| | kWh | kWh/kg |
| Traction motor | | |
| Aluminium die casting | 74.4 | 2.6 |
| Electrical steel production | 46.4 | 0.6 |
| Magnet production | 44.9 | 18.9 |
| Assembly factory | 18.8 | 0.2 |
| Wire enamelling | 3.8 | 0.5 |
| Inverter unit | | |
| Aluminium die casting | 22.0 | 2.6 |
| Assembly factory | 6.8 | 0.4 |
| Power module production | 5.9 | 2.1 |
| Printed circuit board assembly | 1.0 | 5.0 |
| Electroplating | 0.9 | *N/A* |

In both traction motor and inverter unit manufacturing, it was found that most of the electricity usage is attributable to the die casting of aluminium housings. Although it incurs relatively low specific consumptions, the high aluminium mass fraction in the PEMD makes housings the highest $CO_2$ contributor. High absolute values are also associated with the production of electrical steel, due to the high steel mass fraction in the motor and the high scrap rate of punching core laminations. The magnet production chain is by far the most energy intensive process on a specific basis and magnets are the third highest $CO_2$ contributor within the PEMD. With additional energy demands during the mining of rare earth elements, sources estimate the specific energy consumption for virgin magnet production to be as high as 36 kWh/kg. In their 2016 paper, Zakotnik et al. (46) discuss a large scale magnet recycling operation, capable of producing high-temperature magnets suitable for automotive applications with 90% energy savings. Other sources report similar 'magnet-to-magnet' recycling methods can reduce total GHG emissions by 55 to 80% (47, 48). This is an area of interest for future related work as a stable, self-sufficient and environmentally favourable magnet supply chain can be key towards the electrification era the industry is going through.

## 4 RESULTS

### 4.1 Electricity generation

Figure 1 presents the specific GHG emissions of electricity generation in the regions of interest. Apart from nuclear-dominated France, the UK electricity grid had the lowest emissions amongst the regions considered. Compared to coal-based China and Poland, the relatively green UK grid offers a 66% and 71% reduction in electricity related GHG emissions.

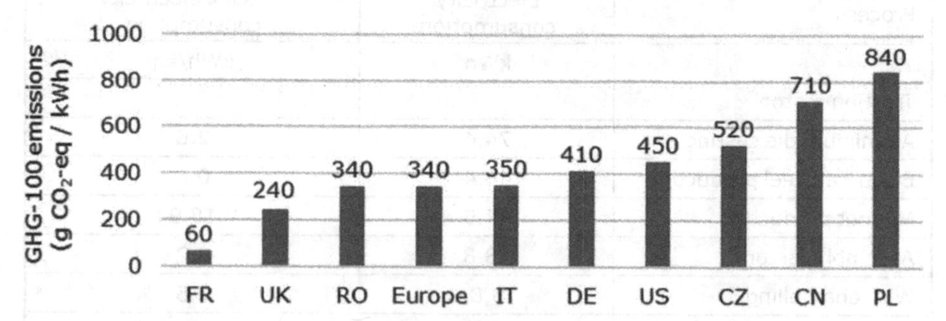


**Figure 1. GWP impact of electricity generation by region.**

### 4.2 Production

Figure 2 presents the GHG emissions of e-powertrain production in the regions of interest. In all cases, the high-voltage battery was responsible for most of the production emissions, with the PEMD accounting for 10-14% of the total. Onshoring all components of the e-powertrain within the UK offers $CO_2$-eq savings ranging from 0.29 (Romania) to 1.78 (Poland) metric tonnes $CO_2$-eq (MT $CO_2$-eq) per vehicle, translating to reductions of 7-32%. On the other hand, e-powertrain manufacturing in the UK predicts a 17% increase to GHG emissions compared to manufacturing in France. Not including the high-voltage battery, onshoring just the PEMD within the UK offers $CO_2$-eq savings ranging from 0.06 (Romania) to 0.35 (Poland) metric tonnes $CO_2$-eq (MT $CO_2$-eq), translating to reductions of 12-46%. PEMD manufacturing in the UK predicts 34% higher GHG emissions compared to manufacturing in France.

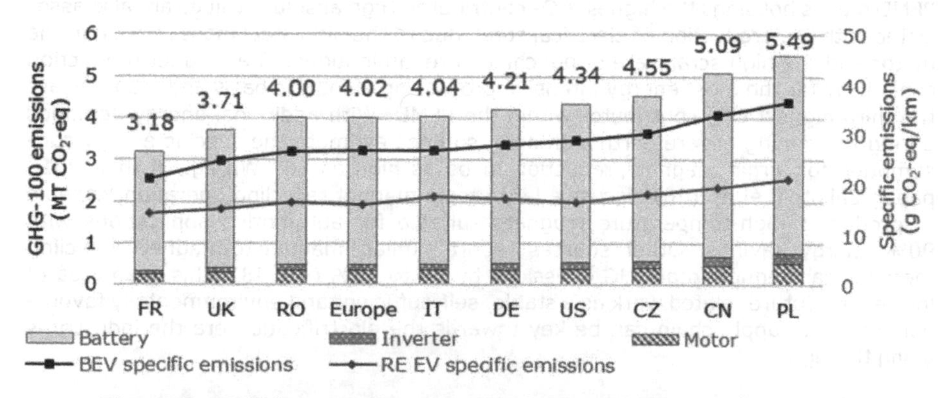


**Figure 2. GWP impact of e-powertrain manufacturing by region.**

The same trend was observed in the other types of powertrains, while the sensitivity to the carbon grid intensity was directly related to the battery capacity. The ICE models had significantly low GHG emissions at 10-12% of the BEV powertrain, ranging from 0.33-0.64 MT $CO_2$-eq (2-4 g $CO_2$-eq/km). Relative to the BEV powertrain, the RE EV reduced the embedded $CO_2$-eq emissions by 32-43% due to the smaller 25 kWh battery, ranging from 2.15-3.18 MT $CO_2$-eq per powertrain (14-21 g $CO_2$-eq/km).

### 4.3 Use phase

Figure 3 illustrates the average European use phase GHG emissions, occurring either through electricity generation for EVs or fuel cycle and tailpipe emissions for ICEVs. The combustion powertrains had the highest carbon footprint while the BEV minimised carbon emissions at 44 g $CO_2$-eq/km, a 64% reduction relative to the SI ICEV. The RE EV with 80% of its lifetime spent in fully electric mode achieved a 52% $CO_2$-eq reduction relative to the SI ICEV and contributed 33% more use phase GHG emissions than the BEV solution.

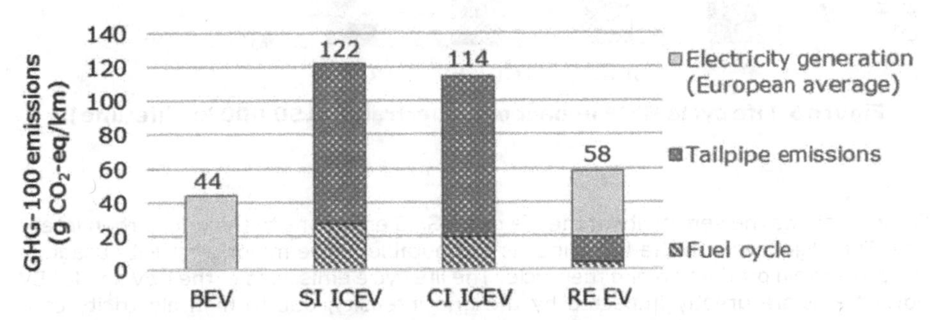


**Figure 3. Use phase GWP impact of powertrains (150,000 km lifetime).**

Figure 4 shows the sensitivity of the use phase GHG emissions to the grid carbon intensity. The slight slope in the ICEV lines is attributable to the minor electricity usage in the fuel cycle. Although the BEV and RE EV use phase emissions are greatly impacted by the grid intensity, even at coal-dependent grids such as Poland the use phase emissions of ICEVs are higher than their electric equivalents.

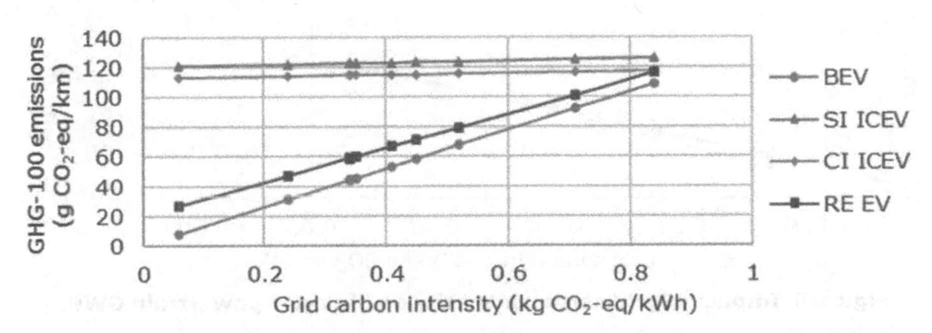


**Figure 4. Impact of grid carbon intensity on use phase powertrain GWP.**

## 4.4 Well-to-Wheel

Figure 5 illustrates the life cycle GHG emissions of the powertrains considered, which are dominated by the use phase. The combustion powertrains had the highest carbon footprint while the BEV minimised carbon emissions at 71 g $CO_2$-eq/km, a 43% reduction relative to the SI ICEV. The RE EV at 74 g $CO_2$-eq/km achieved a 41% $CO_2$-eq reduction relative to the SI ICEV and contributed 4% more GHG emissions than the BEV solution.

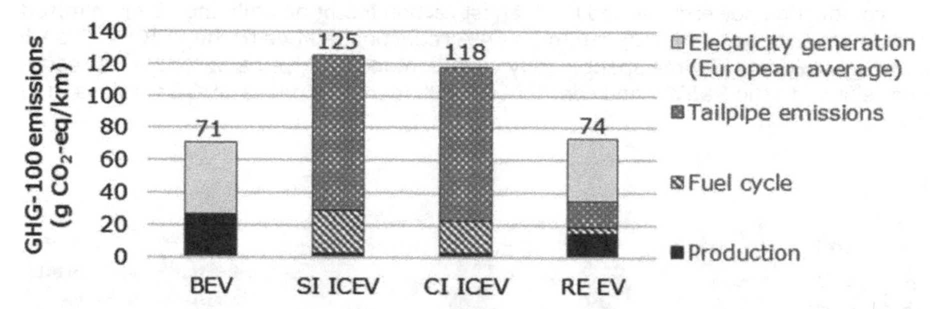


**Figure 5. Life cycle GWP impact of powertrains (150,000 km lifetime).**

Figure 6 shows the sensitivity of the life cycle GHG emissions to the grid carbon intensity. The slight slope in the ICEV lines is attributable to the minor electricity usage in the powertrain production and fuel cycle. The life cycle emissions of the BEV and RE EV powertrains are greatly impacted by the grid intensity, due to high electricity consumption during manufacturing and operation. Even so, the WTW GHG emissions of electrified powertrains only overtake conventional combustion engines in coal-dependent regions, namely Poland in this case. Comparing the two e-powertrains, with low and medium carbon electricity grids BEVs perform better than RE EVs, primarily due to use phase related benefits. In terms of total $CO_2$-eq emissions, BEVs overtake RE EVs at about 500 g $CO_2$-eq/kWh.

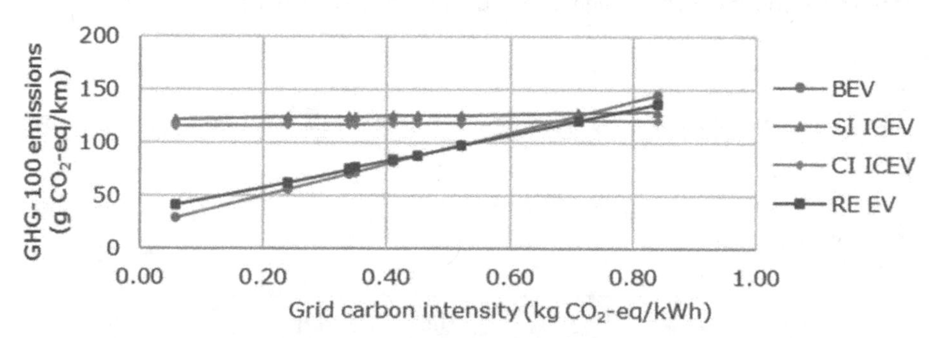


**Figure 6. Impact of grid carbon intensity on life cycle powertrain GWP.**

## 4.5 Carbon break-even analysis

Figures 7 and 8 show the break-even distances of BEV and RE EV powertrains against SI and CI ICEs respectively, using the European average mix for the use phase. The break-even distances are longer for diesel due to the lower fuel consumption of CI ICEs. It was found that a Range-Extender configuration can reduce the carbon break-even distance of EVs by 8 to 24 thousand km. The reduction is proportional to the grid carbon intensity of the manufacturer country, since carbon-heavy grids lead to increased embedded GHG emissions in battery manufacturing and therefore larger break-even distances during the use phase. From an LCA perspective, the use of fuel-efficient compact ICEs utilised for on-board electricity generation seems promising against oversized batteries, as it can significantly reduce the GHG emissions and natural resources associated with their manufacture. This can be explored more extensively by utilising renewable biofuels (e.g. E100), which will further reduce the total GHG deficiency of RE EVs against BEVs.

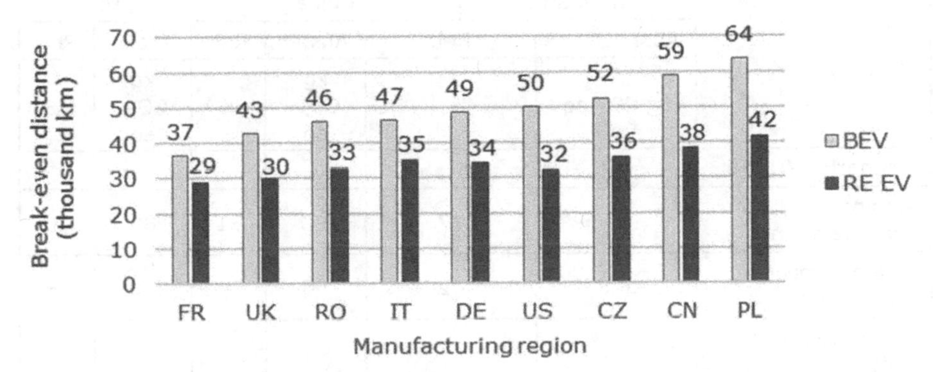


**Figure 7. BEV and RE EV break-even distance against SI ICEV.**

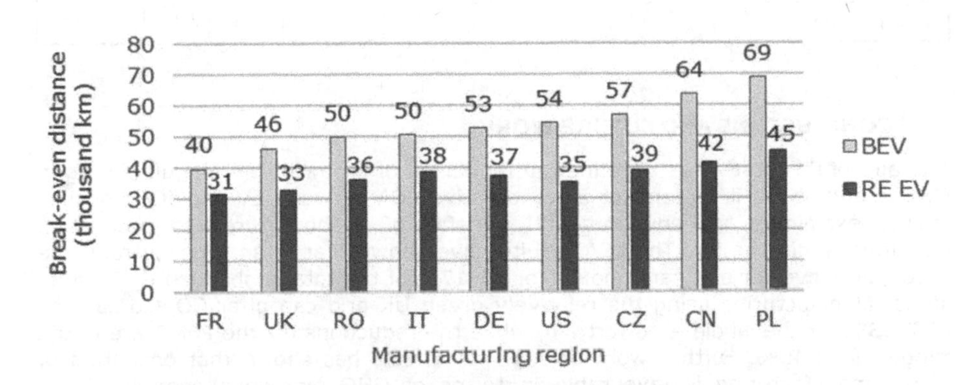


**Figure 8. BEV and RE EV break-even distance against CI ICEV.**

## 4.6 Logistics

Table 6 presents the effects of logistics on the GHG emissions covering the manufacturing and distribution of the PEMD unit for the two scenarios explained earlier. In all cases, it was more beneficial to manufacture the motor and inverter in UK/Italy and export it to Germany, China and the eastern US rather than manufacturing the PEMD locally or importing it from China. The biggest $CO_2$-eq savings against local manufacturing occur when exporting to China despite the large freight distances. Against Chinese manufacturing, the biggest $CO_2$-eq savings occur when exporting to Germany. It is suggested that from a GHG perspective, clean grid manufacturing is favourable even for worldwide distribution. In the future, logistics effects of raw materials will be more extensively studied to assess the impacts of non-European supply chains.

**Table 6. GHG reduction from PEMD manufacturing onshoring.**

| | Export destination | | | | | |
|---|---|---|---|---|---|---|
| | Germany | | China | | US | |
| | Abs. | Rel. | Abs. | Rel. | Abs. | Rel. |
| | kg $CO_2$-eq | % | kg $CO_2$-eq | % | kg $CO_2$-eq | % |
| Scenario 1 (UK) | | | | | | |
| Against local manufacturing | 89 | 17 | 211 | 31 | 86 | 16 |
| Against imports from China | 308 | 42 | - | - | 200 | 26 |
| Scenario 2 (UK/It) | | | | | | |
| Against local manufacturing | 48 | 9 | 174 | 25 | 41 | 8 |
| Against imports from China | 280 | 37 | - | - | 268 | 35 |

## 5 CONCLUSIONS AND FUTURE WORK

The aim of this study was to refine the modelling of e-powertrain manufacturing in GREET and utilise it to conduct a comparative LCA between EV and ICEV powertrains, examining the environmental benefits of onshoring related production activities within the UK. The LCA results have shown that in an e-powertrain, the motor and inverter unit can amount for 12-17% of the total embedded GHG emissions. Manufacturing using the relatively green UK grid can offer $CO_2$-eq savings of 7-33% for the whole e-powertrain, while the reductions for the PEMD are in the range of 12-46%. Further work on logistics effects has shown that onshoring of PEMD manufacturing is favourable in terms of GHG emissions even for long-distance exports. Nevertheless, the life cycle GWP impact is dominated by the use phase GHG emissions, which are lower for electrified vehicles in all cases – even with high-carbon electricity. Using the European average, which is a low to medium carbon intensity generation mix, EVs achieved the lowest life cycle GHG emissions with 66 g $CO_2$-eq/km, against 125 and 118 g $CO_2$-eq/km for SI and CI ICEVs respectively. The RE EV reduced carbon-break even distances by 5-19 thousand km, while having only 8% higher total GHG emissions than the BEV over the lifetime of 150,000 km. Such powertrains will clearly avoid reliance on generally oversized batteries to help reduce remaining issues with battery recycling.

Future work will involve several aspects of the LCA process. Regarding the powertrain life cycles, the logistics effects of raw material sourcing for each of the components will be investigated on a global scale for all supply chains considered. Additionally, the impact of EoL disposal and material recycling will be integrated into the current iteration. Remaining unexplored areas in the manufacturing inventory such as the ICEV transmission, specialised electronics and metalworking processes will be updated using "ecoinvent", a set of dedicated LCI databases developed by the Swiss Centre for Life Cycle Inventories (49). Furthermore, the RE EV case will be revisited to employ E100 fuel and reiterate on the benefits of battery "rightsizing" using range extension via renewable fuels. Lastly, there is ambition to expand the modelling capabilities into other potential e-machine technologies, such as induction motors and Silicon Carbide (SiC) or Gallium Nitride (GaN) MOSFET based inverter units.

## REFERENCES


[1] Mobility 2030: Meeting the challenges to sustainability. Geneva,Switzerland: World Business Council for Sustainable Development (WBCSD); 2004.

[2] Kalghatgi, G. Is it really the end of internal combustion engines and petroleum in transport? Applied Energy. 2018;225: 965–974. doi:10.1016/j.apenergy.2018.05.076

[3] Roadmap to decarbonise European cars. European Federation for Transport and Environment; 2018.

[4] Government takes historic step towards net-zero with end of sale of new petrol and diesel cars by 2030. Last accessed 28 May, 2021 from https://www.gov.uk/government/news.

[5] Electric vehicles from life cycle and circular economy perspectives. European Environmental Agency; 2018.

[6] Gustafsson, T., Johansson, A. Comparison between Battery Electric Vehicles and Internal Combustion Engine Vehicles fueled by Electrofuels - From an energy efficiency and cost perspective; 2015.

[7] Egede, P. Environmental Assessment of Lightweight Electric Vehicles. 1st ed: Springer International Publishing; 2017.

[8] Rangaraju, S., De Vroey, L., Messagie, M., Mertens, J., Van Mierlo, J. Impacts of electricity mix, charging profile, and driving behavior on the emissions performance of battery electric vehicles: A Belgian case study. Applied Energy. 2015;148: 496–505. doi:10.1016/j.apenergy.2015.01.121

[9] Helmers, E., Weiss, M. Advances and critical aspects in the life-cycle assessment of battery electric cars. Energy and Emission Control Technologies. 2017;5: 1–18. doi:10.2147/EECT.S60408

[10] Apostolaki-Iosifidou, E., Codani, P., Kempton, W. Measurement of power loss during electric vehicle charging and discharging. Energy. 2017;127: 730–742. doi:10.1016/j.energy.2017.03.015

[11] Kostopoulos, E.D., Spyropoulos, G.C., Kaldellis, J.K. Real-world study for the optimal charging of electric vehicles. Energy Reports. 2020;6: 418–426. doi:10.1016/j.egyr.2019.12.008

[12] Wang, Q., Santini, D.L. Magnitude and value of electric vehicle emissions reductions for six driving cycles in four US cities with varying air quality problems. IL (United States): Argonne National Lab.; 1992.

[13] Wang, D., Zamel, N., Jiao, K., Zhou, Y., Yu, S., Du, Q., et al. Life cycle analysis of internal combustion engine, electric and fuel cell vehicles for China. Energy. 2013;59: 402–412. doi:10.1016/j.energy.2013.07.035
[14] Wang, M., Elgowainy, A., Lu, Z., Bafana, A., Benavides, P.T., Burnham, A., et al. Greenhouse gases, Regulated Emissions, and Energy use in Technologies Model ® (2020 .Net). 2020. doi:10.11578/GREET-Net-2020/dc.20200913.1
[15] Van den Bossche, P., Vergels, F., Van Mierlo, J., Matheys, J., Autenboer, W. SUBAT: An assessment of sustainable battery technology. Journal of Power Sources. 2006;162: 913–919. doi:10.1016/j.jpowsour.2005.07.039
[16] Matheys, J., Van Mierlo, J., Timmermans, J.-M., Van den Bossche, P. Life-cycle assessment of batteries in the context of the EU Directive on end-of-life vehicles. International Journal of Vehicle Design. 2008;46: 189–203. doi:10.1504/IJVD.2008.017182
[17] Majeau-Bettez, G., Hawkins, T.R., Strømman, A.H. Life Cycle Environmental Assessment of Lithium-Ion and Nickel Metal Hydride Batteries for Plug-In Hybrid and Battery Electric Vehicles. Environmental Science & Technology. 2011;45(10):4548–4554. doi:10.1021/es103607c
[18] Ellingsen, L.A.-W., Majeau-Bettez, G., Singh, B., Srivastava, A.K., Valøen, L.O., Strømman, A.H. Life Cycle Assessment of a Lithium-Ion Battery Vehicle Pack. Journal of Industrial Ecology. 2014;18(1):113–124. doi:10.1111/jiec.12072
[19] Samaras, C., Meisterling, K. Life Cycle Assessment of Greenhouse Gas Emissions from Plug-in Hybrid Vehicles: Implications for Policy. Environmental Science & Technology. 2008;42(9):3170–3176. doi:10.1021/es702178s
[20] Frischknecht, R., Flury, K. Life cycle assessment of electric mobility: answers and challenges. The International Journal of Life Cycle Assessment. 2011;16(7):691–695. doi:10.1007/s11367-011-0306-6
[21] Faria, R., Moura, P., Delgado, J., de Almeida, A. A sustainability assessment of electric vehicles as a personal mobility system. Energy Conversion and Management. 2012;61: 19–30. doi:10.1016/j.enconman.2012.02.023
[22] Faria, R., Marques, P., Moura, P., Freire, F., Delgado, J., de Almeida, A.T. Impact of the electricity mix and use profile in the life-cycle assessment of electric vehicles. Renewable and Sustainable Energy Reviews. 2013;24: 271–287. doi:10.1016/j.rser.2013.03.063
[23] Bartolozzi, I., Rizzi, F., Frey, M. Comparison between hydrogen and electric vehicles by life cycle assessment: A case study in Tuscany, Italy. Applied Energy. 2013;101: 103–111. doi:10.1016/j.apenergy.2012.03.021
[24] Hawkins, T.R., Singh, B., Majeau-Bettez, G., Strømman, A.H. Comparative Environmental Life Cycle Assessment of Conventional and Electric Vehicles. Journal of Industrial Ecology. 2013;17(1):53–64. doi:10.1111/j.1530-9290.2012.00532.x
[25] Elgowainy, A., Burnham, A., Wang, M., Molburg, J., Rousseau, A. Well-to-wheels energy use and greenhouse gas emissions analysis of plug-in hybrid electric vehicles. Argonne National Lab. (ANL), Argonne, IL (United States); 2009. Report No.: ANL/ESD/09-2.
[26] Shiau, C.-S.N., Samaras, C., Hauffe, R., Michalek, J.J. Impact of battery weight and charging patterns on the economic and environmental benefits of plug-in hybrid vehicles. Energy Policy. 2009;37(7):2653–2663. doi:10.1016/j.enpol.2009.02.040
[27] Huo, H., Zhang, Q., Wang, M.Q., Streets, D.G., He, K. Environmental Implication of Electric Vehicles in China. Environmental Science & Technology. 2010;44(13):4856–4861. doi:10.1021/es100520c
[28] Parks, K., Denholm, P., Markel, T. Costs and Emissions Associated with Plug-In Hybrid Electric Vehicle Charging in the Xcel Energy Colorado Service Territory.

National Renewable Energy Lab. (NREL); 2007. Report No.: NREL/TP-640-41410.
[29] Pero, F.D., Delogu, M., Pierini, M. Life Cycle Assessment in the automotive sector: a comparative case study of Internal Combustion Engine (ICE) and electric car. Procedia Structural Integrity. 2018;12: 521–537. doi:10.1016/j.prostr.2018.11.066
[30] Curran, M.A. Environmental life-cycle assessment. The International Journal of Life Cycle Assessment. 1996;1(3):179–179. doi:10.1007/BF02978949
[31] Guinee, J.B. Handbook on Life Cycle Assessment. Operational Guide to the ISO Standards. Netherlands: Kluwer Academic Publishers, Dordrecht (Netherlands); 2002.
[32] Nordelöf, A., Grunditz, E., Tillman, A.-M., Thiringer, T., Alatalo, M. A scalable life cycle inventory of an electrical automotive traction machine—Part I: design and composition. The International Journal of Life Cycle Assessment. 2018;23(1):55–69. doi:10.1007/s11367-017-1308-9
[33] Nordelöf, A., Tillman, A.-M. A scalable life cycle inventory of an electrical automotive traction machine—Part II: manufacturing processes. The International Journal of Life Cycle Assessment. 2018;23(2):295–313. doi:10.1007/s11367-017-1309-8
[34] Nordelöf, A., Alatalo, M., Söderman, M.L. A scalable life cycle inventory of an automotive power electronic inverter unit—part I: design and composition. The International Journal of Life Cycle Assessment. 2019;24(1):78–92. doi:10.1007/s11367-018-1503-3
[35] Nordelöf, A. A scalable life cycle inventory of an automotive power electronic inverter unit—part II: manufacturing processes. The International Journal of Life Cycle Assessment. 2019;24(4):694–711. doi:10.1007/s11367-018-1491-3
[36] International Energy Agency (IEA). Last accessed 22 February 2021, from https://www.iea.org/data-and-statistics.
[37] Winjobi, O., Dai, Q., Kelly, J. Update of Bill-of-Materials and Cathode chemistry addition for Lithium-ion Batteries in the GREET® Model.
[38] Tesla Model 3 Battery Pack & Battery Cell Teardown Highlights Performance Improvements. Last accessed 9 August 2021, from https://cleantechnica.com.
[39] VEH0101: Licensed vehicles by body type (quarterly): Great Britain and United Kingdom. Department for Transport; United Kingdom 2016.
[40] TRA0104: Road traffic (vehicle miles) by vehicle type and road class in Great Britain. Department for Transport; United Kingdom 2018.
[41] NTS0904: Annual mileage band of 4-wheeled cars, England: since 2002. Department for Transport; United Kingdom 2013.
[42] Bassett, M., Hall, J., Warth, M. Development of a dedicated range extender unit and demonstration vehicle. 2013 World Electric Vehicle Symposium and Exhibition (EVS27); 17-20 November 2013.
[43] Impact Report 2019. Tesla, Inc.; 2019.
[44] CO2 Emission Standards for Passenger Cars and Light-Commercial Vehicles in the European Union. International Council on Clean Transportation (ICCT) January 2019.
[45] Sims R., R. Schaeffer, F. Creutzig, X. Cruz-Núñez, M. D'Agosto, D. Dimitriu, M. J. Figueroa Meza, L. Fulton, S. Kobayashi, O. Lah, A. McKinnon, P. Newman, M. Ouyang, J.J. Schauer, D. Sperling, and G. Tiwari, 2014: Transport. In: Climate Change 2014: Mitigation of Climate Change. Contribution of Working Group III to the Fifth Assessment Report of the Intergovernmental Panel on Climate Change [Edenhofer, O., R. Pichs-Madruga, Y. Sokona, E. Farahani, S. Kadner, K. Seyboth, A. Adler, I. Baum, S. Brunner, P. Eickemeier, B. Kriemann, J. Savolainen, S. Schlömer, C. von Stechow, T. Zwickel and J.C. Minx (eds.)]. Cambridge University Press, Cambridge, United Kingdom and New York, NY, USA.

[46] Zakotnik, M., Tudor, C.O., Peiró, L.T., Afiuny, P., Skomski, R., Hatch, G.P. Analysis of energy usage in Nd–Fe–B magnet to magnet recycling. Environmental Technology & Innovation. 2016;5: 117–126. doi:10.1016/j.eti.2016.01.002
[47] Jin, H., Afiuny, P., McIntyre, T., Yih, Y., Sutherland, J.W. Comparative Life Cycle Assessment of NdFeB Magnets: Virgin Production versus Magnet-to-Magnet Recycling. Procedia CIRP. 2016;48: 45–50. doi:10.1016/j.procir.2016.03.013
[48] Jin, H., Afiuny, P., Dove, S., Furlan, G., Zakotnik, M., Yih, Y., et al. Life Cycle Assessment of Neodymium-Iron-Boron Magnet-to-Magnet Recycling for Electric Vehicle Motors. Environmental Science & Technology. 2018;52(6):3796–3802. doi:10.1021/acs.est.7b05442
[49] Weidema, B., Bauer, C., Hischier, R., Mutel, C., T, N., J, R., et al. The ecoinvent database: Overview and methodology, Data quality guideline for the ecoinvent database version 3. Switzerland: The Centre for Life Cycle Inventories; 2013.

## *Session 8: Powertrain development systems for hybrid electric vehicle*

# MAHLE modular hybrid powertrain for large passenger cars and light commercial vehicles

**A. Cooper, M. Bassett, A. Harrington, I. Reynolds, D. Pates**
MAHLE Powertrain Ltd, Costin House, Northampton, NN5 5TZ, UK

## ABSTRACT

For large passenger cars and light commercial vehicles, meeting all the customer requirements with a pure electric vehicle is challenging. Building on work previously presented on the MAHLE modular hybrid powertrain concept, this study examines potential hybrid powertrain configurations for these vehicle applications, based on life-cycle $CO_2$ assessments.

The latest testing results from our 3-cylinder 1.5 litre high-efficiency demonstrator engine are presented, running on both conventional 95 RON gasoline and alternative sustainable low carbon fuels. These results are then used to evaluate the potential weighted $CO_2$ emissions of the proposed plug-in hybrid concept across representative drive cycles for these applications.

## 1 INTRODUCTION

Vehicle manufacturers are experiencing a paradigm shift in legislative requirements and customer attitudes towards powertrain technologies. To support the pathway towards net-zero emissions by 2050, technologies that significantly reduce $CO_2$ emissions will need to be developed that also align with the availability of the supporting infrastructure. Through the introduction of clean air zones, and the UK government's announcement to end the sale of vehicles powered only by petrol or diesel passenger cars by 2030, the requirement for vehicles to drive under electric only propulsion is mandated. In the areas of personal mobility vehicles, compact cars and urban transportation, the adoption of pure battery electric powertrains is expected to accelerate and become the dominant technology. For large passenger cars and light commercial vehicles (LCVs), meeting customer requirements for the combination of factors such as range, payload, towing capability, and purchase cost is challenging. To meet these requirements with a pure electric vehicle requires the use of heavy and expensive battery packs, that have a high embedded $CO_2$ content.

The study builds on the work previously presented on the MAHLE modular hybrid powertrain (MMHP) concept and examines the powertrain requirements to meet the specific needs of large passenger cars and LCVs in the 2030 timescale and beyond. In this concept, the full dynamic performance of the vehicle is provided by the electric traction motor, along with the ability to drive in pure electric mode for a reduced range, using a plug-in capability. On long journeys or when charging is not available, drive can be supplemented, and battery state of charge maintained, through use of a compact, highly efficient dedicated hybrid engine that can run on sustainable low carbon fuel sources.

Once the vehicle has significant electric drive capability, it is possible to remove any dynamic loading from the engine and allow it to operate at much steadier loads, that are now independent from the driver's dynamic performance requirements. The ability

DOI: 10.1201/9781003219217-19

to anticipate starting requirements and control transient operation, particularly during start-up from cold conditions, also significantly reduces the demands placed on the aftertreatment system and allows it to be pre-conditioned prior to starting. When sizing such a dedicated hybrid engine (DHE) the most critical parameter to consider is the steady state cruising conditions that are required for charge sustaining operation. In the study, the optimum powertrain split between the traction drive, battery capacity and DHE specification is explored for example large passenger car and LCV applications, based on life-cycle $CO_2$ assessments.

The paper also presents the latest testing results from our 3-cylinder 1.5 litre high-efficiency demonstrator engine running on both conventional 95 RON gasoline and alternative sustainable low carbon fuels. The hybrid engine concept features a combination of a pre-chamber-based combustion layout, together with high geometric compression ratio, cooled exhaust gas recirculation (EGR) and aggressive Miller-cycle operation, to enable extremely high brake thermal efficiency levels to be achieved. Finally, drive-cycle analysis, based on the engine test results, will also be used to show the efficiency of the entire powertrain system based on the different fuels used.

## 2 MAHLE MODULAR HYBRID POWERTRAIN CONCEPT

The MMHP concept is based around a high-voltage PHEV architecture, with a dedicated hybrid internal combustion engine, featuring the MAHLE jet ignition system, integrated with a dual-mode hybrid electric drive. The concept was developed to showcase MAHLE Powertrain's capabilities through engineering a hybrid technology package optimised for future global automotive markets, targeted to meet emissions and $CO_2$ targets for 2030 and beyond, that is also scalable across a wide range of vehicles. Previous studies have focused on the application of the concept, featuring a 2-cylinder dedicated hybrid engine (DHE), into a compact crossover sports utility vehicle (SUV) (1,2). An image or the MMHP concept in this configuration is shown in Figure 1. These previous studies also illustrated how the modular nature of the MMHP enabled scalable application of the concept into vehicles ranging from a compact car to a large SUV.

**Figure 1. MAHLE modular hybrid powertrain.**

The dual-mode hybrid electric drive, which contains the best features of both series and parallel hybrid arrangements, contains two electric machines, a main traction motor and a smaller generator. The main traction motor can provide the full dynamic performance and maximum speed capability of the vehicle without assistance from the DHE. This, when combined with the plug-in capability, allows the vehicle to operate under electric only propulsion, with zero tail-pipe emissions, in clean-air zones and for the majority of typical use. The electric-only range is determined by the size of the battery pack, the size of which can be optimised to provide different trade-offs between the battery pack and the DHE in terms of both in-use and embedded $CO_2$, as well as the total cost and weight of the system. As the dynamic vehicle performance requirements can be met by the main traction drive the starting, warm-up and transient operation of the DHE can be managed independently of driver demand. This enables improved emissions and reduced after-treatment complexity, by removing the requirement for DHE cold starts with high power demands, which can be a challenge with PHEV systems that rely on the internal combustion engine (ICE) power for full dynamic vehicle performance. Once the battery is depleted, the system can operate as a series hybrid with the DHE driving only the generator at low vehicle speeds, having the NVH and operating flexibility that this arrangement offers. Then at higher vehicle speeds the engine can be connected directly to the wheels, via the dedicated hybrid transmission, with a number of transmission ratios enabling some flexibility in engine operating speed. The traction motor is directly connected to the wheels, thus there is seamless torque delivery, even during a gear-shift event, enabling the use of a simple automated manual transmission.

The MMHP was designed with the intention of being scalable across a broad range of vehicles, as shown in Figure 2. The simplified DHE is designed to be produced as either a 1 litre 2-cylinder with up to 60 kW, or a 1.5 l 3-cylinder unit with up to 90 kW, both variants achieving peak power at 4000 rev/min and a BMEP of 18 bar. Likewise, the transmission can be configured with 1, 2 or 4 gear ratios depending upon application requirements. The architecture has been designed to enable all variants, irrespective of number of ratios, to use common ratios and main transmission casing.

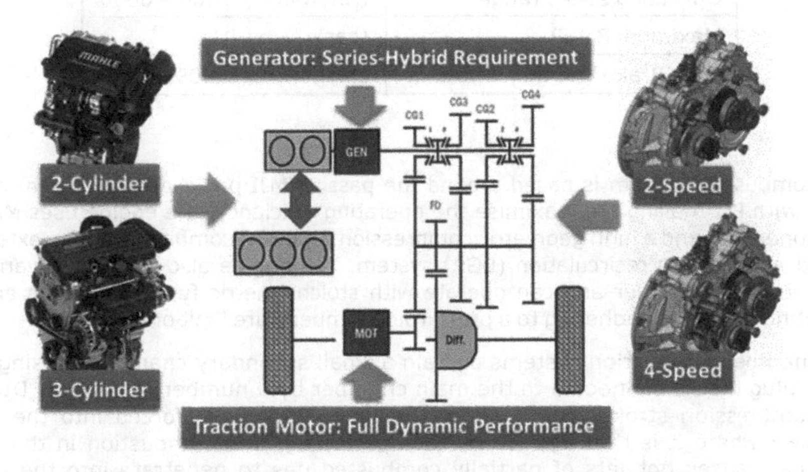


**Figure 2. MAHLE modular hybrid powertrain layout.**

The traction motor can be sized to achieve the desired dynamic performance of the vehicle. The generator specification is set by the road load power requirement at the minimum vehicle speed at which the engine can be switched into direct drive mode. The power requirement of the DHE is then set to enable charge in the battery pack to be sustained under worst case cruising conditions.

This study builds on this previous work and investigates the potential of applying the concept to both a large high-performance SUV and an LCV application, whilst using MAHLE Powertrain's 1.5 litre, 110 kW, high-efficiency demonstrator engine (3,4) as the DHE. This engine differs from the 1.5 litre, 90 kW DHE engine described above by having 4 valves, rather than 2 valves, per cylinder and a lower compression ratio of 12.8:1 instead of 14.7:1. The engine has also been designed to operate at up to 6500 rpm rather than being limited to 4500 rpm.

## 3 1.5 LITRE 110 KW HIGH-EFFICIENCY DEMONSTRATOR ENGINE

MAHLE Powertrain's 1.5l demonstrator engine (3,4) was developed as a low-cost high efficiency engine concept that was suitable for use in applications as both the sole prime-mover as well as being part of a hybridised powertrain. The key specifications of the engine are summarised in Table 1.

**Table 1. MAHLE 1.5l high-efficiency engine specification.**

| Parameter | Units | Value |
|---|---|---|
| Swept volume | (litres) | 1.5 |
| Number of cylinders | (-) | 3 |
| Bore/Stroke | (mm) | 83/92.4 |
| Peak power | (kW) | 110 |
| Compression ratio | (#) | 12.8:1 |
| Operating speed range | (rev/min) | 1000 – 6000 |
| Maximum BMEP | (bar) | 18 |
| Peak Brake Thermal Efficiency | (%) | >40% |

The combustion system is based around the passive MJI pre-chamber system combined with PFI fuelling. To maximise the operating efficiency, the engine uses Miller-cycle operation and a high geometric compression ratio (CR combined with an external cooled exhaust gas recirculation (EGR) system. The engine also features a variable geometry turbocharger and can operate with stoichiometric fuelling over its entire operating map whilst adhering to a pre-turbine temperature limit of 950°C.

Pre-chamber combustion systems contain a small secondary chamber housing the spark plug that is connected to the main chamber by a number of nozzles. During the compression stroke, charge from the main cylinder is forced into the pre-chamber where it is then ignited by the spark plug. The combustion in the pre-chamber causes hot jets of partially combusted gas to penetrate into the main cylinder, through the nozzles, initiating the main cylinder combustion at multiple locations. This gives the benefit of much shorter burn durations, which helps reduce knock, and enhances the ignition of dilute mixtures (either lean or with EGR) that would not be possible with a conventional spark plug (CSP). Figure 3 shows a comparison of the combustion events, at an operating speed of 4000 rpm and an engine load of 18 bar BMEP, between the passive MJI system and a CSP.

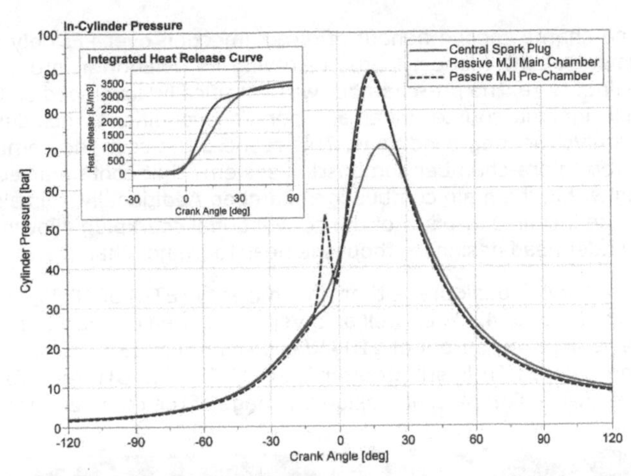


**Figure 3. Comparison of passive MJI combustion to a conventional central spark plug at 4000 rpm and 18 bar BMEP.**

The pre-chamber combustion event can be seen clearly as the spike in the pre-chamber pressure trace just prior to top-dead-centre firing (0° crank angle). This in turn initiates the rapid combustion in the main chamber. Table 2 summarises some of the key combustion metrics for the two cases.

**Table 2. Comparison of combustion metrics at 4000 rpm and 18 bar BMEP between MJI and CSP.**

| Parameter | Units | MJI | CSP |
|---|---|---|---|
| Combustion Phasing (50% Mass Fraction Burned) | °ATDCF | 8.6 | 17.4 |
| Burn Duration (10-90%) | °CA | 15.4 | 25.9 |
| Maximum Cylinder Pressure | bar | 90.9 | 71.2 |
| Maximum Rate of Pressure Rise | bar/°CA | 5.2 | 1.98 |
| Combustion Stability - Coefficient of variance of NMEP (CovNMEP) | % | 0.85 | 2.54 |

At this test condition the 10 to 90 % burn duration is reduced by 40 %, from 25.9 °CA to 15.4 °CA, through use of MJI which then enables the combustion phasing to be advanced by almost 9 °CA before the onset of knock. Because of the faster and more advanced combustion, higher levels of maximum cylinder pressure and the maximum rate of pressure rise are seen. A further benefit of the MJI combustion is the improvement in stability with the combustion stability at this condition improving from a CovNMEP of 2.54 % down to 0.85 %.

For this specification of engine, it was found that a combination of a moderate Miller-cycle operation in addition to use of the moderate levels of EGR (less than 20 %) gave the best compromise between achieving optimum combustion phasing and low pumping work [3]. The addition of EGR was also found to provide a means to control the maximum rates of pressure rise, caused by the fast combustion, to within acceptable limits. Due to the advanced combustion phasing that was achievable with this hardware combination, the charge cooling benefit of direct injection over PFI only gave a small BSFC improvement at the highest loads, justifying its deletion in order to lower the costs of the engine. Here, the pre-chamber was specifically developed so that it can be used as the sole ignition source under all operating conditions, including cold-start, low-load and catalyst heating conditions. This reduces the cost and complexity of the system, compared to pre-chamber combustion systems that contain a second conventional sparkplug within the main combustion chamber. Additionally, this also allows the pre-chamber to be installed in-place of the conventional sparkplug allowing application into existing cylinder head designs without the need for major changes.

The engine can achieve a peak brake thermal efficiency (BTE) of 40 %, or an indicated thermal efficiency (ITE) of 42 % as well as a very wide area of operation with a BTE of over 34 %. The peak power output of 110 kW is delivered at 5500 rpm, under stoichiometric operation where BTE is still greater than 36 %. The BTE map for this engine specification is shown in Figure 4, alongside an image of the complete engine.

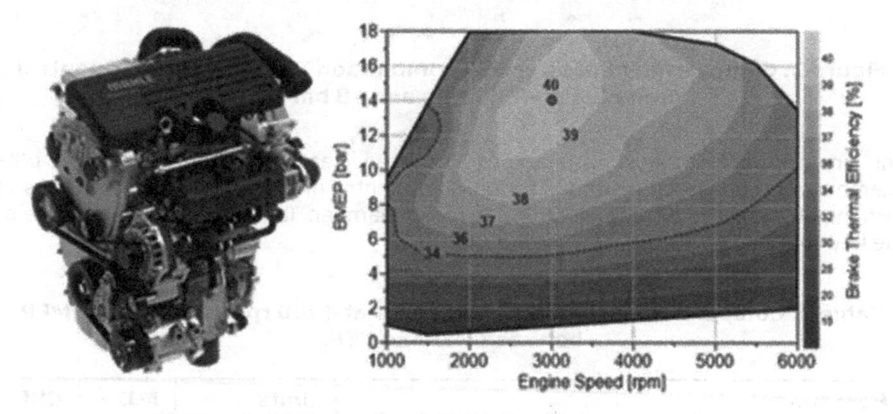


**Figure 4. BTE map for 1.5l high-efficiency engine.**

## 4 VEHICLE APPLICATIONS

In this study the 110 kW high-efficiency MJI version of the MAHLE DI3 engine (3, 4) has been coupled to the dual-mode hybrid transmission and motor arrangement from the MMHP to enable it to support larger vehicles. Two vehicle applications, with quite different requirements, have been considered, a 3.5 tonne truck and a large high-performance 4x4 SUV. The key vehicle parameters for these two vehicles are summarised in Table 3.

**Table 3. Specifications and targets for the two vehicles considered.**

| Vehicle Parameter | Units | Delivery Truck | Large 4x4 SUV |
|---|---|---|---|
| Kerb mass | Kg | 3500 | 2300 |
| Gross vehicle weight | Kg | 5500 | 3100 |
| Frontal area | $m^2$ | 4.9 | 3.0 |
| Aero dynamic drag coefficient | - | 0.47 | 0.35 |
| Wheel rolling radius | M | 0.353 | 0.395 |
| Maximum speed | km/h | 120 | 240 |
| Trailer towing capacity | Kg | N/A | 2500 |

The driveline architecture of the MMHP was shown schematically in Figure 2. The high-voltage traction motor drives directly to the differential. There is also a high-voltage, liquid-cooled generator mounted directly on the engine crankshaft, in the location of a conventional engine flywheel. The transmission input shaft is also directly engaged with the generator. To save cost, weight and package space, there is no clutch within the system, as this is not required because the DHE is not used for vehicle pull away. The engine is decoupled from the driveline, for pure-electric or series-hybrid operation, by selecting neutral in the simple automated manual transmission unit.

The transmission design uses cylindrical helical gears to provide a cost optimised solution and can be tailored to have 1, 2 or 4 ratios depending upon application requirements. The architecture has been designed to enable all variants, irrespective of number of ratios, to use common ratios and main transmission casing. The ratio options selected for the transmission are summarised in Table 4. For the two vehicles considered in this study a common 4-speed transmission was used, but a lower final drive ratio was used for the large delivery truck.

**Table 4. MMHP transmission specification.**

| Parameter | Delivery Truck | Large 4x4 SUV |
| --- | --- | --- |
| 1st gear ratio | 1.48: 1 | |
| 2nd gear ratio | 1.17: 1 | |
| 3rd gear ratio | 0.95: 1 | |
| 4th gear ratio | 0.77: 1 | |
| Traction motor ratio | 3.91: 1 | |
| Final drive ratio | 6.50: 1 | 4.14: 1 |

The transmission ratios were selected to enable the ICE to be operated in direct drive mode over a wide range of vehicle speeds. Figure 5 shows road load curves for both the 3.5 tonne delivery truck (at gross vehicle weight) and the large 4x4 SUV operating on both a level road, a 6 % grade and a 20 % grade. The influence of a 2500 kg trailer on the vehicle road load is also shown for the large SUV on both a 6 % and a 20 % grade.

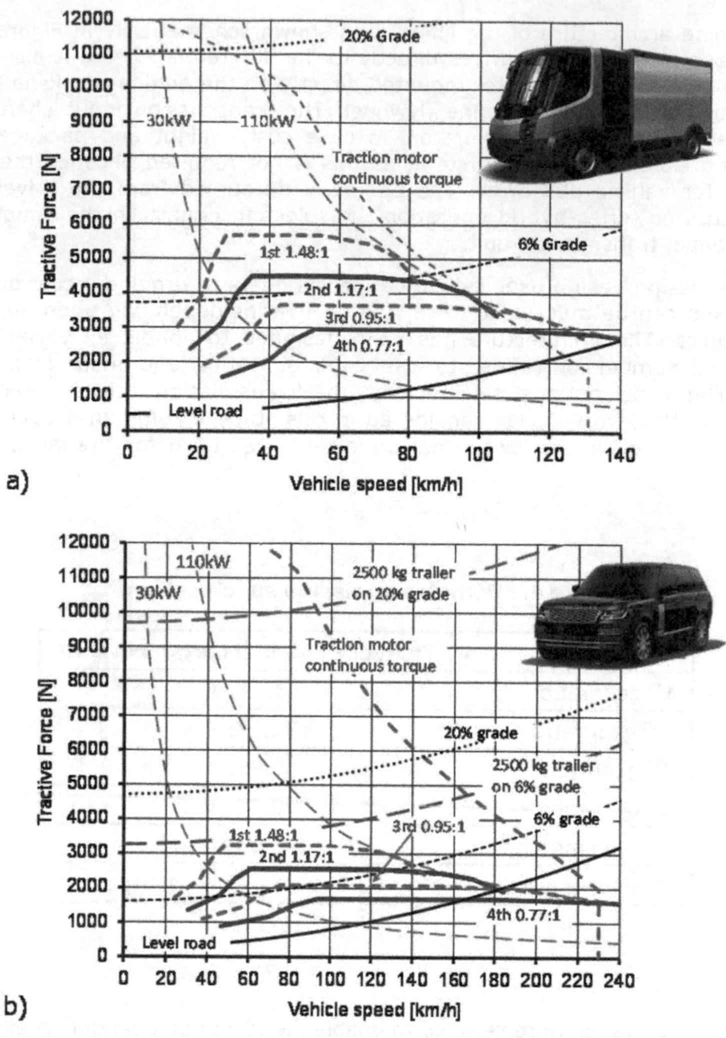


**Figure 5. Cascade Plots; a) Delivery truck; b) Large 4x4 SUV.**

Figure 5 also shows the continuous tractive force available from the traction motors in both vehicles. The delivery truck is powered by a single 135 kW traction motor which is integrated into the MMHP. It can be seen from Figure 5a that this is able to provide sufficient tractive force to enable the fully laden vehicle to cruise at over 120 km/h on a level road and at over 80 km/h on a 6 % grade. To provide the required vehicle performance, and to give full four-wheel drive capability, the large SUV is powered by two 135 kW traction motors, one of which is integrated into the MMHP, the other is incorporated in an electric axle unit (combined motor, inverter and transmission unit) mounted in the rear of the vehicle to drive the rear wheels. With these two electric motors the large 4x4 SUV can achieve a 0-100 km/h time of less than 6.0 seconds. It can cruise at almost 220 km/h on a level road and almost 200 km/h on a 6% grade. Furthermore, the vehicle can tow a 2500 kg trailer on a 20 % grade at 90 km/h and over 160 km/h on a 6 % grade.

The engine tractive wheel force, for the ICE in each of the four transmission ratios listed in Table 4, are also shown in Figure 5. It can be seen from Figure 5a that the ICE can charge sustain the fully laden large delivery truck up to its maximum speed on a level road in $4^{th}$ gear and up to 80 km/h on a 6% grade in either $1^{st}$ or $2^{nd}$ gear. Likewise, from Figure 5b it can be seen that the large 4x4 SUV can charge sustain using the ICE at up to 180 km/h in either $2^{nd}$ or $3^{rd}$ gear (with $4^{th}$ as an overdrive ratio to enable higher ICE efficiency and better NVH to be achieved at lower vehicle speeds). The vehicle can also charge sustain at up to 140 km/h on a 6 % grade. Charge sustaining operation is achieved with the 2500 kg trailer at up to 60 km/h on a 6% grade.

As discussed previously, the MMHP can operate in three discrete modes, pure electric driving, series hybrid and parallel hybrid. In all of the driving modes the traction motor provides the power to meet the instantaneous dynamic performance required. As the traction motor is connected to the wheels by a fixed gear ratio, there is never any torque disruption for gear shifting. When the battery is depleted, but the vehicle speed is too low for the engine to operate above 1500 rev/min in $1^{st}$ gear, the vehicle operates in series hybrid mode. Once the vehicle has reached sufficient speed for the ICE to be directly coupled to the wheels then the vehicle will operate in parallel hybrid mode.

Figure 6 shows a simplistic example of how the engine could be operated. When running, the power output of the engine is modulated as a function of engine speed and load. The engine speed will be determined by vehicle speed and the gear ratio selected. To illustrate the potential of the MMHP, the fuel efficiency has been analysed over the drive cycles summarised in Table 5.

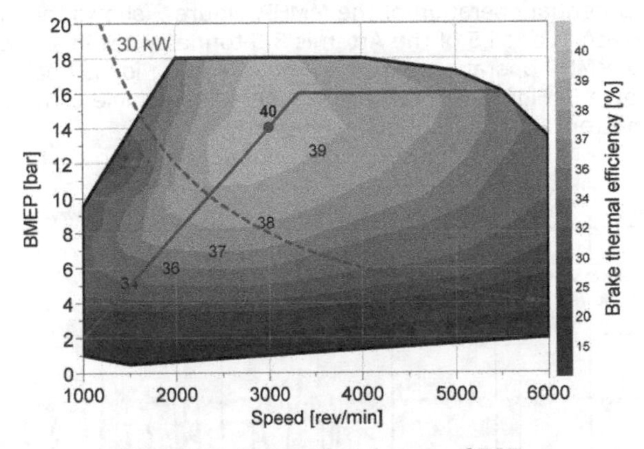


**Figure 6. Operating locus of ICE.**

Both vehicles have been analysed over the NEDC (5) and the WLTP (6) cycles, as these are both well known cycles and provide a good frame of reference. Furthermore, the WLTP is the current legislative test cycle used for passenger car certification in Europe and the NEDC is used as the basis for fleet $CO_2$ limits. The large SUV has also been evaluated over the combined Artemis driving-cycle (CADC) (7), which consists of an urban, rural and motorway phase in a combined cycle, this has been devised to be representative of real driving in Europe and has been shown to compare very well to driving metrics observed for large vehicle fleets (8). The delivery vehicle has also been analysed over the Artemis Cycles for 3.5 tonne vans, which has 5 separate sections, corresponding to slow urban traffic, free-flowing urban traffic, a delivery route, rural road and motorway. The relative applicability of each of these cycles to any delivery van will depend on the type of use and geographic location over which the vehicle is deployed, for this reason these cycles have been analysed and reported separately.

**Table 5. Summary of drive-cycles considered.**

| Drive-cycle | Duration (s) | Distance (km) | Average speed (km/h) | Maximum speed (km/h) |
|---|---|---|---|---|
| NEDC | 1180 | 11.0 | 33.6 | 120.0 |
| WLTP | 1800 | 23.1 | 46.3 | 131.3 |
| Combined Artemis Driving Cycle (CADC) | 3143 | 51.7 | 59.2 | 150.4 |
| Artemis 3.5 Tonne Slow Urban | 649 | 2.2 | 12.1 | 57.9 |
| Artemis 3.5 Tonne Free Flow Urban | 467 | 2.9 | 22.3 | 52.5 |
| Artemis 3,5 Tonne Delivery | 546 | 1.6 | 10.5 | 32.3 |
| Artemis 3.5 Tonne Rural | 819 | 11.5 | 50.4 | 86.2 |
| Artemis 3.5 Tonne Motorway | 1280 | 31.3 | 88.1 | 130.4 |

To illustrate the potential operation of the MMHP, Figure 7 shows the operation of the delivery truck driven over all 5 of the Artemis 3.5 tonne cycles in succession, running at GVW, with the MMHP operation based on the operating locus depicted in Figure 6. The example shown in Figure 7 is for a case where the vehicle commences the cycle with a depleted battery (charge sustaining operation).

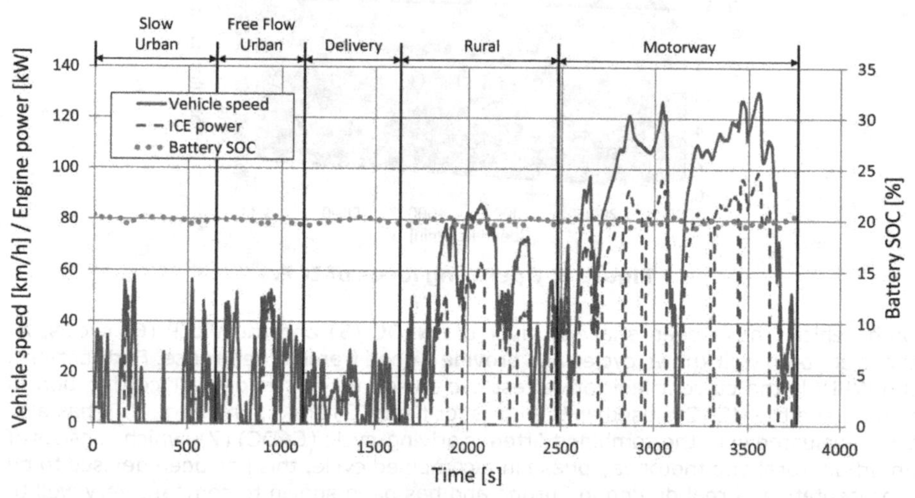


**Figure 7. Operation of delivery truck with the MMHP over the Artemis 3.5 tonne cycles.**

It can be seen from Figure 7 that the engine power output closely follows the vehicle speed. The control strategy implemented operates in a purely reactive manner, based only on vehicle speed and battery SOC. The results presented in Figure 7 correspond to the delivery truck with a 50 kWh battery, which has been sized to give an EV range of between 57 and 90 km (see Table 6). For the initial 60 seconds of the cycle the ICE operates at a fixed speed and load condition (1250 rev/min and 6 kW) to enable the catalyst to light-off. Previous studies on series hybrid strategies identified that low speed and light load operation gave lowest cumulative emissions prior to catalyst light-off (9). The operating strategy for the MMHP has been devised to maintain battery SOC, without charging the battery (as it is most desirable to replenish the battery from an off-board electrical source), and it can be seen that in the example shown in Figure 7 the battery SOC is readily maintained close to 20 % across the entire 49.5 km cycle.

In the example shown in Figure 7 the ICE is activated depending upon battery SOC and the hybrid mode (series or parallel) is determined based on vehicle speed. The ICE simply operates along the operating locus shown in Figure 6. Some modulation about this operating locus could be used if battery SOC departed significantly from the desired value. A more optimised operating strategy could be developed if the speed profile of the route to be driven is known *a priori* from, for example, GPS and traffic information (10). This could also be used for geo-fencing to restrict operation of the engine within clean air zones.

Tables 6 and 7 show the calculated drive cycle fuel consumption figures for the delivery truck and large SUV respectively, based on the control strategy outlined above. All cycles have been analysed with a continuous 1 kW auxiliary electrical load to account for consumption of fans, pumps and electrical control modules. The cycles also include 60 seconds of catalyst light-off operation of the engine. The calculated electrical range for the cycle has been based on full electrical operation (engine off) for the entire cycle. The delivery truck has been analysed at gross vehicle weight and the large SUV at kerb weigh +100 kg. Weighted tailpipe $CO_2$ is also shown in Tables 6 & 7 and has been calculated using the formula for weighting factor for the NEDC regulation 101 (5), using the corresponding electric driving range for each cycle. The average driving distance between recharges ($D_{av}$), in equation, is specified in NEDC regulation 101 (5) as 25 km. The weighting given by equation 1 yields a similar result to that used for weighting plug-in hybrid consumption for the WLTP cycle.

$$M = (D_e \cdot M_1 + D_{av} \cdot M_2)/(D_e + D_{av}) \qquad (1)$$

Where:

- $M$ = weighted mass emission of $CO_2$
- $M_1$ = mass emission of $CO_2$ with a fully charged battery
- $M_2$ = mass emission of $CO_2$ with a fully depleted battery
- $D_e$ = vehicle's electric range (determined for each drive-cycle)
- $D_{av}$ = 25 km (assumed average distance between two battery recharges)

The electric driving range ($D_e$) for the delivery truck has been based on a battery pack capacity of 50 kWh which enables an EV range ($D_e$) of between 57 and 90 km to be achieved, depending upon the cycle. The large SUV analysis has been based on a battery pack capacity of 25 kWh, which enables it to achieve an EV range of between 67 and 98 km for the three cycles analysed for this vehicle.

**Table 6. Calculated drive-cycle consumptions and $CO_2$ emissions for the delivery truck at gross vehicle weight with a 50 kWh battery pack.**

| Drive-cycle | Fuel Used (l/100km) | $CO_2$ (g/km) | Electrical Consumption (kWh/km) | Electric Range (km) | Weighted $CO_2$ (g/km) |
|---|---|---|---|---|---|
| NEDC | 13.26 | 313 | 0.48 | 83.9 | 72 |
| WLTP | 15.26 | 361 | 0.57 | 70.3 | 95 |
| CADC | - | - | - | - | - |
| Artemis Slow Urban | 20.67 | 488 | 0.67 | 60.0 | 144 |
| Artemis Free Flow Urban | 15.28 | 361 | 0.50 | 80.8 | 85 |
| Artemis Delivery | 13.02 | 308 | 0.42 | 94.4 | 64 |
| Artemis Rural | 12.25 | 289 | 0.44 | 90.2 | 63 |
| Artemis Tonne Motor-way | 17.97 | 424 | 0.70 | 57.1 | 129 |

**Table 7. Calculated drive-cycle consumptions and $CO_2$ emissions for the large SUV at kerb weight +100kg with a 25 kWh battery pack.**

| Drive-cycle | Fuel Used (l/100km) | $CO_2$ (g/km) | Electrical Consumption (kWh/km) | Electric Range (km) | Weighted $CO_2$ (g/km) |
|---|---|---|---|---|---|
| NEDC | 6.32 | 149 | 0.20 | 98.5 | 30 |
| WLTP | 7.12 | 168 | 0.24 | 83.9 | 39 |
| CADC | 8.70 | 205 | 0.30 | 67.3 | 56 |

## 5 BIO-GASOLINE

As part of the testing program completed on the 1.5 litre high-efficiency engine specification an assessment was made on the effect of using three different bio-gasoline blends as a low-carbon alternative to conventional fossil fuels. The three bio-gasoline blends were provided by Coryton and were all compliant with the EN228 fuel standard. The use of these bio-gasoline blends offers the potential for up to 80 % savings in life-cycle greenhouse gas emissions, based on the European Commission's Renewable Energy Directive (RED II). Table 8 summarises the key properties of the bio-gasoline fuels that were assessed against a standard pump grade 95 RON gasoline with a 10 % ethanol content.

**Table 8. Comparison of EN228 drop-in bio-gasoline properties.**

| Property | Baseline | 95RON E5 | 98RON E5 | 98RON E10 |
|---|---|---|---|---|
| Total Bio-Content [%] | 10 | 88.8 | 71.8 | 73.5 |
| Alcohol Content [% v/v] | 10 | 5 | 5 | 10 |
| RON | 95.0 | 95.4 | 97.7 | 99.0 |
| MON | 85.0 | 85.2 | 86.5 | 85.3 |
| Net Cal. Value [MJ/kg] | 41.14 | 42 | 41.9 | 41.3 |
| Aromatics Level [% v/v] | - | 35.5 | 34.6 | 24.5 |
| GHG Savings (RED II) (% basis 94.0g CO2e/ MJ) | - | ~80 | ~65 | ~66 |

All three bio-gasoline fuels were compared to the baseline fossil based 95 RON gasoline over a suite of tests representative of a range of typical engine operating conditions. These tests included operating conditions representative of cold start catalyst heating, idle, high residency drive-cycle points, a 3000 rev/min load sweep and a high-speed power curve and they also assessed the effects on EGR dilution tolerance at the peak efficiency operating point. Figure 8 shows the results of these tests in the form of scatter bands generated from the ensemble of all the test points, for the three bio-gasolines, as a function power output compared the baseline fossil fuel. At the top of Figure 7 the key efficiency and combustion metrics of; brake thermal efficiency, 10 to 90 % mass fraction burned duration, combustion phasing and combustion stability are compared in the four graphs.

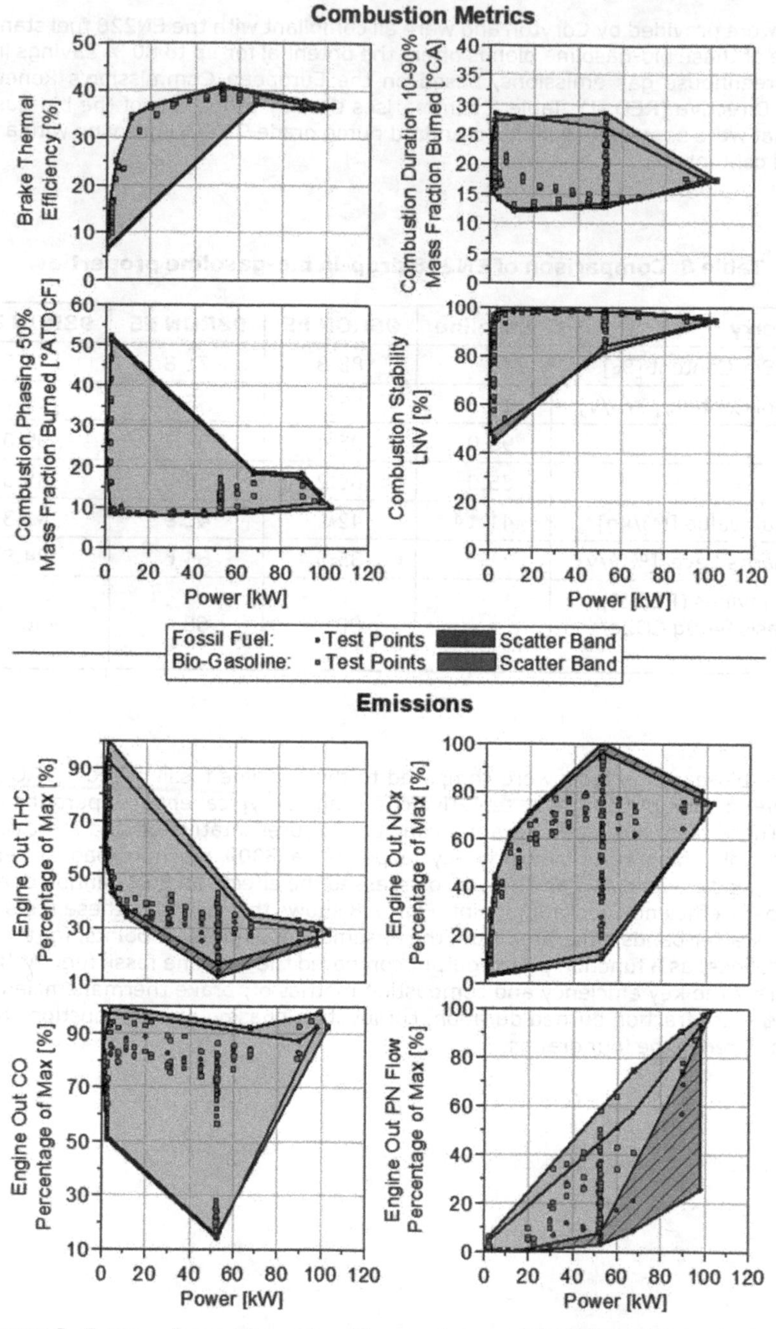


**Figure 8. Comparison of combustion metrics and engine out emissions between conventional fossil fuel and bio-gasoline.**

As was expected, given that all three bio-gasoline blends comply with the EN228 fuel standard, the combustion metrics show very similar performance across the whole suite of tests. A small improvement in combustion phasing, leading to a corresponding improvement in brake thermal efficiency, can be seen with bio-gasoline due to the higher RON of two of the fuel samples. A small increase in the 10 to 90 % burn durations was also identified for all three of the bio-gasoline blends. At the bottom of Figure 7, a comparison of raw engine out emissions of total hydrocarbons, NOx, CO, and Particulate flow are also presented. Again, the raw engine-out emissions and trends were found to be generally comparable between the bio-gasoline blends and the baseline fossil fuel. Due to the higher proportion of heavier fuel fractions in the bio-gasoline blends engine-out emissions of total hydrocarbons and particulate matter are slightly higher at all operating points. At higher engine loads and power outputs, the highest levels of NOx emissions were seen with the fossil fuels, however at 100 kW it can be seen that the NOx emissions of the different bio-gasoline blends fall within the scatter band of the fossil fuel results. The effect of these differences on tail-pipe emissions with a fully warm aftertreatment system in expected to be negligible. Further testing is planned to assess this is detail and to also evaluate the distribution of particulate sizing with the different fuels. This study will include the evaluation of a 95 RON E10 bio-gasoline and a fuel with 100 % bio-content.

If the 80% savings in $CO_2$ enabled by use of the bio-gasoline are applied to the charge sustaining NEDC tailpipe and weighted NEDC tailpipe figures quoted in Tables 6 and 7, the g/km values would be reduced to those shown in Table 9.

**Table 9. Calculated drive-cycle $CO_2$ emissions with 95RON E5 bio-gasoline.**

| Vehicle Parameter | Units | Delivery Truck | Large 4x4 SUV |
|---|---|---|---|
| Charge sustaining NEDC tail-pipe $CO_2$ | g/km | 63 | 30 |
| Weighted NEDC tail-pipe $CO_2$ (100 km EV range) | g/km | 14 | 6 |

## 6 LIFECYCLE AND FLEET AVERAGE $CO_2$ COMPARISONS

Currently vehicles are assessed on a tailpipe $CO_2$ emissions basis, and not on a total life-cycle $CO_2$ equivalent basis. In future the total life-cycle impact of the vehicle may come under scrutiny, and it is important (from an environmental as well as legislative perspective) that we consider the total life-cycle impact of a vehicle – otherwise it is very difficult to compare ICE, PHEV and BEVs on an even basis. Of note with a BEV or PHEV is the equivalent $CO_2$ impact of the battery generated during its production. For a conventional mid-sized vehicle, the embedded $CO_2$ from production is about 5.6 tonnes equivalent $CO_2$ (11). For our large 4x4 SUV, scaling by vehicle mass gives an embedded $CO_2$, to produce a conventional ICE variant, of 8.7 tonnes. Assuming a useful vehicle life of 200,000 km, this gives an embedded vehicle $CO_2$ equivalent to 43.5 g/km. A comprehensive life-cycle study of the Polestar 2 vehicle (12) suggests that 95 g/kWh is a reasonable figure for the embedded $CO_2$ generated in producing a battery pack, although it should be noted that this is likely to reduce as the electricity grid is decarbonised. The carbon intensity of the electrical grid, used for recharging EVs and PHEVs, varies significantly from country to country. For this simplified analysis a grid $CO_2$ intensity of 275 g/kWh has been used, which represents the EU average for

2019 (13). Again, it should be noted that this will reduce over the vehicle lifetime as the grid generating mix becomes increasingly supplied from renewable sources. Additionally, the marginal grid carbon intensity, which is arguably more applicable to the recharging of electric vehicles, is also likely to be lower than this figure.

The large 4x4 SUV using the 1.5 litre engine alone can achieve ~153 g/km $CO_2$ emissions over the NEDC (although it would not be able to meet the vehicle performance targets with just this engine). If we accept that the tail-pipe weighting factor for UN/ECE Regulation 101 (5) produces an appropriate weighting for PHEVs based on typical usage profiles, it can be estimated how the tail-pipe $CO_2$ of a PHEV varies with electric driving range, as shown in Figure 9, where it can be seen to vary from 153 g/km at zero electric range and drop to 14 g/km for a 250 km electric driving range.

We can also calculate the well to tank contribution of the fuel used by the vehicle as a function of electric driving range; again, this is shown in Figure 9. Additionally, we can compute the electrical energy required by the vehicle as the electric driving range increases, based on the inverse of the weighting factor of NEDC regulation 101 (5). We can then convert this into an equivalent $CO_2$ based on the electricity grid carbon intensity; again, shown in Figure 9. By assuming that the battery pack embedded $CO_2$ is simply supplemental to the embedded $CO_2$ from the conventional vehicle construction (assuming any addition for electric machines is offset by a reduction in ICE requirement), we can then consider the life-cycle $CO_2$ burden of vehicle production, as shown in Figure 9 where it increases from 43.5 g/km at zero electric driving range to 77 g/km for a 250 km electric driving range, again based on a 200,000 km vehicle life.

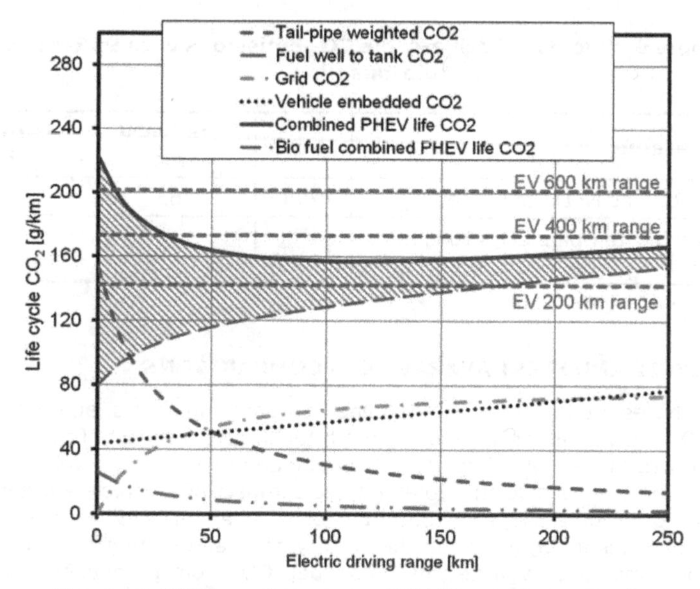


**Figure 9. High performance SUV life-cycle analysis.**

If we sum all of these contributions, we get a curve that shows the life cycle $CO_2$ for a PHEV for varying electric driving range, again this is shown in Figure 9. Interestingly, the life cycle $CO_2$ for the PHEV exhibits a minimum for an electric driving range of around 100 km. For comparison life cycle $CO_2$ values, for BEVs with three differing driving range capabilities, based on the same assumptions outlined for the PHEV, are also shown in Figure 9. From this analysis the PHEV, with a 100 km electric driving range results in a very similar life cycle $CO_2$ to a BEV with a ~300 km driving range. As the BEV driving range is increased the embedded $CO_2$ in the battery pack increases the vehicle life cycle $CO_2$ proportionally. Therefore, based on these simplistic assumptions, for most users (dependent on local electricity grid carbon intensity), a PHEV with a 100 km range represents a good solution for minimising environmental impact.

The influence of using a fully bio-gasoline (95 RON E5 in Table 8) upon the life cycle $CO_2$ for the large 4x4 SUV is also shown in Figure 9. In this case the analysis that was used to calculate the combined PHEV life cycle $CO_2$ has been used, but in this case the contribution of the tail-pipe weighted $CO_2$ has been reduced by 80 %, to account for the GHG savings offered by the bio-gasoline. With this fuel the combined life cycle $CO_2$ for the vehicle is lowest at zero electric driving range (i.e., with no plug-in hybridisation). However, the PHEV system has the benefit of enabling zero tail-pipe emissions during use when operating using battery electric power, which is the preferable option for use in densely populated areas. Additionally, depending upon the proportion of biomass in the fuel mix used to operate the vehicle during its lifetime, the actual life cycle $CO_2$ impact will sit somewhere in the shaded region of Figure 9, between the conventional gasoline and fully bio-gasoline life cycle $CO_2$ lines.

## 7 CONCLUSIONS

The MAHLE Modular Hybrid Powertrain (MMHP) has been based around a high-voltage PHEV architecture, with a dedicated hybrid internal combustion engine (DHE) integrated with a dual-mode hybrid electric drive, featuring two electric machines. The traction motor provides full vehicle dynamic performance and vehicle maximum speed without assistance from the DHE. This enables improved emissions and reduced after treatment complexity. Additionally, the direct drive arrangement enables seamless torque delivery, enabling the use of a simple automated manual transmission.

In this study the MMHP concept has been combined with MAHLE's 1.5 litre high-efficiency demonstrator engine to enable the concept to be applied to even larger vehicles. The engine features the MAHLE Jet Ignition (MJI) pre-chamber system, which produces jets of partially combusted species that induce ignition in the main combustion chamber enabling rapid, stable combustion. The system has been developed to operate over the entire engine range without the requirement for a second igniter. This has been combined with a high geometric compression ratio, Miller cycle operation and exhaust gas recirculation to produce an engine that achieves a peak brake thermal efficiency of over 40 %, whilst also achieving a peak power output of 110 kW under stoichiometric operation. Testing on this engine has also been performed to confirm the suitability of using bio-gasoline blends that are EN228 compatible as a drop-in replacement for conventional fossil based 95RON E10. Use of the bio-gasolines tested offers the potential for an 80 % reduction in green-house gases (based on RED II guidelines) compared to fossil-based fuels and showed no significant effect on key combustion metrics and raw engine out emissions.

The study has shown how this combination of the MMHP and the 1.5 litre high-efficiency engine could be applied to meet the needs of both a large high-performance SUV, using two 135 kW traction motors to provide four-wheel drive capability, and a large delivery truck with a single integrated 135 kW traction motor. Through use of a 25 kWh battery pack for the large SUV and a 50 kWh battery pack for the delivery truck, both would be capable of around 60 to 100 km pure electric driving range, depending upon usage. For the SUV this is sufficient to cover most typical urban daily usage without the need to recharge and coincident with the minimum combined PHEV life cycle $CO_2$ for the SUV.

Based on these proposed configurations, the large delivery truck has the potential to achieve a weighted tail-pipe CO2 emissions value of between 63 to 144 g/km, depending upon the usage cycle. These calculations have been conducted using a measured BSFC map from the demonstrator engine. These figures could be reduced further through use of bio-gasoline (based on RED II GHG savings). For the high performance 4x4 SUV the proposed powertrain configuration has the potential to achieve weighted vehicle fuel consumption of 1.3 litres/100 km over the NEDC. This equates to a weighted tail pipe $CO_2$ value of 30 g/km with conventional fuels, that could be reduced to 6 g/km through use of bio-gasoline over the NEDC cycle, whist still offering the performance and towing capability expected from this class of vehicle.

Vehicle manufacturers are facing increasing pressure by legislation and economics to reduce vehicle emissions and deliver improved fuel economy. MAHLE Powertrain have proposed a possible hybrid technology pathway for such powertrains that offers a balanced use of battery and fuel resources to minimise life-cycle $CO_2$ emissions

## ACKNOWLEDGEMENTS

The authors would like to thank David Richardson at Coryton and Steve Sapsford for the supply of the bio-gasoline fuels used in this study.

## REFERENCES

[1] Bassett, M., Reynolds, I., Cooper, A., Reader, S., Berger, M., "MAHLE Modular Hybrid Powertrain," Proceedings from the 28th Aachen Colloquium, 2019.

[2] Bassett, M., Cooper, A., Reynolds, I., Hall, J., Reader, S., and Berger, M., "MAHLE Modular Hybrid Powertrain," IMechE, Internal Combustion Engines and Powertrain Systems for Future Transport 2019.

[3] Cooper, A., Harrington, A., Bassett, M., Reader, S. et al., "Application of the Passive MAHLE Jet Ignition System and Synergies with Miller Cycle and Exhaust Gas Recirculation," SAE Technical Paper 2020-01-0283, 2020.

[4] Bassett, M., Cooper, A., Harrington, A., Pates, D. et al., "Passive MAHLE Jet Ignition System Demonstrator" Proceedings from the 28th Aachen Colloquium, 2020.

[5] UNECE Addendum 100: Regulation No. 101, Revision 3, 1995. (E/ECE/324/Rev.2/Add.100/Rev.3), 16th October 1995. http://www.unece.org/fileadmin/DAM/trans/main/wp29/wp29regs/2015/R101r3e.pdf, (accessed 29th April 2021).

[6] UNECE Addendum 153: Regulation No. 154, Revision 3, 2017. (E/ECE/TRANS/505/Rev.3/Add.151), https://unece.org/sites/default/files/2021-08/R154e.pdf, (accessed 19th Aug 2021).

[7] André, M., Keller, M., Sjödin, Å., Gadrat, M., and McCrae, I., "The Artemis European Tools for Estimating the Pollutant Emissions from Road Transport and their Application in Sweden and France", 17th International Conference Transport and Air Pollution, Graz, 2008.

[8] Bassett, M., Brooks, T., Fraser, N., Hall, J., Thatcher, I., Taylor, G. "A Study of Fuel Converter Requirements for an Extended-Range Electric Vehicle",SAE Technical Paper 2010-01-0832, 2010.

[9] Warth, M., Bassett, M., Hall, J., Taylor, G., and Mahr, B., "Development of a Compact-class Range Extended Electric Vehicle".21st Aachen Colloquium Automobile and Engine Technology, 2012.

[10] Bassett, M., Brods, B., Hall, J., Borman, S., Grove, M., and Reader, S., "GPS Based Energy Management Control for Plug-in Hybrid Vehicles". SAE Technical Paper 2015-01-1226, 2015. https://doi.org/10.4271/2015-01-1226.

[11] Patterson, J., Alexander, M., and Gurr, A., 'Preparing for a Life Cycle CO2 Measure'. Low Carbon Vehicle Partnership Report RD.11/124801.5, 20th May 2011. http://www.lowcvp.org.uk/assets/reports/RD11_124801_4%20-%20LowCVP%20-%20Life%20Cycle%20CO2%20Measure%20-%20Final%20Report.pdf, (accessed 29th April 2021).
[12] Bolin, L., 'Life cycle assessment – carbon footprint of the Polestar 2', Polestar, 2020. https://www.polestar.com/dato-assets/11286/1600176185-20200915polestarlcafinala.pdf, (accessed 29th April 2021).
[13] https://www.eea.europa.eu/data-and-maps/daviz/co2-emission-intensity-6#tab-googlechartid_googlechartid_googlechartid_googlechartid_chart_11111 (accessed 29th April 2021).

*International Conference on Powertrain Systems for Net-Zero Transport*
*Institution of Mechanical Engineers, ISBN 978-1-032-11281-7*

# Numerical evaluation and optimization of a hybrid vehicle employing a hydrogen internal combustion engine as a range extender

**R.T. Wragge-Morley[1], G. Vorraro[2], J.W.G. Turner[2], C.J. Brace[3], Xingyu Xue[4], Jihad Badra[5], Amir Abdulmanan[4]**

[1]University of Bristol
[2]King Abdullah University of Science and Technology
[3]Institute of Advanced Automotive Propulsion Systems, University of Bath
[4]Strategic Transport Analysis Team, Aramco Asia, Shanghai, China
[5]Transport Technologies Division, R&DC, Saudi Aramco, Dhahran, Eastern Province, Saudi Arabia

## ABSTRACT

During the last decade internal combustion engines have become deemed to be one of the main cause of pollution. As a consequence, governments and policy makers are strongly pushing towards electric technologies and in particular battery electric vehicles (BEVs), despite the huge efforts of scientists and researchers in designing cleaner and more efficient hybrid powertrains employing both electric motors and thermal machines. Hybrid electric vehicles (HEVs) have demonstrated capability of delivering low emissions while reducing costs compared to the BEVs. This work is focussed on the study of a series hybrid vehicle (also commonly called a Range Extended Electric Vehicle – REEV) equipped with a range extender running on hydrogen fuel. The study has been conducted mainly at a numerical level by simulating the performance of the commercial BMW i3 over the common NEDC and WLTP drive cycles and then comparing the results with the original powertrain in terms of energy, fuel consumption and emissions. All the modelling activities and optimisations have been carried out using the commercial software GT-Suite by Gamma Technologies.

The vehicle model has been validated against the experimental data provided in literature and other previous works from the authors, while the internal combustion engine model setup relies on experimental data coming from a previous project carried out at Argonne National Laboratory. Interestingly, the present study also takes into account the different fuel storage system employed for the hydrogen, characterized by a high ratio between its total weight and the weight of fuel contained. In that respect an evaluation of the differences in weight and tank volume, performance and energy consumption of both the original and the hydrogen vehicles have been carried out.

Finally, the simulations demonstrate that the increased weight due to the different fuel storage system does not heavily affect the performance and the energy consumption of the entire vehicle. In addition the improved results in terms of fuel consumption establish that hydrogen internal combustion engines, with their higher efficiency (higher than 44%), are a viable solution for range extender applications while giving the opportunity of a thorough optimisation on the choice of the battery size in order to maximise range and minimise weight and cost while maintaining the desired driveability.

DOI: 10.1201/9781003219217-20

## 1 INTRODUCTION

The automotive sector and the technology of personal transportation are in a state of flux. The levels of change and uncertainty are at their highest for a century and the established, $20^{th}$ century conventions of propulsion are being challenged. In a world after the Paris Climate Agreement in 2015 [1], many governments have introduced more and more stringent regulations on tailpipe emissions from road vehicles. In particular, the reduction of $CO_2$ emissions, and by analogy fuel consumption, has been targeted in order to reduce the impact of the transportation sector on greenhouse gases and global warming. The EU, US and China each have sets of regulations that between them cover a significant portion of the global vehicle market, and for the purposes of this paper we shall focus on how the development of EU emissions regulations is influencing propulsion technology.

The over-arching trend resulting from the tightening of emissions regulations for road transport, both with respect to greenhouse gas and harmful pollutant emissions, is towards electrification of powertrains. Many Western governments have committed to roadmaps for the removal from sale of vehicles propelled purely by internal combustion within the next two decades [2]. However, pure battery electric vehicles (BEVs) still present a number of widely discussed practical and environmental challenges. One of the most pertinent questions surrounds readiness for *en masse* domestic charging of EVs, and although certain countries with access to renewables and relatively small populations, such as Sweden, Denmark or Norway, are well placed to switch to relatively clean charging, larger countries, more reliant on conventional power generation, such as Germany and the UK, currently lack the infrastructure to support a behavioural switch to BEVs [3, 4]. In recent years, automotive OEMs have been guided by a requirement for the European fleet average $CO_2$ production to be reduced 15% from 2021 to 2025, and 37.5% from the same baseline by 2030; this also has the assumption that the 95g/km fleet average target for new vehicles is met in 2021. However, in July 2021 the European Commission approved the Green Deal adopting a set of proposals relating to climate change, energy, environment and industry for reducing $CO_2$ emissions by at least 55% by 2030 [5]. Specifically the Green Deal set the objectives to reduce by 55% the $CO_2$ emissions from cars by 2030, by 50% those from vans by the same year and finally achieve zero emissions from new cars by 2035 [6]. Obviously, the adoption of internal combustion engines fuelled by hydrogen is in line with all the aforementioned Green Deal objectives, given that $H_2$ ICEs produce zero tailpipe $CO_2$ emissions and contribute in reducing the average $CO_2$ emissions of the entire vehicle fleet. A scenario based on the adoption of pure $H_2$ ICE cars instead of the current battery technology may even give a net benefit if the $H_2$ is produced by green energy (Green Hydrogen).

There are already many forms of partially-electrified vehicle powertrain, and since the turn of the millennium, the most commonplace types have been variations on a parallel hybrid architecture. These are constrained by the direct coupling of the ICE to the transmission to harness maximum system power, and thus require the ICE to still be designed for some level of dynamic performance to retain drivability. To move from this state of affairs, to something closer to a pure BEV, the natural solution is the Range Extender (REx) or plug-in series hybrid vehicle. In this topology, the ICE and associated hardware can be significantly downsized and optimised for a relatively small operating range, as it will never need to answer the transient demands of an electric final drive directly. In this paper, we seek to examine a variety of potential range extender technologies, including hydrogen combustion, by benchmarking against one of the few currently available series hybrid vehicles, the BMW i3.

Compared to the preceding methodology used in Europe, the Worldwide harmonized Light vehicles Test Cycle (WLTC) presents a more dynamic and aggressive test cycle and it has been designed to adhere more to real driving conditions and provide more accurate fuel consumption and emissions. There are four different variants of the WLTC based on the vehicle Power-to-Mass Ratio (PMR) and top speed. An example is reported in the figure below for vehicles with maximum speed higher than 120km/h.

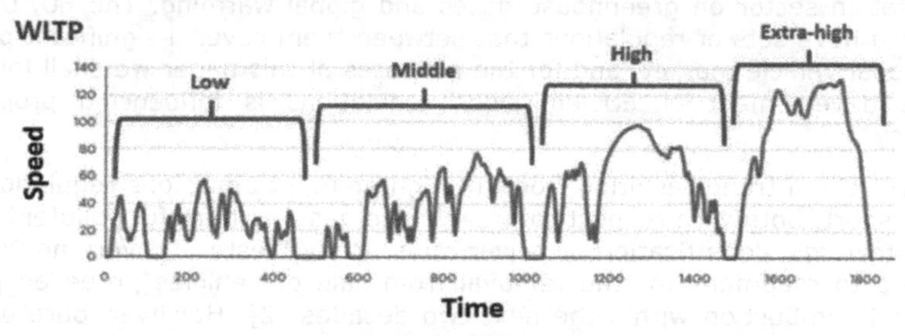


**Figure 1. WLTC test cycle [7].**

The Worldwide harmonized Light vehicles Test Protocol (WLTP) testing protocol for plug-in HEVs, including REEVs, is arranged to bias strongly in favour of those vehicles with a longer electric range. The vehicles are first tested with the battery at the maximum allowable SOC (Test cycle n-1 in Figure 2), and they are then run on repeated cycles (n, n+1...) until the minimum allowable state of charge (SOC) is reached, before then being tested in 'charge sustaining' operation. This is defined as when the net energy change in the battery is less than 4% of the energy measured at the wheels over a cycle and is illustrated in cycle n+1 in Figure 2.

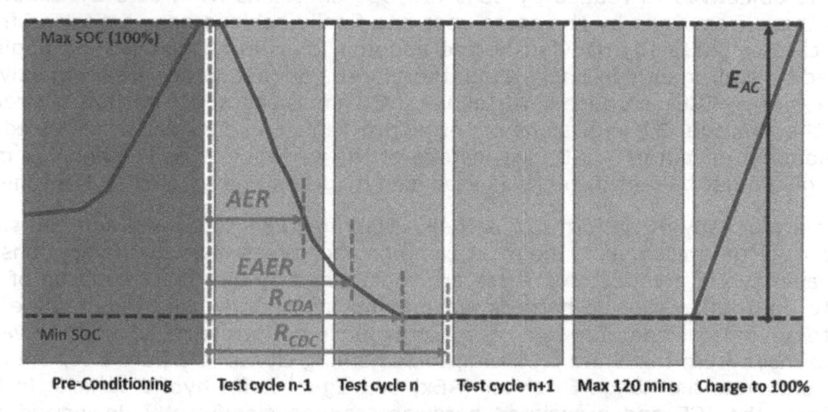


**Figure 2. WLTP for HEVs (modified from [7]).**

Practically this means that the vehicles are run as EVs until the system uses the REx to maintain some kind of charge-sustaining regime. It is accepted that specific driving events may cause the ICE to fire earlier on in the testing in order to boost system power output. Once charge sustaining operation is reached, at least one more whole cycle must be run to confirm charge sustaining fuel economy.

Utility Factor

For the WLTP, the range and charge depleting and charge sustaining fuel economy values are combined to arrive at the quoted fuel consumption. The weighted fuel consumption C in l/100km is:

$$C = UF \cdot C1 + (1 - UF) \cdot C2 \tag{1}$$

where:

$C$ = weighted fuel consumption [l/100km]

$C_1$ = fuel consumption [l/100] in Charge Depleting (CD) mode

$C_2$ = fuel consumption [l/100] in Charge Sustaining (CS) mode

$UF$ = Utility Factor as a function of the electric range $R_{CDC}$, defined as the distance driven up to and including the transition cycle. The Utility factor is illustrated in Figure 3 below.

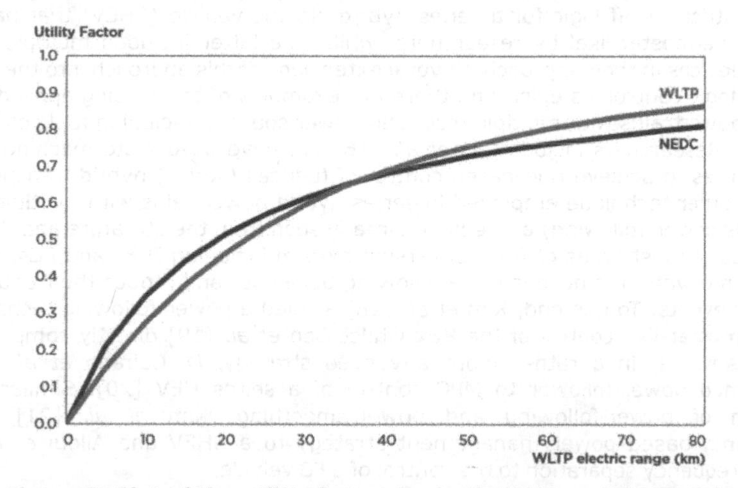


**Figure 3. Plot of the utility factor for both NEDC and WLTP [7].**

The testing procedure also seeks to determine the electrical charge energy used, so as to quote an energy consumption in Wh/km. The recharge energy is determined by fully recharging the vehicle after no more than 120 minutes after the end of the final charge sustaining cycle. This value $E_{AC}$ is converted into a quoted energy consumption simply by dividing by equivalent all-electric range (EAER).

$$EC = \frac{E_{AC}}{EAER} \tag{2}$$

The EAER is determined by subtracting the distance driven with the combustion engine in operation from the total distance up to and including the transition cycle (cycle n). This is calculated using the $CO_2$ emissions as per equation 3.

$$EAER = \left(\frac{M_{CO_2,CS} - M_{CO_2,CD,avg}}{M_{CO_2,CS}}\right) \times R_{CDC} \tag{3}$$

where:

$R_{CDC}$ = the charge depleting range up to and including the transition cycle,

$M_{CO_2,CS}$ = charge sustaining $CO_2$ emissions in g/km.

$M_{CO_2,CD,avg}$ = average charge depleting $CO_2$ emissions in g/km.

The control strategy and optimisation of REEV powertrains is more simple than that for other classes of hybrid and electrified powertrain, because the ICE or other REx component is fully decoupled from the final drive. There is a partial exception to this rule which leads to the documented loss of performance for REEV vehicles in charge sustaining mode [8], where there is insufficient battery energy available to safely operate the electric drive components at full power and the REx is required to supplement the available electrical power. Nonetheless a body of work exists proposing both online and offline methods of achieving close to optimal control.

It is widely recognised that one of the key benefits of a REEV architecture is that it is possible to operate an ICE at steady-state, optimal, operating points. Barsali *et al.* [9] and Konev *et al.* [10] both proposed power management control algorithms reflecting this. Jalil *et al.* [11], Mohammadian and Bathahee [12] and Yoo *et al.* [13] both proposed a strict on-off logic for a series hybrid electric vehicle (SHEV) that has been termed 'thermostat-like' by researchers, whilst the latter included multiple energy storage devices in their approach. Several extensions of this approach into the domain of fuzzy-logic controllers exist, and there are examples of them being applied effectively to powertrains with multiple electrical power sources, including fuel cell hybrids [14, 15]. Researchers including Li *et al.* [16, 17] have used state-machine control architectures to achieve rule-based control of fuel cell (series) hybrid transport systems. Another technique employed in series hybrid powertrains with multiple power sources is power-following; as we have already seen from the literature and the basic properties of most types of fuel conversion system including ICEs and FCs, there is a strong motivation to decouple the following behaviour and reduce the frequency of transient events. To this end, Kim *et al.* [18] applied a power-following technique to the thermostat-like control of the REx whilst Gao *et al.* [19] directly compared the two. Meanwhile, in a rather more advanced strategy, Di Cairano *et al.* applied a smoothed power-follower to MPC control of a series HEV [20]. Similar to the approach of power-following and power-smoothing, Kim *et al.* [21] applied a frequency-based power management strategy to a SHEV and Alloui *et al.* [22] applied frequency separation to the control of a FC vehicle.

The most commonplace online form of controller for parallel HEVs is based on the Equivalent Consumption Minimisation Strategy (ECMS), introduced as the concept of 'Equivalent Fuel Consumption' by Kim *et al.* [M ref 14] and fully developed by Paganelli *et al.* [23] . Serrao *et al.* [24] demonstrated that ECMS was an online realisation of Pontryagin's Minimum Principle for optimisation. Although less commonplace, ECMS also has a relevance for the on-line control of series-HEVs as demonstrated by Pisu and Rizzoni [25] and Sezer *et al.* [26]; Geng *et al.* [27] used ECMS for energy management of a gas turbine SHEV, and there have been many applications of ECMS to FC vehicles, with varying degrees of complexity including García *et al.* [28], Hemi *et al.* [29], Fu *et al.* [30], Rodatz *et al.* [31] and Li *et al.* [32].

All of the literature referred to thus far deals with the direct, online control of a REEV or SHEV, which is the main focus of this research. However, it is important to briefly mention the main techniques used for off-line optimisation of energy management strategies (EMS). In this context, computational overhead is less onerous, and some of the methods introduced will be of relevance for the reader to understand the automated optimisation process used in the present work.

Generally, Dynamic Programming (DP) [33, 34], with its basis in Bellman's Principle of Optimality is regarded as the gold standard for determining the optimal control strategy for complex systems. It has been applied to many hybrid electric vehicle control problems, including the optimisation of SHEV energy management by Brahma *et al.* [35], Pérez *et al.* [36] and Peng *et al.* [37]. Dynamic programming, like most of the optimisation methods reserved for off-line use is computationally very intensive. Other methods that have been used offline include Convex Optimisation [38], Simulated Annealing [39, 40], Support Vector machines [41] and strategies derived directly from Pontryagin's Minimum Principle, [42]. In recent years, as high power computation has increased in accessibility, the use of Evolutionary Algorithms (EAs, of which a good overview is given in [43]), Genetic Algorithms (GAs) [44] and Particle Swarm Optimisation (PSO) [45], has gained traction for solving complex optimisation problems. These are more efficient than a brute-force Monte Carlo approach and are able to work in multi-dimensional search spaces with many optimisation parameters. An automated GA is built into the GT-Suite optimisation tool, GT-IDO, and this was used to fine-tune parameters for the optimisation of the different RExs studied in this research.

The typology and technology of internal combustion engines used for range extender applications is of foremost importance for the power and emissions performance of hybrid electric vehicles. Ideally, a range extender should be lightweight, powerful and able to operate at high efficiency in a reasonable range of power [46]. Apart from the well established and reliable 4-stroke technology, a huge amount of research has been carrying out on different engine solutions by several research groups. Interestingly, amongst the different internal combustion engines, the rotary Wankel engine has been demonstrated to have high potential for range extender applications fulfilling the aforementioned requirements of advantageous power to weight ratio [47, 48] together with advanced control strategies for efficiency enhancement and emissions reductions [49] [50] or particular configurations for energy recovery [51, 52]. Alternatively, the opposed piston technology also proved to be a valid solution with its impressive efficiency due to the favourable configuration of the combustion chambers reducing heat loss and the high compression ratio [53].

## 2 THE METHODOLOGY AND THE MODEL DESCRIPTION

In order to examine the most effective future range extender technology, a simulation-based approach was taken. A vehicle model was based closely around the BMW i3 94Ah REx vehicle, which is one of a handful of currently available series hybrid REEV cars and has been the subject of examination by Argonne National Laboratory (ANL) [54], who have produced a publicly available dataset allowing for straightforward validation of a model and the examination of baseline control strategies. To effect a fair comparison of the different Range Extender units (REx) under consideration, the remainder of the powertrain, as described in subsequent sections, is kept very close to the BMW i3 baseline.

The entire powertrain and vehicle model was implemented in GT power, and the REx models were based on available data either from the literature or from prior experimental work by the authors. The comparison of RExs is based around the way in which hybrid vehicles are tested for emissions in Europe. Referring back to the discussion around Figure 3, we note that there is a significant reduction in homologated $CO_2$ output from increased electric range, and the regulations are formulated around the assumption that vehicles will operate in a 'charge depleting' fashion until it is no longer viable and then switch to a 'charge sustaining' mode. In order to compare RExs, it is assumed that the 'charge depleting' regime is exactly equivalent between the different proposed vehicle configurations and therefore only the 'charge sustaining' performance of the vehicle will be examined. As per the regulations, the duty cycle used to determine vehicle performance is the WLTC, while the validation was carried out using the duty cycles created by ANL for their dynamometer tests.

The different RExs are compared using several strategies. The first makes the assumption that the most efficient way to operate the REx is to run the ICE at its minimum BSFC operating point in order to generate the energy required for a charge-sustaining duty cycle, and then operate as an EV drawing down on this energy. This type of strategy, of course, requires *a priori* knowledge of the duty cycle that needs to be driven once the battery is sufficiently depleted to enter 'charge sustaining' mode. Given that in the real world, the operating strategy cannot depend on *a priori* knowledge, an operating strategy was devised for the REx based on instantaneous power demand and optimised for each REx under examination using a variety of tuneable parameters.

## 2.1 The vehicle and engine model

In addition to the ANL downloadable dynamometer database values, other public domain data, reported in Table 1, were used in the modelling work [55, 56]. Despite the fact that most of the data reported in Table 1 are related to the New European Drive Cycle (NEDC), the numerical models were validated against different drive cycles, most of them more representative of real driving conditions, and thus of the real fuel consumption and emissions, such as the WLTC.

**Table 1. BMW i3 technical data from [55, 56].**

| Technical Data | |
|---|---|
| **Electric motor and Range Extender** | |
| Electric motor maximum power | 125 kW |
| Range extender displacement | 647 cc |
| Range extender maximum power | 28 kW |
| Fuel tank | 9 litres |
| Electric Range | |
| NEDC | 239.79 km |
| Real | 168.98 km |
| Fuel consumption and $CO_2$ emissions | |
| NEDC | 0.49 liters/ 100km |
| NEDC $CO_2$ | 14 g/km |
| WLTP $CO_2$ | 5 g/km |

(*Continued*)

The engine and vehicle models were developed in the commercially available GT-Suite and GT-Power software produced by Gamma Technologies [57, 58]. The initial model was inherited from previous research work carried out at the Institute for Advanced Automotive Propulsion Systems (IAAPS) at the University of Bath, using the BMW i3 as a reference vehicle [59]. The modelling activities carried out for that work took into account the entire vehicle system (including the wheels, suspensions, chassis, etc.) and included the range extender, the electric machine and the battery pack. The layout of the aforementioned model is reported in Figure 4.

For the current work many new features have been introduced in order to validate the model against the ANL data. Firstly, all the most relevant parameters related to the mass of the vehicle, the frontal area, the drag coefficient *Cd* and the tires were analysed for the evaluation of the total resistance

**Table 1. (*Continued*)**

| Technical Data | |
|---|---|
| **Electric motor and Range Extender** | |
| NEDC empty battery fuel consumption | 5.32 litres/ 100km |
| NEDC $CO_2$ empty battery | 124 g/km |
| Real empty battery fuel consumption | 6.89 litres/ 100km |
| Real $CO_2$ empty battery | 162 g/km |
| NEDC Fuel range | 170.59 km |
| Real fuel range | 130.36 km |

applied to the vehicle in static and dynamic conditions. Furthermore, the ANL experimental work used specific coastdown coefficients to emulate the road load by means of the formula:

$$F_r = a + bV + cV^2 \tag{4}$$

where:

$F_r$ = Road load resistance [N]

a, b, c = Coastdown coefficients [N], [N/ m/s] and [N/ $m^2/s^2$] respectively

V = Vehicle speed [m/s]

The coefficients used by ANL were slightly different from the ones reported on the extensive United States Environmental Protection Agency (EPA) database. In the current work yet different ones were used to reach a good validation with the ANL experimental data. All of the above-mentioned coastdown coefficients are reported in the Table 2. The differences are likely to result from subtleties of modelling techniques used for rolling resistance and aerodynamic effects in the different modelling software employed by the research teams.

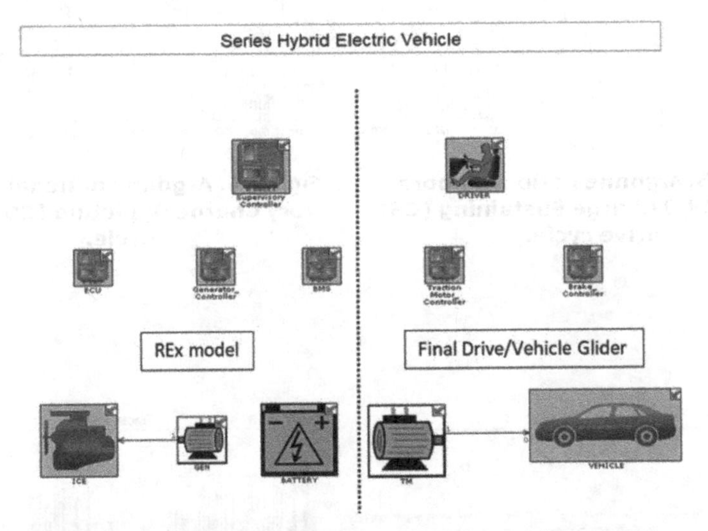


**Figure 4. BMW I3 Complete model layout, from [59].**

**Table 2. Coastdown coefficients used in the different research works [54 60].**

| | a [N] | b [N/m/s] | c [$N/m^2/s^2$] |
|---|---|---|---|
| EPA | 134.781 | 2.992 | 0.402 |
| ANL | 105.943 | 6.426 | 0.264 |
| Current work | 105.42 | 3.443 | 0.373 |

In addition, the numerical model for the current work was mainly developed for fuel consumption and emissions optimisations. As a consequence all the aforementioned and redundant details regarding the suspension, vehicle body, etc. were compiled into a single power demand in order to reduce the variables analysed by the internal GT-Suite optimiser and thus speed up the simulations. The reduced model structure is reported in the left hand side of Figure 4, with the electrical system traction power demand implemented as a lookup table in the BMS model.

While the data for the REx and the electric machines were relatively easy to find in the literature or to estimate, work on the REx strategy was carried out to match the experimental results from the ANL work. The strategy elaborated was then implemented in the "Supervisory control" block in Figure 4. Validation against the ANL experimental data was carried out using a Charge Sustaining (CS) and Charge Depleting (CD) drive cycle, reported in Figure 5 and Figure 6. A comparison of electrical flows and ICE speed between the ANL data and the simulated control strategy is given in Figure 7 and Figure 8. As is possible to appreciate there are some detail differences, as may be expected without access to the actual BMW i3 controllers, but as may be seen, the overall trends validate closely with the data.

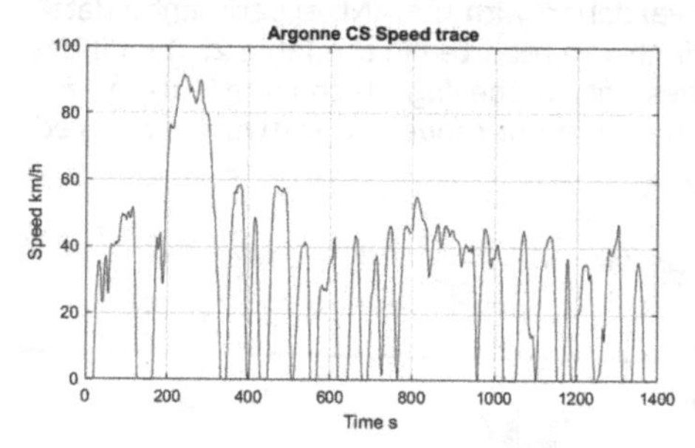


**Figure 5. Argonne national laboratory BMW i3 Charge Sustaining (CS) drive cycle.**

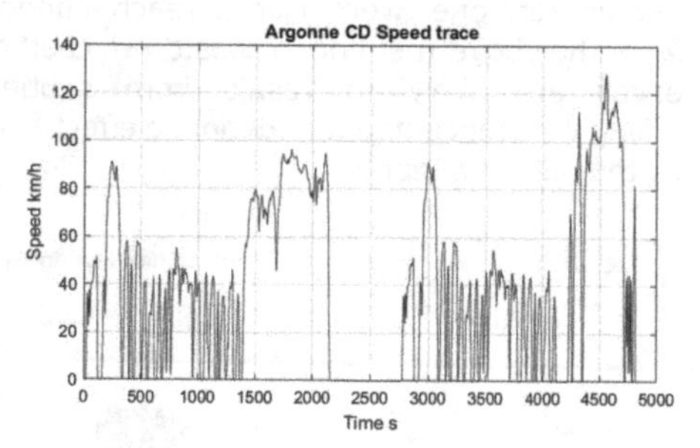


**Figure 6. Argonne national laboratory Charge Depleting (CD) drive cycle.**

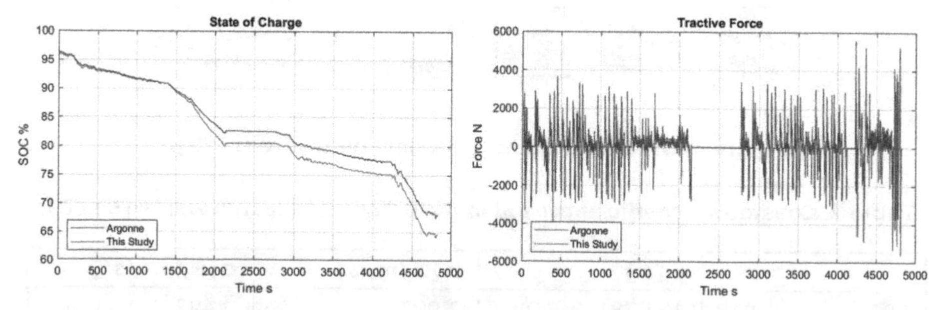


**Figure 7. Comparative results for BMW i3 model over ANL Charge Depleting Duty Cycle.**

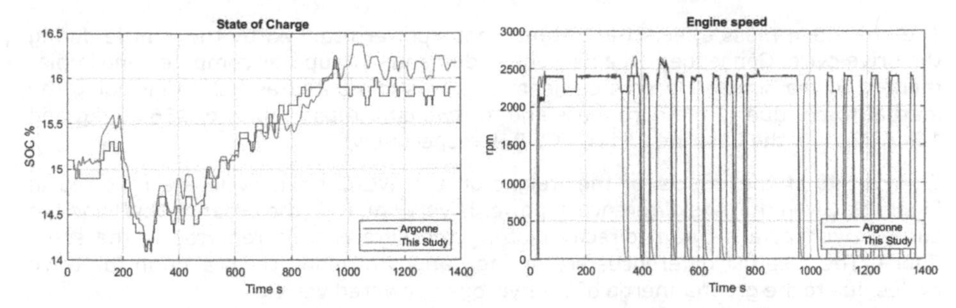


**Figure 8. Comparative results for BMW i3 model over ANL Charge Sustaining Duty Cycle.**

Since this work focussed on the comparison of different REx technologies including an hydrogen engine, some further considerations on the effects of a completely different fuel storage system need to be reported when considering the hydrogen solution. The storage of hydrogen at high pressure and low temperatures requires large and heavy tanks. Neglecting the consequences of a larger tank volume at this stage, the effects of the larger vehicle mass need to be considered for a correct evaluation of the instantaneous power demand and total energy consumed by the vehicle. BMW faced that issue already with its Hydrogen 7 model and the declared data [61] on that storage system were used to inform the computations in this work.

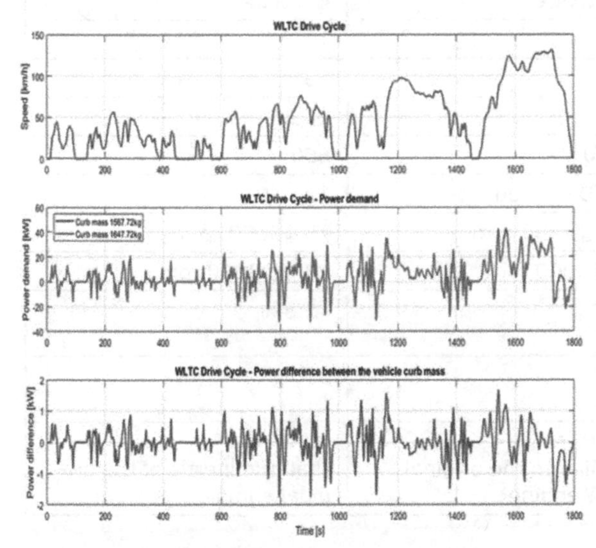


**Figure 9. WLTC Drive cycle and its power demand for two different curb masses and the resulting difference.**

In the literature related to the BMW Hydrogen 7 vehicle it is reported that the percentage in weight of the stored hydrogen (8kg) is the 3.2% of the total storage system (250 kg). Given that hydrogen has nearly three times the lower heating value (LHV) of gasoline (120 MJ/kg and 43 MJ/kg respectively), and taking into account the same amount of energy stored in the i3 for a fair comparison, it is assumed that 2.5 kg of hydrogen should be stored on the i3 equipped with the hydrogen engine (instead of ~7.5 kg of gasoline). Using the same ratio between fuel and storage system weight the hypothetical i3 $H_2$ storage system weight is 78.125kg for a total weight (fuel and storage system) of ~80kg. This is considered to be a conservative estimate, i.e. a more modern 700 bar pressurized storage system would be expected to be lighter than this; the value given dies, however, allow for a practical consideration of the increase in mass due to having to accommodate an actual hydrogen tank system.

The change in mass affects the instantaneous power required by the vehicle during the drive cycle. Consequently, a new power demand look-up was computed and implemented in the simulation environment. This was computed as before including the inertial forces due to the curb mass and for two curb mass values of 1567.72kg and 1647.72kg for the baseline and $H_2$-ICE REx respectively.

The results of the effects of the weight on the WLTC drive cycle are reported in Figure 9. Given that the difference in power is very small (in the order of 1kW) and the curves overlap, a diagram directly plotting the difference is reported in the same figure. The greatest differences are during periods of high acceleration in the drive cycles due to the greater inertia of the hydrogen powered vehicle.

In the current work two different 4-stroke solutions were implemented in the validated simulation model and the results compared subsequently. Instead of the original BMW engine, a 4-stroke 2-cylinder Mahle "Dedicated Hybrid Internal Combustion Engine" (DHICE) [62] and a hydrogen 4-stroke single cylinder $H_2$-ICE tested at Argonne National Laboratory [63]. A summary of the most important engines characteristics are reported in Table 3.

**Table 3. Mahle DHICE and argonne national laboratory $H_2$-ICE engines characteristics.**

| | **Mahle DHICE** | **ANL $H_2$-ICE** |
|---|---|---|
| **Type** | **4-Stroke** | **4-Stroke** |
| Number of cylinders [/] | 2 | 1 |
| Bore [mm] | 83 | 83 |
| Stroke [mm] | 92.4 | 105.8 |
| Displacement [cc] | 1000 | 660 |
| Compression Ratio [/] | Not Disclosed | 12.9:1 |
| Maximum torque [Nm] | 143@2500 rpm | 75@3000 rpm |
| Maximum power [kW] | 60@4000 rpm | 23.6@3000 rpm |
| Peak brake thermal efficiency [/] | 40% | 45.5% |
| Minimum operating speed [rpm] | 500 | 350 |
| Engine inertia [$kgm^2$] | 0.04 (similar to the original BMW engine) | 0.02 (half the inertia of the original engine) |

In order to give a thorough insight on the inputted models, it is worth reporting here the power and torque curves together with the brake specific fuel consumption (BSFC), brake mean effective pressure (BMEP) and friction mean effective pressure (FMEP) (where available or otherwise estimated) for both of the engines.

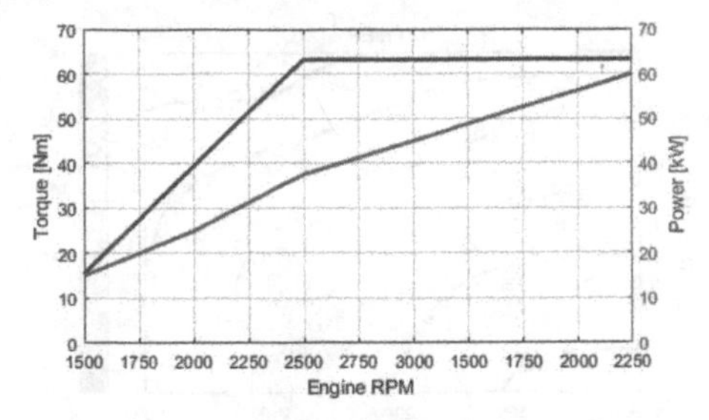


**Figure 10. MAHLE DHICE torque and power curves.**

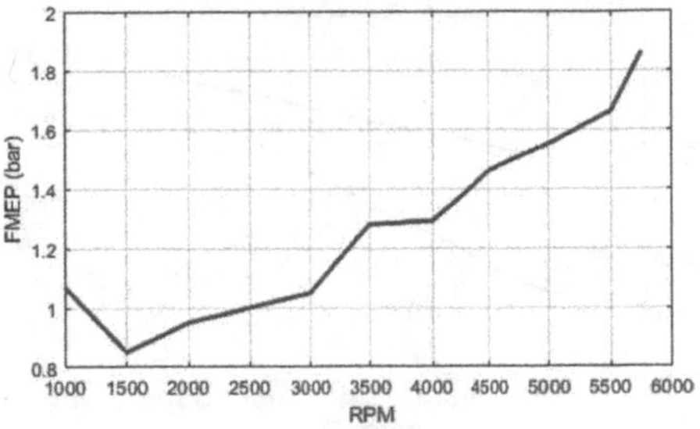

**Figure 11. Mahle DHICE estimated FMEP [bar].**

The power and torque curves for the Mahle engine are reported in Figure 10. As can be appreciated it produces a constant and maximum torque from 2500 rpm. This is also reflected in the BMEP map with an extended range at the maximum value of 18bar (see Figure 12). More importantly, the engine presents a large operating range at high efficiency with its peak at 3000rpm and 13bar BMEP as reported in Figure 12.

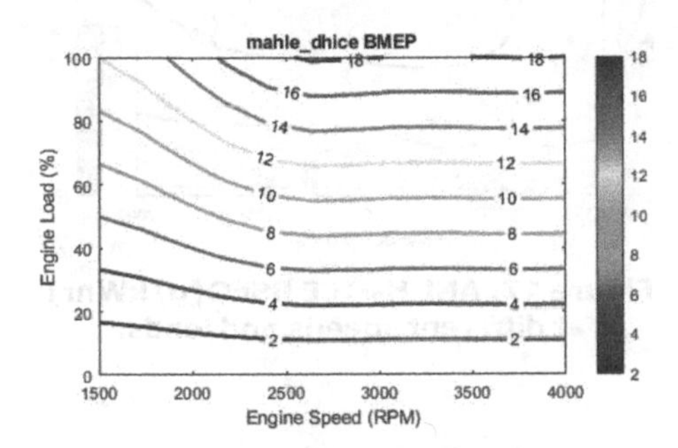


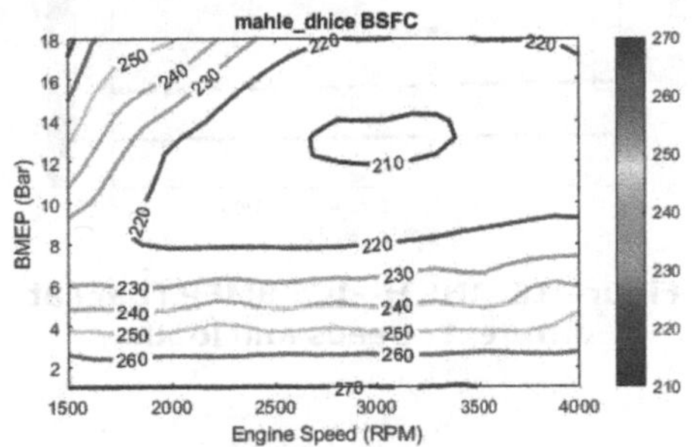


**Figure 12. Mahle DHICE BMEP [bar] at different speeds and loads.**

**Figure 13. Mahle DHICE BSFC [g/kWhr] at different speeds and loads.**

For the FMEP some assumptions had to be made since no data were presented in the literature from Mahle. It was assumed that the FMEP was equal to that of a similar engine (both in terms of displacement and configuration, 4-stroke 2-cylinder inline) tested at the University of Bath. The other assumption was that the FMEP was the same at all the loads thus reducing the map to a single curve as reported in Figure 11.

Conversely, the ANL $H_2$-ICE engine presented peculiar curves of torque and power as represented in Figure 14. Given the monotonic trend of the aforementioned curves, it seems that the engine was limited to run up to 3000rpm while being able to run faster than the maximum speed presented in the plot. Regardless of this, for an engine employed as a range extender it is more important that it is able to deliver the desired power at the maximum efficiency possible. The ANL $H_2$-ICE was capable of achieving a maximum efficiency of 45.5% at 2000 rpm and maximum load, with BSFC lower than 70g/kWhr (considering the LHV of the hydrogen of 120MJ/kg) as reported in Figure 17. Unlike the Mahle engine literature, ANL reported the FMEP map for their $H_2$-ICE at different speeds and loads as represented in Figure 15. Finally, all the above reported maps were implemented in the Mean Value Engine Model (MVEM) block in the GT-Power model.

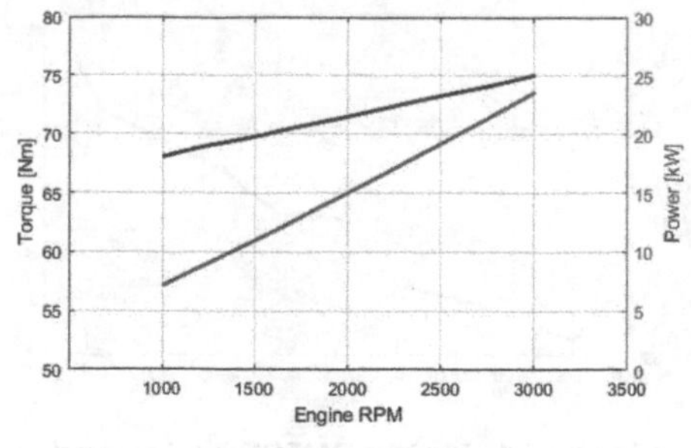


**Figure 14. ANL $H_2$-ICE torque and power curves.**

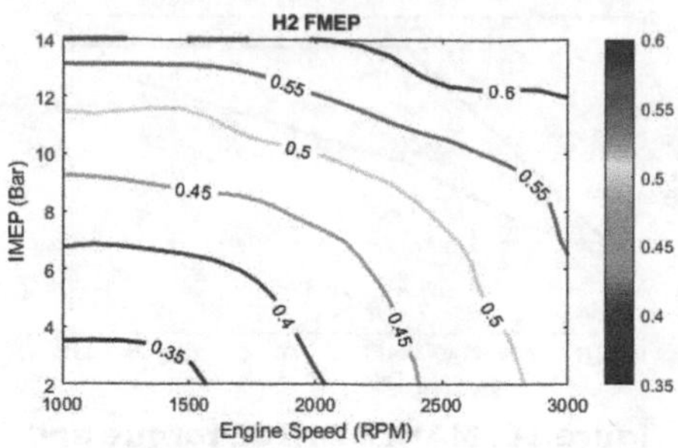

**Figure 15. ANL $H_2$-ICE FMEP [bar].**

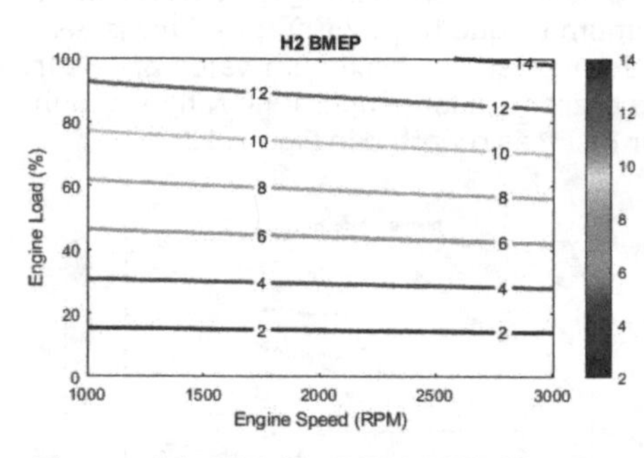


**Figure 16. ANL $H_2$-ICE BMEP [bar] at different speeds and loads.**

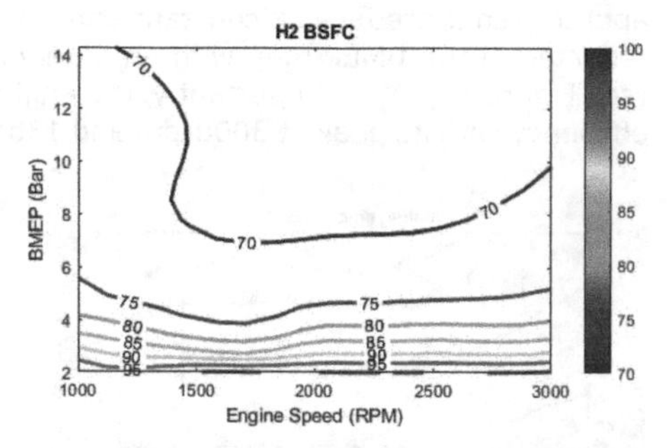


**Figure 17. ANL $H_2$-ICE BSFC [g/kWhr] at different speeds and loads.**

## 2.2 The range extender and battery management system model

The other elements of the powertrain are modelled based on the BMW i3 baseline vehicle, but some modifications are made for each REx to ensure that the generator machine and battery charge/discharge power remain rightsized for that ICE. The battery sub-model emulates the behaviour of the battery system subject to the fluxes of electrical energy and taking into account different important parameters such as the SOC, the geometry of the cells, the internal resistance, the open-circuit voltage, the thermal behaviour and all their inter-dependencies.

## 2.3 REx control strategy

Two control strategies were employed to compare the performance of the RExs. One of these simply involved examining the fuel and $CO_2$ cost (where applicable) of generating the necessary electrical energy to drive the duty cycle at the best-BSFC operating point, with some heuristics included for catalyst light-off and coolant and lubricant warm-up. The other was a time variant energy management strategy, dependent on instantaneous power demand and current SOC was implemented to provide a more realistic operating regime. This strategy is designed to maintain SOC between predefined bounds (13.5% and 16.5% in this particular case), and once again includes some modelling of a catalyst light-off and warm-up procedure which is normalised across the RExs under consideration. The strategy is described in Table 4.

**Table 4. REx hybrid rule-based strategy.**

| | SOC<Lower Threshold | | Lower Threshold < SOC < Upper Threshold | | Upper Threshold < SOC | |
|---|---|---|---|---|---|---|
| | **REx Warmup Incomplete** | **REx Warmup Complete** | **REx Warmup Incomplete** | **REx Warmup Complete** | **REx Warmup Incomplete** | **REx Warmup Complete** |
| Conditioned P demand < Low Power Threshold | Light-off /Warmup setpoint | REx demand == Low Power | Light-off /Warmup setpoint | REx demand == Low Power | Light-off /Warmup setpoint | ICE Off |
| Low Power Threshold < Conditioned P demand < High Power Threshold | Light-off /Warmup setpoint | REx demand == Conditioned Power | Light-off /Warmup setpoint | REx demand == Conditioned Power | Light-off /Warmup setpoint | ICE Off |
| High Power Threshold > Conditioned P demand | Light-off /Warmup setpoint | REx demand == High Power | Light-off /Warmup setpoint | REx demand == High Power | Light-off /Warmup setpoint | ICE Off |

The warm-up routine consists of two stages, one at lower power to achieve catalyst light-off from cold, and one at higher power to rapidly warm coolant and lubricant in the ICE. The setpoints for this operation are selected as the best BSFC point for the higher power operation and the point on the best BSFC curve representing approximately 1/3 of that power for the light-off operation. Each of these setpoints is run for 45 seconds; under the current EU6 regulations, catalyst light-off must be achieved in 20 seconds from cold start.

The conditioned power referred to in Table 4 is computed from the instantaneous power demand. The instantaneous demand is smoothed by passing through a moving average filter – the time window of this filter is one of the tuneable parameters of the strategy, making it more or less reactive to the dynamics of the power demand signal. Additionally, an offset and multiplier are used to alter the scaling of the REx demand relative to the instantaneous demand. This is an important piece of functionality because the ICE may not be right-sized for the power demand of a duty cycle, and forcing it to operate at higher or lower powers will have an efficiency benefit. The scaling is also carried out with respect to the instantaneous error of the SOC from its target, in order to force the system towards that target SOC.

$$P_{conditioned} = P_{smoothed} \times \left(\widehat{SOC} + SOC_{const}\right) \times P_{factor} \quad (5)$$

Where $SOC_{const}$ and $P_{factor}$ are the tuneable parameters and $\widehat{SOC} = SOC_{target} - SOC$ is the SOC error, which is scaled to be reported as a percentage. Upper and lower SOC thresholds are used to determine the hysteresis of the engine start/stop logic. Additional thresholds are used to bookend the ICE power demand and prevent operation in excessively high or low power regimes where efficiency is less good. All four of these thresholds are tuneable parameters. The tuning of this control strategy is facilitated by the use of GT-Power's built-in optimisation suite, GT-IDO, which uses a genetic algorithm within a user-designed search space.

## 3 NUMERICAL RESULTS

In order to directly compare the efficiency of the different REx ICEs in the study, it is important to consider that one of them uses a different fuel. Thus the comparison is made with energy contained within the fuel used, rather than the volume of fuel itself. The results of comparing an optimal steady-state charge sustain strategy are given in Table 5. However, for the hydrogen combustion engine, an equivalent volume of gasoline and associated actual and certification fuel consumption and emissions values have been computed. The reader should note that there are in reality no tailpipe $CO_2$ emissions from the $H_2$-ICE powertrain (disregarding any lubricating oil combustion, which will be very small in a modern-technology ICE). In each case, the engines are simulated at a steady state and the selected pareto-optimal result is close to perfectly charge sustaining. The utility factor for the BMW i3 baseline vehicle is beyond the range normally considered or presented, for example in Figure 3, however by substituting the quoted legislative values and real world values from the literature (presented in Table 1) into equation 1, an effective utility factor of 0.968 is arrived at. This value and a duty cycle length of 23.25km are used to compute the WLTP equivalent $CO_2$ and fuel consumption values presented in Table 5.

**Table 5. Comparison of steady state energy conversion.**

| | Latent Fuel Energy Consumed | Fuel Mass | Fuel Volume (gasoline equivalent) | Actual Charge Sustaining FC | WLTP FC | Actual Charge Sustaining $CO_2$ | WLTP $CO_2$ |
|---|---|---|---|---|---|---|---|
| **Units** | **kJ** | **g** | **l** | **l/100km** | **l/ 100km** | **g/km** | **g/km** |
| Baseline i3 | 39082 | 885 | 1.17 | 5.03 | 0.176 | 116 | 4.07 |
| Mahle DHICE | 37139 | 841 | 1.11 | 4.78 | 0.167 | 110 | 3.87 |
| $H_2$-ICE | 32256 | 252 | 0.73 | 3.14 | 0.110 | 72.5 | 2.54 |

The other trade-off that is presented is the result tuning of the realistic bounded strategy to achieve charge sustaining performance. Once again there is an entirely expected direct proportionality between fuel energy consumed and battery energy generated. The result for the ICEs under consideration is given in Table 5. Using the described strategy, the larger ICEs show a tendency to overcharge, but the spread of results and axis crossings highlight that the hydrogen ICE requires less fuel energy than the gasoline ICEs to maintain an equivalent level of charge sustenance. The gradient of the trend in the point clouds, as well as the crossing points, suggests that the Mahle DHICE engine is more efficient than the baseline i3 at converting stored energy to a different form, but not as efficient as the $H_2$-ICE.

Given the greater energy conversion efficiency of the $H_2$-ICE, combined with the ability of the an REx system to completely decouple ICE operating point from instantaneous power demand, it is only natural that this is the option that consumes the least energy.

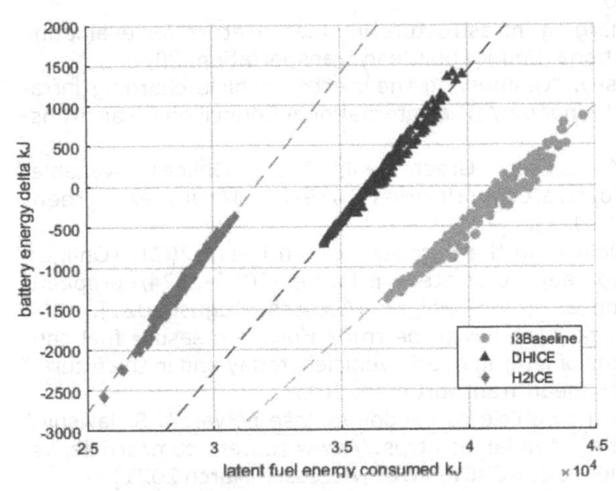


**Figure 18. Comparison of stored energy consumed by each type of engine during multiple simulations to search for optimal control strategy tuning.**

## 4 CONCLUSIONS

From this study of range extender technologies and control strategies for series hybrid vehicles, the authors conclude that there are still significant improvements in system efficiency to be had by careful design of dedicated ICEs. The work of Mahle on improving the efficiency of conventional combustion systems presents a significant improvement over the baseline BMW i3 engine.

The modelling carried out as part of this study confirms that hydrogen combustion offers a tangible benefit in terms of efficiency of combustion, compared by using latent energy consumed between fuels of very different mass. It also produces no tailpipe $CO_2$ from its combustion. This is offset against the additional mass of the pressurised fuel system. Part of the reason for this result is that only relatively small amounts of fuel need to be carried, and the engine is optimised for a particular operating point, unlike a conventional propulsion application where a large amount of fuel is consumed away from the best BSFC region of the engine.

The ICEs were compared with reference to the testing methodologies defined under the NEDC and WLTP regulations, and the necessity to operate in 'charge' sustaining mode, after a long period of depleting electric range. The authors note that this may not be the most effective use of the available battery capacity and that entering a charge sustain mode often brings limitations in drivability. It was also noted that requiring particularly short periods of ICE operation was a problem for achieving catalyst light-off and coolant and lubricant warm up. Heuristics were introduced to the trialled control strategies to address this, and were made equivalent across the board in the interest of making a fair comparison between the different ICE RExs. As a means of demonstrating feasible performance if the duty cycle was known *a priori* the amount of fuel (and the latent energy contained therein) required to generate one WLTC's worth of energy from a low initial SOC was computed for steady-state, best BSFC operation.

In a more realistic strategy, it was noted that between the ICEs compared, the hydrogen combustion engine offers a 22% saving on latent energy consumed, for an equivalent delta-SOC compared to the baseline. It also offers the steepest gradient of energy released vs energy contained, for any of the engines under consideration, implying the greatest overall efficiency. Further improvements to real-world performance are likely to be possible using more advanced control strategies that are less closely related to the structure of the current legislative framework for emissions.

## REFERENCES


[1] United Nations, "Paris Agreement," 2015.

[2] 3Sandra Wappelhorst, "The end of the road? An overview of combustion-engine car phase-out announcements across Europe," The International Council on Clean Transportation, 2020.

[3] Dale Hall, Nic Lutsey, "Charging infrastructure in cities: Metrics for evaluating future needs," The International Council on Clean Transportation, 2020.

[4] Michael Nicholas, Nic Lutsey, "Quantifying the electric vehicle charging infrastructure gap in the United Kingdom," The International Council on Clean Transportation, 2020.

[5] European Commission, "A European Green Deal," 2021. [Online]. Available: https://ec.europa.eu/info/strategy/priorities-2019-2024/european-green-deal_en. [Accessed August 2021].

[6] European Commission, "Delivering the European Green Deal," 2021. [Online]. Available: https://ec.europa.eu/info/strategy/priorities-2019-2024/european-green-deal/delivering-european-green-deal_en. [Accessed August 2021].

[7] Iddo Riemersma, Peter Mock, "Too low to be true? How to measure fuel consumption and CO2 emissions of plug-in hybrid vehicles, today and in the future," The International Council on Clean Transportation, 2017.

[8] J. Stempel, "BMW's i3 electric vehicle can suddenly lose power: U.S. lawsuit," Reuters, May 2016. [Online]. Available: https://www.reuters.com/article/us-bmw-lawsuit-electricvehicles-idUSKCN0Y91WU. [Accessed March 2021].

[9] S Barsali, C Miulli, A Possenti, "A control strategy to minimize fuel consumption of series hybrid electric vehicles," *IEEE Transactions on Energy Conversion*, vol. 19, no. 1, pp. 187–195, 2004.

[10] A. Konev, L. Lezhnev, I. Kolmanovsky, "Control Strategy Optimization for a Series Hybrid Vehicle," in *SAE 2006 World Congress & Exhibition*, Detroit MI, 2006.

[11] NJalil, N A Kheir, M Salman, "A rule-based energy management strategy for a series hybrid vehicle," in *American Control Conference*, Alberquerque, NM, 1997.

[12] M Mohammadian, M T Bathaee, "Motion control for hybrid electric vehicle," in *The 4th International Power Electronics and Motion Control Conference*, Xi'an, China, 2004.

[13] H Yoo, S-K Sul, Y Park, J Jeong, "System Integration and Power-Flow Management for a Series Hybrid Electric Vehicle Using Supercapacitors and Batteries," *IEEE Transactions on Industry Applications*, vol. 44, no. 1, pp. 108–114, 2008.

[14] H Hemi, J Ghouili, A Cheriti, "A real time fuzzy logic power management strategy for a fuel cell vehicle," *Energy Conversion and Management*, vol. 80, pp. 63–70, 2014.

[15] A. Melero-Perez, W. Gao, J. Jesus Fernadez-Lzano, "Fuzzy Logic energy management strategy for Fuel Cell/Ultracapacitor/Battery hybrid vehicle with Multiple-Input DC/DC converter," in *2009 IEEE Vehicle Power and Propulsion Conference*, Dearborn, MI, 2009.

[16] Q Li, B Su, Y Pu, Y Han, T Wang, L Yin, W Chen, "A State Machine Control Based on Equivalent Consumption Minimization for Fuel Cell/ Supercapacitor Hybrid Tramway," *IEEE Transactions on Transportation Electrification*, vol. 5, no. 2, pp. 552–564, 2019.

[17] Q Li, H Yang, Y Han, M Li, W Chen, "A state machine strategy based on droop control for an energy management system of PEMFC-battery-supercapacitor hybrid tramway," *International Journal of Hydrogen Energy*, vol. 41, no. 36, pp. 16148–16159, 2016.

[18] M Kim, D Jung, K Min, "Hybrid Thermostat Strategy for Enhancing Fuel Economy of Series Hybrid Intracity Bus," *IEEE Transactions on Vehicular Technology*, vol. 63, no. 8, pp. 3569–3579, 2014.

[19] J-P Gao, G-M G Zhu, E G Strangas, "Equivalent fuel consumption optimal control of a series hybrid electric vehicle," *Proceedings of the Institution of Mechanical Engineers, Part D: Journal of Automobile Engineering*, vol. 223, no. 8, pp. 1003–1018, 2009.

[20] S Di Cairano, W Liang, I V Kolmanovsky, M L Kuang, A M Phillips, "Power Smoothing Energy Management and Its Application to a Series Hybrid Powertrain," *IEEE Transactions on Control Systems Technology*, vol. 21, no. 6, pp. 2091–2103, 2013.

[21] Y Kim, A Salvi, J B Siegel, Z S Filipi, A G Stefanopoulou, T Ersal, "Hardware-in-the-loop validation of a power management strategy for hybrid powertrains," *Control Engineering Practice*, vol. 29, pp. 277–286, 2014.

[22] H Alloui, M Becherif, K Marouani, "Modelling and frequency separation energy management of fuel Cell-Battery Hybrid sources system for Hybrid Electric Vehicle," in *21st Mediterranean Conference on Control and Automation*, Platanias, Greece, 2013.

[23] G Paganelli, T M Guerra, S Delprat, J-J Santin, M Delholm, E Combes, "Simulation and assessment of power control strategies for a parallel hybrid car," *Proceedings of the Institution of Mechanical Engineers, Part D: Journal of Automobile Engineering*, vol. 214, no. 7, pp. 705–717, 2000.

[24] L Serrao, S Onori, G Rizzoni, "ECMS as a realization of Pontryagin's minimum principle for HEV control," in *Proceeding of the American Control Conference*, St. Louis, MO, 2009.

[25] P Pisu, G Rizzoni, "A supervisory control strategy for series hybrid electric vehicles with two energy storage systems," in *IEEE Vehicle Power and Propulsion Conference*, Chicago, IL, 2005.

[26] V Sezer, M Gokasan, S Bogosyan, "A Novel ECMS and Combined Cost Map Approach for High-Efficiency Series Hybrid Electric Vehicles," *IEEE Transactions on Vehicular Technology*, vol. 60, no. 8, pp. 3557–3570, 2011.

[27] B. Geng, J. K. Mills, D. Sun, "Energy Management Control of Microturbine-Powered Plug-In Hybrid Electric Vehicles Using the Telemetry Equivalent Consumption Minimization Strategy," *IEEE Tranactions on Vehicular Technology*, vol. 60, no. 9, 2011.

[28] P García, J P Torreglosa, L M Fernández, F Jurado, "Energy Management Control of Microturbine-Powered Plug-In Hybrid Electric Vehicles Using the Telemetry Equivalent Consumption Minimization Strategy," *International Journal of Hydrogen Energy*, vol. 37, no. 11, pp. 9368–9382, 2011.

[29] H Hemi, J Ghouili, Ahmed Chenti, "A real time energy management for electrical vehicle using combination of rule-based and ECMS," in *IEEE Electrical Power & Energy Conference*, Halifax, NS, Canada, 2013.

[30] Z Fu, Z Li, P Si, F Tao, "A hierarchical energy management strategy for fuel cell/battery/supercapacitor hybrid electric vehicles," *International Journal of Hydrogen Energy*, vol. 44, no. 39, pp. 22146–22159, 2019.

[31] P Rodatz, G Paganelli, A Sciarretta, L Guzzella, "Optimal power management of an experimental fuel cell/supercapacitor-powered hybrid vehicle," *Control Engineering Practice*, vol. 13, no. 1, pp. 41–53, 2005.

[32] H Li, A Ravey, A N'Diaye, A Djerdir, "Equivalent consumption minimization strategy for fuel cell hybrid electric vehicle considering fuel cell degradation," in *IEEE Transportation Electrification Conference and Expo (ITEC)*, Chicago, IL, 2017.

[33] R. Bellman, Dynamic Programming, Princeton, NJ, USA: Princeton University Press, 1957.

[34] R. Bellman, The Theory of Dynamic Programming, Santa Monica, CA: The Rand Corporation, 1954.

[35] A. Brahma, Y. Guezennec, G. Rizzoni, "Optimal energy management in series hybrid electric vehicles," in *Proceedings of the American Control Conference*, Chicago, IL, USA, 2000.

[36] L V Pérez, G R Bossio, D Moitre, G O García, "Optimization of power management in an hybrid electric vehicle using dynamic programming," *Mathematics and Computers in Simulation*, vol. 73, no. 1-4, pp. 244–254, 2006.

[37] J Peng, H He, R Xiong, "Rule based energy management strategy for a series–parallel plug-in hybrid electric bus optimized by dynamic programming," *Applied Energy*, vol. 185, no. 2, pp. 1633–1643, 2017.

[38] X Hu, N Murgovski, L Johanneson, B Egardt, "Energy efficiency analysis of a series plug-in hybrid electric bus with different energy management strategies and battery sizes," *Applied Energy*, vol. 111, pp. 1001–1009, 2013.

[39] Z Wang, B Huang, Y Xu, W Li, "Optimization of Series Hybrid Electric Vehicle Operational Parameters By Simulated Annealing Algorithm," in *IEEE International Conference on Control and Automation*, Guangzhou, China, 2007.

[40] K Chen, Y Deng, F Zhou, G Sun, Y Yuan, "Control strategy optimization for hybrid electric vehicle based on particle swarm and simulated annealing algorithm," in *International Conference on Electric Information and Control Engineering*, Wuhan, China, 2011.

[41] Wang, Z.; Li, W.; Xu, Y., "A novel power control strategy of series hybrid electric vehicle," in *Proceedings of the IEEE International Conference on Intelligent Robotics Systems*, 2007.

[42] L Serrao, G Rizzoni, "Optimal control of power split for a hybrid electric refuse vehicle," in *American Control Conference*, Seattle, WA, USA, 2008.

[43] T Bäck, H-P Schwefel, "An Overview of Evolutionary Algorithms for Parameter Optimisation," *Evolutionary Computation*, vol. 1, no. 1, 1993.

[44] J. H. Holland, "Genetic Algorithms and the Optimal Allocation of Trials," *Journal of Computing*, vol. 2, no. 2, 1973.

[45] R Eberhart, J Kennedy, "A New Optimiser Using Particle Swarm Theory," in *Sixth International Symposium on Micro Machine and Human Science*, 1995.

[46] Turner, J., Blake, D., Moore, J., Burke, P. et al., "The Lotus Range Extender Engine," *SAE International Journal of Engines*, vol. 3, no. 2, pp. 318–351, 2010.

[47] Vorraro, G., Turner, M., and Turner, J., "Testing of a Modern Wankel Rotary Engine - Part I: Experimental Plan, Development of the Software Tools and Measurement Systems," *SAE Technical Paper 2019-01-0075*, 2019.

[48] Pennycott, A. et al., "Modelling and simulation of a rotary-powered range extender engine," in *AVL International Simulation Conference*, 2019.

[49] A. S. Chen et al., "Nonlinear Observer-Based Air-Fuel Ratio Control for Port Fuel Injected Wankel Engines," in *2018 UKACC 12th International Conference on Control (CONTROL)*, 2018.

[50] Chen, A. et al., "Control-Oriented Modelling of a Wankel Rotary Engine: A Synthesis Approach of State Space and Neural Networks," *SAE Technical Paper 2020-01-0253*, 2020.

[51] G. Vorraro et al., "Application of a rotary expander as an energy recovery system for a modern Wankel engine," in *ImechE Internal Combustion Engines and Powertrain Systems for Future Transport 2019*, Birmingham, 2019.

[52] Turner, J. et al., "Initial Investigations into the Benefits and Challenges of Eliminating Port Overlap in Wankel Rotary Engines,", 2020," *SAE Int. J. Adv. & Curr. Prac. in Mobility*, vol. 2, no. 4, pp. 1800–1817, 2020.

[53] Turner, J., Vorraro, G., "The Opposed-Piston Engine: Studies on efficiency and geometry and some new concepts to further improve its thermodynamics in concert with increased powertrain electrification," in *Direct-Injection Two-Stroke Engines - International Workshop and Conference*, Rueil-Malmaison, France, 2020.

[54] Argonne National Laboratory, "Downloadable Dynamometer Database 2014 BMW i3-REX," [Online]. Available: https://www.anl.gov/es/energy-systems-d3-2014-bmw-i3rex. [Accessed May 2021].
[55] EV-Database, "BMW i3 94Ah," [Online]. Available: https://ev-database.uk/car/1068/BMW-i3-94-Ah. [Accessed May 2021].
[56] European Environment Agency, "Monitoring of CO2 emissions from passenger cars – Regulation (EU) 2019/631," [Online]. Available: https://www.eea.europa.eu/data-and-maps/data/co2-cars-emission-18. [Accessed May 2021].
[57] Gamma Technologies, "GT Suite Overview," [Online]. Available: https://www.gtisoft.com/gt-suite/gt-suite-overview/. [Accessed May 2021].
[58] Gamma Technologies, "GT-POWER Engine Simulation Software," [Online]. Available: https://www.gtisoft.com/gt-suite-applications/propulsion-systems/gt-power-engine-simulation-software/. [Accessed May 2021].
[59] Turner, M., Turner, J., and Vorraro, G., "Mass Benefit Analysis of 4-Stroke and Wankel Range Extenders in an Electric Vehicle over a Defined Drive Cycle with Respect to Vehicle Range and Fuel Consumption," *SAE Technical Paper 2019-01-1282*, 2019.
[60] EPA, "Data on Cars used for Testing Fuel Economy," [Online]. Available: https://www.epa.gov/compliance-and-fuel-economy-data/data-cars-used-testing-fuel-economy. [Accessed June 2021].
[61] wikipedia, "BMW Hydrogen 7," [Online]. Available: https://en.wikipedia.org/wiki/BMW_Hydrogen_7. [Accessed June 2021].
[62] Bassett, M., et al., Mahle Powertrain, "Mahle modular hybrid powertrain," in *Internal combustion engines and powertrain systems for future transport 2019, IMechE 2019*, Birmingham, UK.
[63] N S Matthias, T Wallner, R Scarcelli, "A Hydrogen Direct Injection Engine Concept that Exceeds U.S. DOE Light-Duty Efficiency Targets," *SAE International Journal of Engines*, vol. 5, no. 3, pp. 838–849, 2012.

# Author Index